SIGILL : COLL : CARLETON
DECLARATIO SERMONUM TUORUM ILLUMINAT
NORTHFIELD, MINN. A.D. 1866

NEMESIS DIVINA

ARCHIVES INTERNATIONALES D'HISTOIRE DES IDÉES

INTERNATIONAL ARCHIVES OF THE HISTORY OF IDEAS

177

NEMESIS DIVINA

by
CARL VON LINNÉ

Introduced and translated by
M.J. PETRY

CARL VON LINNÉ

NEMESIS DIVINA

Edited and translated
with an introduction and explanatory notes

by

M.J. PETRY
Fellow of the Linnean Society of London
Emeritus Professor of the History of Philosophy,
Erasmus University Rotterdam

KLUWER ACADEMIC PUBLISHERS
DORDRECHT / BOSTON / LONDON

A C.I.P. Catalogue record for this book is available from the Library of Congress.

ISBN 0-7923-6820-7

Published by Kluwer Academic Publishers,
P.O. Box 17, 3300 AA Dordrecht, The Netherlands.

Sold and distributed in North, Central and South America
by Kluwer Academic Publishers,
101 Philip Drive, Norwell, MA 02061, U.S.A.

In all other countries, sold and distributed
by Kluwer Academic Publishers,
P.O. Box 322, 3300 AH Dordrecht, The Netherlands.

Dutch non-Scientific version published in 1996 at Agora:
Carl von Linné: Nemesis Divina,
bezorgd door T. de Vlaming-van Santen en M.J. Petry

Printed on acid-free paper

Printed in the Netherlands.

To Adam and his mother, —
both much concerned with gardens.

FOREWORD

The main autograph text of the *Nemesis Divina* is in the University Library at Uppsala, and there is also important manuscript material in the libraries of the Linnean Society of London and the Caroline Institute in Stockholm. Making proper use of these sources in preparing this edition has been a challenging task, and the whole project might well have been abandoned at a very early stage had the archivists in charge of the material not gone out of their way in providing help and encouragement. Carl-Otto von Sydow, Keeper of Manuscripts at Uppsala, responded to my initial enquiry by calling attention to the way in which the most recent Swedish edition of the manuscript had been criticized, sending on a complete microfilm, and adding that I was free to make what use I wanted of it. When I ran into difficulties and came back to him with queries and suggestions, he was always ready to consult the original and give his opinion on readings and interpretations. Gina Douglas, Librarian and Archivist of the Linnean Society and the Librarians of the Caroline Institute have been no less helpful, not only in answering questions concerning the condition and the sequence of the sheets, but also in sending on extra copies of them whenever I have had difficulty in deciphering the text. Very little progress could have been made without help of this kind, and I am most grateful for the friendly and forthcoming manner in which it has been given.

Of recent years there has been a great surge of popular interest throughout the whole of Scandinavia in the activities of local history societies and the study of genealogy, and it is largely on account of this that it has been possible to investigate the background to Linnaeus' case-studies in such depth and detail. My indebtness to others in compiling the commentary will be apparent to all, and has been acknowledged in many of the separate notes. Contacting a local library or society has nearly always brought me into contact with a network of researchers deeply informed concerning the careers of the persons I have been attempting to trace, and I have often found that they have enabled me to throw new light on Linnaeus' own sources of information. In this connection I feel that special mention should be made of Kerstin Lindblom, Secretary of the *Linné's Råshult Trust*, who has not only documented all the material relating to Linnaeus' home village, but has also taken the initiative in tackling what I had found to be completely intractable problems concerning his knowledge of events in northern Sweden and Finland. I shall never cease to wonder at her persistence in badgering librarians and archivists, her genius in ferreting out information, the ultimate success with which she has managed to piece together the background to so many of the case-histories.

The *Nemesis Divina* provides an ideal basis for exploring many of the leading themes of the European enlightenment, – the general reaction against the facile rationalism of the later Cartesians, Spinozists and Leibnizians, the wider implications of Newtonian physics, the growing importance of the organic sciences in determining philosophical reflection. In the courses on the enlightenment given at the Erasmus University Rotterdam over the last decade or so it has proved to be immensely popular with the students, one of the results of this being the Dutch translation of the text published by Trudi de Vlaming in 1996. It is all the more surprising, therefore, that despite the way in which Levertin, Malmeström and Hagberg managed to bring out the broader significance of the work during the early years of the twentieth century, later Swedish scholarship in the field seems to have lost direction. It is to be hoped that this edition will help to give a new impetus to Swedish scholars, not only in investigating the historical and theological background to the work, but also in opening up its potential as an introduction to mainstream developments in the history of philosophy.

M.J.P.

Rotterdam
October 2000

TABLE OF CONTENTS

PART ONE

INTRODUCTION

INTRODUCTION

Linnaeus' contemporaries were well aware that the roots of his contributions to natural history lay in something more than a rigorous empiricism and an effective taxonomy. The earliest statement of his general programme of scientific research, the 1735 Leiden edition of the *System of Nature*, consists of no more than seven folio sheets, but the whole of the opening section of it is devoted to raising theoretical issues with implications far beyond the bounds of natural history. Since many of these issues are still relevant to reaching a balanced assessment of his general scientific achievement, it may be useful to begin this study of the inter-relationship between his scientific and his ethical and religious thinking by calling attention to three of the most central of them. – If we agree to follow Linnaeus in deriving the necessity of classification from the very nature of human knowledge, what status are we to ascribe to whatever it is that is being classified? How does what is mental or artificial relate to what is natural or actual? If we accept that it is reasonable to classify man with animals, as part of the natural world, how do we account for our being able to do so? How do what is mental and what is physical inter-relate within man himself? If we agree that it was reasonable for Linnaeus to take the apparent fixity of the species as evidence of Divine creation, how are we to regard this proposition now that its basic premise is no longer so plausible? How do man's assumptions about the natural world relate to his moral and religious awareness?[1]

Eleven further editions of the *System of Nature* appeared during Linnaeus' lifetime, four of which were extensively revised and updated. The opening section was extended and reformulated, the natural history sections grew into a veritable encyclopedia of the three realms of nature. His contemporaries therefore had opportunities enough for considering both aspects of his work, and the issues we have just outlined fuelled a widespread and often polemical debate. Since he was keen on publicizing his scientific investigations and furthering the acceptance of his views, we have a wealth of source material for assessing the ways in which his ideas were mulled over by the learned world of the eighteenth century.[2]

It is all the more remarkable, therefore, that in respect of the keystone to his intellectual convictions, the moral and religious matters nearest and dearest to his heart, he should have remained so uncommunicative. Since this is in

---

[1] Linnaeus, C., 1735a, observations 10, 8, 4; Hofsten, N. von, 1935; Cain, A.J., 1958.

[2] Linnaeus, C., 1735b; Larson, J.L., 1967b, 1971; Stafleu, F.A., 1971; Anderson, L., 1976; Sloan, P. R., 1976; Duris, P., 1993.

any case a field in which it is difficult to pinpoint determining influences, research into the matter has tended to lack any proper bearings. As far as we can make out, it must have been the experiences of his earliest youth, life in his father's rectory at Stenbrohult, the customs and beliefs of the country people of Småland, which first led him to distinguish between fate and Providence, between the all too evident inscrutability of fortune and misfortune, and the light thrown on the way of the world by conscience and faith. It was probably as a result of the Swedish legal system of the time, together with his reading of the Old Testament, that he first became aware of the significance of talion or retributive justice, the principle of an eye for an eye and a tooth for a tooth, of punishing an offender by making him undergo the same injuries or harms as those he has inflicted on his victim. It looks as though his mature conception of the central issue raised here developed out of his realizing the importance of collecting concrete case-history evidence for the fourth main feature of his keystone – that of Nemesis, the Greek goddess personifying the final and effective interweaving of fate and Providence, the actual meting out of Divine justice in accordance with talion. He seems to have grasped the importance of this very early on, evidently as a result of the way in which Jacob Flachsenius (1683–1733) was teaching systematic theology at the cathedral school he attended in Växjö. He may well have been encouraged to make such a collection by his patron Nils Reuterholm (1676–1756), while he was staying with him at Falun in Dalecarlia during the summer of 1734. We know that he shared an interest in such cases with the most powerful and influential of his later patrons Carl Gustav Tessin (1695–1770), leader of the Hat Party, with whom he first came into contact toward the end of 1738. Yet the earliest concrete evidence of his having actually invoked the goddess dates from the summer of 1746, when he was journeying through Västergötland. On July 24th, while crossing the heathland near Brålanda, not far from the western shore of Lake Vänern, he came upon the remains of a farmhand who had been broken on the wheel for committing incest with his step-daughter. In the published account of the journey he comments as follows:

> *Nemesis* divina, experimentally demonstrated by an unfortunate who lay broken on a wheel, by the wayside on Brålanda Heath; the locals said that he was a farmhand who had been in service with a farmer, and who had usurped his master's sole right, that the master eventually pined away and died, the farmhand marrying the widow. A rash moment had, however, made him the father of his own step-daughter's child. Woe betide any who fail to remember that God in His judgement is both patient and just.[3]

Linnaeus continued to collect similarly specific instances of this evident interweaving of fate and Providence over a period of some thirty years. His

---

[3] Wallerström, I., 1974; Ehnmark, E., 1941, 1944, 1951/52; Munktell, H., 1936; Koch, K., 1972; Hederich, B., 1770, 1701–1707; *Introduction*, III.iii.3; Uggla, A.Hj., 1961; *Nemesis Divina*, III.i.1.5; III.i.3.14; III.iii.1.10; IV.ii.2.7; Linnaeus, C., 1747, 227; *Bibliothèque*, 1747, 39: 267–281; Gillby, J., 1961.

reasons for doing so seem to have been partly purely private, and partly the desire to have to hand a range of useful examples for warning and edifying his son. In any case, none of this material was published until long after his death, and only the very closest of his friends even knew that it existed.[4]

I. The Text

i. *Manuscripts*

1. *The London manuscript*

There are two quite distinct manuscript versions of the material Linnaeus collected on the theme of Nemesis. The earlier of these, which would appear to have been written out and re-worked between about 1758 and 1765, was purchased in 1784, together with the rest of Linnaeus' papers and collections, by the English botanist James Edward Smith (1759–1828), and is now in the keeping of the Linnean Society of London. In its present state it consists of thirty folio and four folded or quarto pages: certain cross-references in the text seem to indicate that two of the original folio pages must now be missing.

These early papers cannot be regarded as of any particular importance on account of the material they contain, since this is nearly always to be found in a more satisfactory form in the later version. They do, however, provide us with a unique insight into the way in which Linnaeus attempted to organize his basic material, which he quite evidently acquired in an essentially haphazard manner.

The main body of the text – twelve of the folio pages – consists of a numbered series of some seventy-nine case-studies. There are three further short-title listings of parts of the same material, in which the attempt is evidently being made to define and inter-relate general themes. Certain of the case-studies are grouped under the headings of *Nemesis* (nos. 23–37), *Fate* (nos. 38–48) and *Sin* (nos. 73–79), sins are distinguished from vices, and there is a separate list devoted to sin and punishment. One of the folio pages contains a series of nine cases taken from a recently translated Danish book on talion, which is also mentioned in the later manuscript.[5]

Although this preliminary version provides no real evidence that Linnaeus had managed to work out any satisfactory classification of his basic material, it does provide us with a number of important leads for understanding the general principles he employed in attempting to do so. The focus of his attention, his point of departure, was quite clearly the central theme of Nemesis itself. Fate and sin were considered because they were closely related to it. It

---

[4] Malmeström, E., 1926.

[5] Walker, M., 1985; Uggla, A.Hj., 1967; Friess, F.C., 1758, 1763; Linnean Society of London, Linn. Pat. Mss.: Annotations on *Nemesis Divina*.

is almost certainly the case, therefore, that this original version of the *Nemesis Divina* has affiliations with another set of notes, dating from about the same time, and also to be found among the Smith acquisitions now in the keeping of the Linnean Society of London. In these notes, which were first deciphered and translated from Latin into Swedish some sixty years ago, Linnaeus' point of departure is the injunction that man should understand himself. He then goes on to expound in considerable detail the exact sequence of issues raised once one analyzes out the presuppositions of this general premise. Summarizing the train of his thought in very broad terms, one might say that theological issues are conceived of as having their immediate presuppositions in ethical issues, and these in their turn as having arisen out of laws, commandments and the subconscious, which have their roots in vices and passions, as well as even more basic psycho-somatic states. This is by no means an isolated or uncorroborated exposition, for the categorization here is quite clearly based on the treatment of the same subject matter outlined in the sixth edition of the *System of Nature*, published at Stockholm in 1748. In earlier versions, Linnaeus had simply appended the injunction "know thyself" to the denomination "man". Probably as a result of his preoccupation with Nemesis and related themes throughout the 1740s, he now expanded this simple coupling of concepts into a lengthy footnote, specifying in detail the precise sequence of its theological, moral, natural, physiological, dietetic and pathological implications. It is evident enough, that the attempts at systematization apparent in the original version of the *Nemesis Divina*, must have been the outcome of the same line of thinking.[6]

2. *The Uppsala manuscript*

Given this background to the London text, it is important to note that in the tenth edition of the *System of Nature*, that published at Stockholm in 1758/1759, this footnote dealing with the mental as distinct from the physical nature of man, is not only expanded and revised, but also reversed. Instead of beginning with the theological implications of self-knowledge, Linnaeus ends with them. One might say that instead of approaching the central issue analytically, he does so synthetically, progressing from what is more basic and physical in man – his physiological, dietetic and pathological characteristics – to his psychology, and only then to his political, moral and theological capabilities. The general idea of a rational sequence remains the same, but the order within it is reversed.

It is almost certainly the case, therefore, that when he began work on the original drafting of the *Nemesis Divina*, Linnaeus already had his doubts concerning the viability of starting with a religious or theological premise, and treating it analytically. It is in any case clear, that by 1758 he had definitely

---

[6] Malmeström, E., 1939, Linnaeus, C., 1735b (1748[6]): 3.

opted for expounding things synthetically, and that it was this decision which determined the whole development of his subsequent thinking on the subject.[7]

In the twelfth edition of the *System of Nature*, published at Stockholm in 1766/1768, the last he prepared for the press, Linnaeus keeps to the categorization he had worked out a decade or so earlier, and in six of the seven subsections of the note in question, adds references to his wholly private collection of material on *Nemesis Divina*. To the paragraph on physiology, he adds the observation that we should "dwell on what is our own", to that on pathology that we should "remember our mortality", to that on psychology that we should "live irreproachably, God being near". Politics is said to find its focal-point in "a man of old-fashioned virtue and trust", morality in "doing good and being glad", while theology is made to turn on "living morally under God's rule, convinced of the complete justice of His Nemesis", and "remembering our Creator". There can be little doubt, therefore, that in the form in which it was published in 1766, the note reflects the general framework Linnaeus had devised for the material he had assembled in the revised version of the *Nemesis Divina*.[8]

In the manuscript material available, there is some evidence of the way in which Linnaeus went about the preparation of this revised version. The folio pages of the London manuscript are concerned almost exclusively with the ordering of what he would have referred to as the "experimental" basis of his work, the case-histories illustrating the general themes. The four folded pages, however, which are probably of a slightly later date, provide evidence of the way in which he began to give substance to the themes by citing the Bible, and to highlight what is essential in them by means of literary quotations. In this particular case the theme was Nemesis itself, the Biblical quotations tally fairly closely with those subsequently employed in the Uppsala version, and the fine lines addressed to "the mighty Ruler" who holds sway over "the work of human hands", were used in order to highlight the matter.

In the library of the Caroline Institute in Stockholm, there are two folded folio sheets, four quarto pages, in Linnaeus' hand, dating from 1762/1765, in which once again we find him attempting to define the central theme of Nemesis by listing Biblical and literary quotations, and in which he also gives a fairly complete account of five of the case-studies subsequently included in the Uppsala version.[9]

Despite its outmoded structure, Linnaeus must have continued to add to the London version at least until the summer of 1765, since like its Uppsala counterpart it also contains an account of the mysterious happenings at Ham-

---

[7] Linnaeus, C., 1735b (1758/1759[10]) 1: 20–22.

[8] Linnaeus, C., 1735b (1766/1768[12]) I: 28–32; Appendix C; IV.iii.2.1; II.i.3.1; IV.iii.1.1; III.iii.8.1; IV.iii.3.2; IV.iv.3.1.

[9] *Nemesis Divina*, I.v; IV.iv.1.1; Karolinska Institutets Bibliotek och Informationscentral, Nemesis anteckningar, signature 44: 6: 4.

marby during the night of July 22/23. By December 1765, however, he had transcribed all the material he then had available onto loose octavo sheets, and it is these sheets, together with those he added later, which constitute the final revised version of the *Nemesis Divina*. There were eventually two hundred and three of them, and they were evidently kept in a box similar to the ones then used for hymn books. So far as we know, this was the only occasion on which Linnaeus used octavo paper for his notes, and kept them together in this way. The reasons for his having done so in this case would appear to have been partly the nature of the subject matter he was dealing with, and partly the practical problems he faced in ordering it. As we have seen, a radical reversal of his general conception of the subject matter had taken place between 1748 and 1758, and a significant development in his manner of ordering it after about 1762. There was every likelihood that the details of the rational sequence established would continue to need revision. It was therefore desirable that he should be able to insert new "experimental" material easily, and shift things around to fit in with revised perspectives. Simply listing and numbering the cases on folio sheets made it difficult to carry out effective revision. Keeping octavo sheets in a box was quite clearly a much more convenient arrangement.[10]

Although Linnaeus drew up a title-page for the work, evidently put a lot of effort into structuring it properly, and certainly gave careful thought to providing it with a suitable conclusion, there is no evidence that he ever seriously considered publishing it. As is evident from the dedication, he envisaged it as being useful mainly as a source of instruction for his son. Many of the case-histories present people in ways which contrast so sharply with the social images they were cultivating for themselves, that if what Linnaeus had written had ever become public property, he would soon have regretted it. He knew this only too well:

> I should have named no names, set none of them down,
> Had I not to convince you of the truth.
> Guard the names as your own, as you do your sight and heart.
> Trust no one in the world, all can turn enemy.
> Should families, kin and relatives learn of it,
> You would be pursued as long as you live,
> And even, perhaps, to the grave.

One should never forget, moreover, the intensely personal motivation at the very heart of the work. In the last instance, the wholly universal message it conveys is also purely and utterly private:

---

[10] *Nemesis Divina* II.iii.2; Uppsala Universitetsbibliotek: Ms. Linnaeus, *Nemesis Divina*, Cod. Ups. X. 232; Hedin, S., 1808, II: 63: "i ett foderal som liknade dem i hvilka Psalmböcker förvaras"; Uggla, A.Hj., 1967, 14.

Thanks be to Thee, great and almighty God,
For all the goodness shown me in the world.[11]

3. *Ownership*

This is not to say that no one but his son knew anything about the existence of these papers. Glimpses of them were evidently given to a few select friends. In the December of 1765, for example, Linnaeus received a visit from Johann Beckmann (1739–1811), soon to be appointed to a chair at Göttingen, and subsequently to gain international fame on account of his *History of Inventions*. Beckmann kept a journal of his travels in Sweden, and recorded the following of his visit to Linnaeus:

On a certain day in December, Court Physician Linnaeus told me that it was not only in the things of nature that he had found so many proofs of God's Providence, that for many years now, he had also been collecting instances of it from people's lives, and that the collection had now grown beyond the scope of a thick quarto, despite his having confined himself to cases of which he had no doubt. He illustrated this with a great many instances of Divine retribution, which if it had been possible I would have liked to note down in full, since I was exceedingly moved by them, and his having collected them reflected so well on him. ... He assured me that there were cases of this kind in the present Parliament. ... He is so convinced of the Divine governance of events as a result of cases such as these, that over the door of his country residence he has had inscribed the words of the poet: *Live irreproachably, God is near*.

The two case-histories Beckmann did manage to note down are those of Hauswolff, who gerrymanders himself into a position in the judiciary and dies of starvation while attempting to keep himself alive by consuming his letter of appointment, and that of Provost Boëthius, who was imprisoned for criticizing the government, and lived to see the person who had had him imprisoned incarcerated in the same dungeon.

The next recorded sighting of the manuscript took place some ten years later, when Linnaeus received a visit from his fellow Smålander Sven Hedin (1750–1821), later to become well known for his work in medicine, and at that time studying the subject at Uppsala. The visit can be dated from the fact that Hedin was chairman of the Småland Students' Union between May 1774 and December 1775, and that it was in that capacity that he visited Linnaeus:

One day, when *von Linné* had occasion to discuss certain business with me in my capacity as Chairman of the Småland Union, the conversation turned to one of our fellow countrymen who had offended in that he had acted deceitfully in a matter of the heart. Linné took down this box, and allowed me to select at random two separate case-histories. One of them was that of Kihlmark. The other was that of a man who deceived and dishonoured a young woman after having promised her marriage, and who subsequently paid for this by experiencing constant misfortune. Referring back again to the fellow countryman we had just been discussing, von Linné spoke as follows, emphasizing the plural you, as he always did when addressing anyone: "You will live longer than I shall. This person has behaved basely, and you will see him come to a miserable end". – And twenty-two years later he was in fact hauled up out of his well, having committed suicide by throwing himself into it.

---

[11] *Nemesis Divina* I.i; IV.iv.4.7.

It is rather curious that one of the case-histories Hedin selected – evidently that of Captain Sten Ribbing – should have been so directly relevant to the situation being discussed. It could be, however, that the incident should be regarded as indicating that Linnaeus had in fact sorted his material into a suitable sequence, and that the papers he allowed Hedin to select from were those containing material relevant to the case they were discussing.[12]

It was probably as a result of incidents and conversations such as those recorded by Beckmann and Hedin, that despite Linnaeus' essentially private reasons for collecting this material, the fact that he had done so became fairly common knowledge. Linnaeus died in Uppsala on January 10th 1778, and his best friend the Court Physician Abraham Bäck (1713–1795), in a memorial address delivered to the Academy of Science, made mention of his having "noted down a large number of occurrences ... under the title of Nemesis Divina". In April 1781 his son Carl travelled to London, where he stayed for some fifteen months, helping his father's pupil Daniel Solander (1736–1782) to classify the palms and lilies in Sir Joseph Banks' collections. While he was there, he was approached by a Mr. Romer with regard to the possibility of preparing an edition of his father's writings, which Romer evidently knew included the *Nemesis Divina*. Carl died in Uppsala on November 1st 1783, and James Edward Smith purchased the papers and collections not long afterwards. They were loaded onto an English brig in Stockholm on September 17th 1784, and arrived safely in England a few weeks later. While they were being sorted through, Jonas Dryander (1748–1810), another of Linnaeus' pupils then working on the Banks collections, wrote to his friend Carl Thunberg (1743–1823) in Uppsala, that so far as he could make out, the *Nemesis Divina* had not been shipped across.[13]

It was known in Uppsala that on account of the purely private nature and the scandalous contents of the *Nemesis Divina* papers, they had been removed from the material due to be shipped to England. Johan Gustav Acrel (1741–1801), member of the medical faculty at the University and executor of the estate of Linnaeus the younger, had exercised his discretion and retained them in his own hands. Their whereabouts became unknown, and they seem after a while to have been forgotten. In due course they passed into the possession of Acrel's son Olof, an army surgeon, who died at Kalmar in 1844. At his death they were acquired by a Dr. O.C. Ekman of Kalmar, who gave the sheets their present numbering, evidently keeping them in the order in which they were when they came into his possession. Ekman contacted the distinguished anatomist Professor Anders Retzius (1796–1860) of the Caroline Institute in Stockholm about his acquisition, and Retzius suggested that it would be fitting

[12] Beckmann, J., 1911, 112–113; Hedin, S., 1808, II:63; *Nemesis Divina*, III.iii.2.4: IV.ii.2.10; IV.ii.2.11; III.ii.1.2.

[13] Bäck, A., 1779; Uggla, A.Hj., 1967, 14–15; Walker, M., 1985.

to offer the papers to the University Library at Uppsala. This was done, and on March 3rd 1845 the University purchased them for a hundred rix-dollars.[14]

ii. *Editions*

Since there is no knowing for certain how Ekman's numbering relates to the original sequence of the sheets, the Uppsala text does not enable us to draw any hard and fast conclusions concerning the way in which Linnaeus might have ordered his material had he intended to publish it. The younger Linnaeus added a short observation at the end of one of the case-histories, but apart from that, he seems to have left the manuscript much as it was when it came into his possession. Considered simply by themselves, therefore, divorced from the evidence available in the London manuscript and the very basic ordering indicated in the *System of Nature*, these Uppsala papers certainly present any would-be editor with some pretty challenging problems.

It is not that there is any doubt about the broad outline of things. The first sixty-three sheets are clearly concerned with an attempt to define general categories, and as in the London papers, the main focus of interest is Nemesis itself, the defining of which involves an extensive deployment of Biblical quotations and references to classical literature (pp. 26–40). The problems arise once one considers the detailed working out of the main line of argument. In the case of Nemesis, for example, some consideration is certainly given to the essential ancillary distinction between Fate (p. 14) and Fortune (p. 63). There is, however, no evidence in the present ordering of the papers that this distinction is being brought to bear in any constructively systematic manner upon the central issue of Divine retribution.

In the final editions of the *System of Nature*, the physiological, dietetic and pathological aspects of man, the foundations of his subconscious, are concisely defined in their natural inter-relatedness. In the first main sections of these Uppsala papers, however, although there is certainly an attempt to present various related categories – rutting (p. 57), death (p. 54), haunting (p. 54), manes (p. 51), spell-binding (p. 15), fortune-telling (p. 44), portent (p. 48) – one does not get the impression that Linnaeus is working in the light of any systematically coherent approach.

The same is true of his treatment of sexual and psychological matters. In the final editions of the *System of Nature*, these aspects of human behaviour are shown as having their foundations in physiology and the subconscious, and as being the immediate foundations of man's social and political life. As presented in these papers, however, they might well seem to have no such systematic context – one is simply confronted with an apparently random listing of onanism (p. 172), incest (p. 62), adultery (p. 10), wifehood (p. 11),

---

[14] Hulth, J.M., 1921.

parenthood (p. 60), greed (p. 19), envy (p. 61), ingratitude (p. 60), malice (p. 56) and friendship (p. 20).

As has already been noticed, in the note on man in the final edition of the *System of Nature*, theology is made to turn on "living morally under God's rule, convinced of the complete justice of His Nemesis", and the final sequence of political, moral and theological issues is shown as having a direct bearing on this ultimate focal point. In the first main section of these Uppsala papers, however, although one certainly encounters categories such as diligence (p. 69), law (p. 70), poverty (p. 73), riches (p. 75), parliamentarianism (p. 91), imperialism (p. 99), naturalism (p. 120), gladness (p. 126), blessing (p. 139), there is little indication of how they are conceived of as inter-relating, or of the precise bearing they are supposed to have upon the main theme of the work.

The second main section, which runs to one hundred and forty pages, consists of Linnaeus' famous collection of case-histories. Although it is clear that these constitute the empirical or "experimental" material from which his general categories have been abstracted, he gives little indication of any very precise correlation. Some connections and inter-relationships are obvious enough, others are by no means so, and quite a number of the cases can be regarded as illustrating two or even more distinct categories.

In fact it looks as though it may have been an awareness of this problem which led Linnaeus to arrange this section alphabetically. It could be that as in the London text, he envisaged bringing out the more natural relationships by means of enumerative cross-referencing. In actual practice, however, after putting the name of the person in question at the head of each sheet, he seems simply to have ordered the material in accordance with the letters of the alphabet. This is not immediately apparent from Ekman's pagination. There is every likelihood, however, that the present order is not completely identical with that of Linnaeus, and this second section can be brought into a fairly consistent alphabetical sequence simply by dividing it into seven subsections, and re-arranging them.[15]

1. *Fries, Barr, Hagberg*

The first editor was a distinguished botanist, Elias Magnus Fries (1794–1878), professor of botany and of practical economics at Uppsala. He saw the curiosity value of the manuscript acquired by the university, and in 1848 presented a selection of extracts from it in a University publication, subsequently reprinting the material in his collected papers in 1852. Linnaeus' language was standardized and modernized. The broad distinction between categories and case-histories was indicated, but no attempt was made to throw any light upon the ways in which they might be regarded as inter-relating. A fair selection of case-histories was presented, but few scandalous or sexually explicit narrat-

[15] Uppsala Universitetsbibliotek: Ms. Linnaeus, *Nemesis Divina*; the observation by the younger Linnaeus concludes the case-history of Anders Nordenflycht, III.iv.3.13.

ives were included, and the material relating to the subconscious was simply treated as a curious extra.

The edition did arouse interest, however, and towards the end of his life Fries co-operated with his son Thore Magnus Fries (1832–1913), also professor of botany at Uppsala, in preparing a somewhat fuller and more accurate version of it, published in 1878. In this revised edition, Linnaeus' language is presented more in its original form, the quotations from classical literature by means of which he defines his basic categories are given in full, and the corresponding Biblical references are indicated.

In 1923 the journalist Knut August Barr (1871–1929) published what was essentially little more than a slightly revised version of the Fries edition, making the text more accessible to the general public by modernizing the Swedish and translating a little more of the Latin. In 1960 this popular edition was succeeded by that prepared by the Carlyle-expert and literary critic Knut Hjalmar Hagberg (1900–1975), and published in a well-known series of Swedish literary classics. Twenty years earlier, in an extended monograph on Linnaeus, Hagberg had made the important point that "the *Nemesis Divina* is to be regarded as a defective supplement to the *System of Nature*", but in his edition of the work he made no real attempt to substantiate this. Barr's text was reprinted, together with a few extra case-histories, allowed in on account of the increased broadmindedness of the age.

By the 1960s, therefore, the popular conception of the work was still much the same as it had been during the 1850s. There was a general awareness of the distinction between broad categories and concrete case-histories, but little understanding of the way in which Nemesis fitted into the wider structure of Linnaeus' systematic thinking, and no real appreciation of the ethical and religious issues being raised.[16]

2. *Reception*

This is not to say that during this period there was no worthwhile criticism of the work. Its mere existence helped to fix in people's minds the manysidedness of Linnaeus' overall accomplishment. It was no longer possible to regard him as simply a natural scientist whose avowedly artificial system of classification had been rendered obsolete by the scientific advances that had taken place since his death. The theoretical and theological issues raised in the opening section of the *System of Nature* were looked on in a new light. How rewarding was it to see Linnaeus as the complement to Newton? One of the earliest foreign reviews of the elder Fries's edition, that published by Mathieu Geffroy (1820–1895) in 1861, gave some prominence to the question, noting the similarity between the contemplation of the Deity which concludes the *Principia*, and the theological overtones of Linnaeus' treatment of the three

[16] Fries, E.M., 1848, 1852, II. no. 12: 299–344; Linné, C. von, 1878, 72 pp.; 1923, 162 pp.; 1960, 121 pp.; Hagberg, K., 1939; 1940, 263.

realms of organic nature. For Newton, it was inconceivable that "this most beautiful system of the sun, planets and comets" could have proceeded from anything but "the counsel and dominion of an intelligent and powerful Being". For Linnaeus, the empire of nature was not simply under the sway of the "Lord God Pantokrator", it also culminated in an anthropology over which Nemesis reigns supreme. For the nineteenth century, the new evidence that this was in fact the case, helped to throw a flood of light on the further implications of the famous passage which Linnaeus had first inserted in the tenth edition of the *System of Nature* (1758):

Roused as I was, I saw from the back, as He went forth, the *everlasting, all-knowing, almighty God*, and I reeled! I tracked His footsteps throughout the field of nature, and I found in each of them, even in those I could scarcely make out, an infinite wisdom and power, an unfathomable perfection! I saw there how *animals* were sustained by plants, *plants* by the soil, the *soil* by the Earth; how night and day the *Earth* revolved about the Sun, which gave it life; how the *Sun* and the *Planets*, together with the fixed *Stars*, turned as it were upon their axis, in inconceivable numbers within infinite space; and that all was sustained within this empty nothingness by the incomprehensible *First Mover*, the *Being of all Beings*, the *Cause and Steersman of all Causes*, the Lord and Master of this World. To say that He is *Fate* is not to be mistaken, for everything hangs upon His finger; nor is it wrong to say that He is *Nature*, for all things are born of Him; and it is also right if we say that He is *Providence*, for everything happens in accordance with His will. He is wholly and completely *Sense*, wholly and completely *Sight*, wholly and completely *Hearing*; and although both *Soul* and sense, He is also solely *Himself.* No human conjecture can discover His *Form*: suffice it that He is a *Divine Being*, eternal and unchanging, neither created nor begotten, an *Essence* outside of which nothing made has being, which although It has founded and built all that shimmers here before our eyes, can Itself be seen only in thought; for so sublime a Majesty invests so holy a throne, that only the soul can have access to It.

The Swedish literary world of the period responded to this newly-discovered aspect of Linnaeus' intellectual significance and explored its contemporary relevance. August Strindberg (1849–1912), for example, had a high opinion of Linnaeus, and in his correspondence and his *Blue Book* one comes across a whole series of acute observations on the philosophical significance of his scientific work. He also casts a curiously suggestive light on Nemesis, taking Linnaeus' main point as made, and going on to bring out the complementary paradox of its more purely human aspect:

Nemesis Humana. The disciple spoke: We know Nemesis divina: it is the immanent execution of divine justice which each person carries out within himself, the checking procedure which restrains evil; human justice, however, works in another way. I have already told how it came about that I once made an enemy by refusing to allow someone to rule over me, to decide my views and my company as well as my finances. I was in the right in defending my personal freedom, and he was in the wrong; but because he was unable to do me an injustice, he took his revenge on me.

Thore Fries's magisterial biography of Linnaeus, published in 1903, opened up the wider cultural setting of his writings, and before long it was not so much the scientific accomplishments as the literary and philosophical significance of his works which was attracting public attention. The uncompleted account of Linnaeus by the literary historian and critic Oscar Levertin (1862–1906), first published in 1906, is still the finest work on the subject in Swedish, and it concentrates not upon his contributions to natural science, but upon his significance as a creative writer – the poetry and lyricism of his travel-books, his story-telling, his religious convictions, and the compelling vision of his work as a moralist – the wider implications of his conception of Nemesis and talion, the ways in which his case-histories are to be interpreted. It was Levertin who first brought out the sharp contrast between the evident disorder of the Uppsala papers and the equally evident coherence of the universal message they convey:

> There is a concentric system of laws around man, and they enclose each other within invisible rings. The most inward of them is the individual's conscience, which defines the limits of his being. This is enclosed by the wider ring of the written and unwritten laws, the regulative patterns of society, the norms of general opinion, of inherited and developing doctrines, the moral consciousness of the country to which the individual belongs. Beyond that moves the widest ring of all, within which the superhuman powers have enclosed the existence of both individuals and peoples. Nemesis is reverence for this system of laws, reluctance to transgress any aspect of these boundary lines.[17]

3. *Malmeström and Fredbärj*

It was evidently the climate of opinion created by Fries and Levertin which encouraged Elis Malmeström (1895–1977) to investigate in depth the cultural background to the *Nemesis Divina* papers, and eventually to prepare the first complete edition of them. In his doctoral dissertation on Linnaeus' religious views, defended at Uppsala on May 12th 1926, Malmeström drew attention to the whole tradition of theodicy as initiated by Bayle and developed by Leibniz, and showed that its central problem, that of reconciling the postulation of an all-knowing and all-powerful God with the existence of evil, was also the prime reason for Linnaeus' having fastened upon the significance of Nemesis and talion. In a highly instructive and illuminating manner, he mapped out the ground Linnaeus shared with the physico-theologians of his day, and brought to light the common factors in the ways in which the concept of natural harmony was then being invoked in such diverse fields as ecology, economics and political theory. He analyzed the contents of the Uppsala papers in detail, attempted to establish their exact chronology, and showed how many of the themes they give expression to are also to be found in Linnaeus' other writings, notably the lectures on diet, the funeral oration on his friend Andreas Neander (1714–1765), and the correspondence.

---

[17] Geffroy, M.A., 1861; Newton, I., 1729, bk. 3, general scholium; Linnaeus, C., 1735b (1766/1768[12]) I: 10 (the text translated here is Linnaeus' Swedish version); Fries, T.M., 1903, II: 361; Malmeström, E., 1925, 16–17; 1954/1955; Strindberg, A., 1948/1976, X: 33, 219, 269, 335, 343, 345, 370, 371 (March 1894–November 1896), XII: 10 (December 1896), XV: 115, 358 (March 1905–April 1907); 1907/1912, I: 155–156; Levertin, O., 1906, 75.

In both his doctorate, and in the major study of Linnaeus' "struggle for clarity" with which he crowned his career in such studies some forty years later, Malmeström managed to make out a persuasive case for relating the preoccupation with the ethico-religious principles of Nemesis to two particularly disturbing personal crises in Linnaeus' professional career. It is certainly the case that under the heading of *Nemesis* in the Uppsala manuscript, Linnaeus does in fact insert a purely personal observation:

When I contemplated revenge everything went wrong for me. When I changed and put everything in the hands of God (1734), everything went well.

This is a reference to his much-discussed differences with Nils Rosén (1706–1773), his colleague at Uppsala. Rosén was subsequently appointed to one of the chairs of medicine at the University and Linnaeus to the other, and they were to co-operate amicably throughout the greater part of their academic careers. In 1734, however, they were still potential rivals for a permanent position. Linnaeus had been lecturing successfully without being fully qualified to do so, and in the May, faced with the necessity of giving a course on botany, Rosén asked him if he might make use of his notes. When Linnaeus hesitated, Rosén reminded him of his lack of formal qualifications, and the notes were made available. A few weeks later, however, when Linnaeus discovered that Rosén had made a copy of them, he refused any further co-operation, and was evidently so disturbed that he even contemplated murder and suicide. Before the end of the year he had left the University, and during the following months he obtained his formal qualifications at Harderwijk and began to make his mark in Amsterdam and Leiden.

It is also certainly the case that from 1748 onwards Linnaeus began to concern himself increasingly with the issues documented in the Uppsala papers, and that this too was a year of particularly aggravating upsets and disturbances. During the summer he was pulled up by the Faculty of Theology at the University for overstepping the mark in dealing with theological issues, and by the University authorities for the excessive noise and ostentation of his botanizing expeditions with the students. At about the same time he heard that the government was putting restrictions on publishing abroad, a move which he felt was aimed particularly at him. These domestic irritations happened to come together with the final disruption of his friendship with the famous Swiss botanist and anatomist Albrecht von Haller (1708–1777), and his becoming aware of the way in which his work was being satirized by the French materialist Julien de Lamettrie (1709–1751).

It is, however, difficult to follow Malmeström in concluding from this evidence, that Linnaeus' whole preoccupation with Nemesis was essentially a matter of his coming to terms with these crises, that in the last instance what we have in the Uppsala papers is the documentation of an ageing man's disillusionment with the ways of the world. Malmeström's own pioneering researches into the theodicy and physico-theology traditions of the time, into the widely-used concept of natural harmony, into Linnaeus' obsessive preoccupation with taxonomic clarity, ought to have enabled him to bring into perspective the more purely personal motivation, and give due prominence to the undoubted connections between the Nemesis papers and the *System of Nature*.[18]

The edition of the *Nemesis Divina* which Malmeström published together with Telemak Fredbärj in 1968, excellent though it is in so many respects, contributes very little indeed to any rational re-arranging of the manuscript material. It is important, since it makes the complete contents of the Uppsala papers available in a reliable form. The printed text enables one to see at a glance exactly what Linnaeus dashed down onto his octavo sheets. Most of the references are identified, and there is a useful index. Given the fact that no attempt is made to take the London papers and the *System of Nature* into consideration, the re-arrangement of the basic material has to be regarded as reasonable enough – both the exposition of the general categories and the case-histories are simply ordered alphabetically. All the Latin is given in full, just as it was noted down. It is also translated into Swedish, although not always in a wholly satisfactory manner.

Although this edition has certainly enabled the general reader to get to grips with what Linnaeus actually wrote, it has not thrown much light on the wider intellectual context of the work, nor has it brought about any clearer understanding of the more purely systematic significance of his treatment of ethics and theology. This becomes apparent if we take a look at the ways in which recent Swedish scholarship has tended to assess the work. Rather than any readiness to acknowledge the significance of Linnaeus' conception of it as the keystone to his intellectual achievement, the final bedrock and basis of his accomplishments in natural science, we find a marked tendency to psychologize and marginalize it, to regard it as radically anachronistic, wholly unworthy of one of the great luminaries of the enlightenment. Sten Hjalmar Lindroth (1914–1980), for example, both in his standard survey of Swedish learning during the eighteenth century, and in a fine essay on Linnaeus published in English soon after his death, makes much of "the aged Linnaeus, as the years slid by, brooding upon the enigma of human life", and of the *Nemesis Divina* as "the strangest of all his writings". He notes the "savage impact" made by the work, its "mixture of high and low, of simple superstition and chilling fatalism", the unique insight it provides into an otherwise unknown Linnaeus – unexpectedly rigid and inaccessible: "We glimpse an old-fashioned costume, almost a clergyman's gown, a Linnaeus who is a complete stranger to us, a preacher and a scholastic, caught up in dogmas and prejudices". Gunnar Eriksson, Lindroth's successor in the chair of the History of Ideas at Uppsala,

---

[18] Malmeström, E., 1926, 255, 1964, 390; *Nemesis Divina*, III.ii.4.1; IV.iii.3.3; IV.iv.3.3; Pehrsson, A.-L., 1965; Fries, Th.M., 1903, I: 176–185; Blunt, W., 1984, 76; Lamettrie, J.O. de, 1748/1750; Duris, P., 1993.

in an otherwise exhaustive article on Linnaeus' life and scientific accomplishments, published in the Swedish Dictionary of National Biography, sets the "bitter doctrine" of the *Nemesis Divina* on a par with Linnaeus' belief that swallows spend the winter at the bottom of lakes, and associates it with the other "speculations which became particularly prominent towards the close of his life".[19]

iii. *This Edition*

In the main, this edition may be regarded as building on that of Malmeström and Fredbärj, since it makes available in translation all the material contained in the Uppsala papers. Like its predecessor, it also includes the account of Olof Renhorn (1706–1764), Mayor of Arboga, the original version of which was on the only page of the Uppsala manuscript which is now missing. It is not known what happened to this sheet (no. 112): it evidently went astray sometime between Uppsala University's acquisition of the papers in 1845 and the publication of Elias Fries's selection in 1848, since Fries informed his readers that he had had to prepare his version of the Renhorn case from a copy. Unfortunately, this copy too has now been lost.[20]

Since the whole of the Uppsala manuscript is in Linnaeus' own hand, it has to be regarded as authoritative. He had not prepared it for the press, however, and it is clear not only from the lay-out of the pages but also from the handwriting, that when he jotted the great bulk of this material down, he was working at speed. In fact it is part of the charm and interest of the *Nemesis Divina* that it is not a carefully revised and finished text. Linnaeus certainly put a lot of effort into collecting the Biblical references and literary passages by means of which he defines his basic categories. When he came across the specific instances illustrating his general themes, however, he seems to have responded in much the same way as he did when recording his impressions of stones, plants and animals – he noted them down there and then. If we are to judge from the state of the manuscript, it must be this spontaneous transferring of thought onto paper which accounts for the vigour and vivacity of much of his writing, especially in the presentation of the case-histories.

1. *Errors*

Although the Malmeström–Fredbärj edition provides us with a reasonable version of all the material contained in the Uppsala papers, there are certain respects in which it has had to be changed and modified when preparing the basic text for translation.

---

[19] Linné, C. von, 1968, 1981; Håkanson, L., 1982/1983; Lindroth, S., 1978, 146–296, 1983; Eriksson, G., 1980/1981.

[20] *Nemesis Divina* IV.i.2.7; Linné, C. von, 1968, 220; Fries, E.M., 1848.

It is a rather curious fact that although Linnaeus' main contribution to natural science is his standardization of the way in which we name species, genera, orders, classes and kingdoms, in the *Nemesis Divina* he is irritatingly careless about the spelling of people's names. Both the manuscript and this standard Swedish edition of it contain references to people we shall never be able to trace in any biographical dictionary. We may well want to check up on the extraordinary stories he has to tell about so many of them, find out if these instances of Divine retribution can be squared with the information we have to hand as the result of ordinary historical research. But who were the "Stuerts" who once ruled over Scotland and England? Who was the "Marleborock" whose wife managed to get on the wrong side of Queen Anne? What are we to make of Linnaeus' account of the career of the "Blatckwell" who was executed in Stockholm as an English spy? It might be interesting to search out further details concerning the "Heydecoper" who nearly managed to kidnap Louis XIV, if we could only identify such a Dutch family, or look further into the activities of the "Giörts" and the "Polheim" who served as ministers under Charles XII, if we had any chance of finding them in the ordinary history books.

In some cases, the Swedish editors have done Linnaeus an injustice in this respect. It looks very much as though he did in fact refer to Soldau, the town in East Prussia, when jotting down his account of the "old irreligious linen-weaver" Georg Gallus, but this is not how the name is spelt in the Swedish edition. Gabriel Mathesius (1705–1772), professor of theology at Uppsala, married into the Uddbom family, and Linnaeus wrote "Uttbom" when referring to the fact, providing us thereby with a fine illustration of Grimm's law. In the standard Swedish edition, however, the name appears as "Ullbom".

Given the confusion that would certainly have followed from these misspellings, had they been reproduced in the translation, it has been thought advisable to standardize all the proper names, to present them in the way that enables further research to be carried out in the most convenient manner.[21]

Latin was Linnaeus' only foreign language, and he uses it fairly frequently in this text, both when quoting and when discussing particularly delicate matters. He does not use it carefully, however – the quotations are often inaccurate or even garbled – and his vocabulary, grammar and phrasing leave much to be desired. The Swedish editors have attempted to reproduce his Latin precisely as it stands in the manuscript, and to provide an intelligible translation of it. It is not very surprising, therefore, that their effort should have come in for criticism. – Hence the decision, when preparing this edition, to translate all the Latin quotations from the original texts, in so far as it has been possible to identify them, and simply to disregard blemishes in the original when

---

[21] *Nemesis Divina*, IV.i.5.12 (Stuarts), IV.i.4.1 (Marlborough), III.iv.3.18 (Blackwell), IV.i.5.13 (Huydecoper), IV.i.3.4 (Görtz, Polhem), IV.iv.3.4 (Soldau), IV.ii.2.4 (Uddbom).

translating the rest of the Latin.

In the Malmeström–Fredbärj edition, there are various instances of obvious lapses in the manuscript being either noted and reproduced or overlooked; for example, Linnaeus writes "Russians" when he should have written "Prussians", he leaves a gap for a date or a name and forgets to fill it in, he notes that he has given the wrong year but fails to correct the main text, he scribbles in an addition so carelessly that it is difficult to make out what he has written. In cases such as these, all of which are essentially trivial, the original is simply corrected in the translation.[22]

2. *Biblical quotations*

Linnaeus turns readily to Biblical quotation when he is attempting to define his general categories or indicate how a particular case-history is to be understood and interpreted. He was evidently enough of a scholar in this field to be appointed member of Gustav III's Bible Commission in 1773, although after his death his colleagues complained that: "they did not find his membership much of a help, since for the most part he spent his time plaguing them with hypotheses and paradoxes". The Bible he quotes from is that authorized by Charles XII and published in 1703. It was based, somewhat uncritically, on the earlier authorized Swedish translations of 1618 and 1541, and by the middle of the eighteenth century was universally recognized as being in need of radical revision. Like the English King James's version of 1611, however, and the Dutch Estates version of 1637, it had become such a loved and respected part of the culture which had given rise to it, its cadences and turns of phrase had so entered into the heart of the nation, that nearly everyone was reluctant to see it superseded. It seems only right and proper, therefore, that in preparing this translation of Linnaeus' text, we should make use of the corresponding King James's version in rendering his Biblical quotations.

Considering that Elis Malmeström was bishop of Växjö (1950–1962), and a Biblical scholar in his own right, it is rather surprising that the Swedish edition should be somewhat lax in this particular respect. There are several instances of its accepting uncritically the erroneous numeration Linnaeus ascribes to the verses he is quoting, and of its not identifying correctly Biblical quotations for which he fails to give chapter and verse. These lapses have been set right in the translation.

There are many instances in the *Nemesis Divina* of Linnaeus' referring to a passage in the Bible, and then not quoting but summarizing it. If it is clear that this is what he is doing, the general policy in the translation has been to

[22] Håkanson, L., 1982/1983; *Nemesis Divina*, IV.i.5.17 (Prussians), III.iii.5.1 (1755), III.ii.2.2 (Charlotta Gerner), II.ii.3 (Parliament 1738), IV.ii.2.4 (Uppsala ms. 198, 23) – reading: "som horade till pengar även medan Axberg lefwer", i.e., "who had made money by fornicating, even when Axberg was still alive".

translate his actual words rather than supply the official version of the passage being referred to.

Care has to be taken to distinguish such summarizing from simple error, however, as is evident from Linnaeus' reference to *Tobit* 4: 13 in his general definition of Pride. Tobit is advising his son Tobias to beware of whoredom, not to despise his brethren, but to take a wife of the seed of his fathers. In the King James's version he justifies his point by observing that: "in pride is destruction and much trouble and in lewdness is decay and great want: for lewdness is the mother of famine". Linnaeus, who quite clearly has in mind the 1703 Swedish translation, which includes a cross-reference to *Genesis* 3: 5 and is the exact counterpart of Luther's, quotes as follows: "Let not pride rule, since it is an initiation of punishment". He wrote carelessly, however, and in the Malmeström–Fredbärj edition the second word of the quotation is erroneously emended, transforming it from a perfectly straightforward admonition into the somewhat enigmatic statement that: "Pride will take no advice, since pride is an initiation of punishment". One might very well ask what reasons Linnaeus could possibly have had for entertaining the quite evidently perverse and paradoxical idea that pride refuses advice because it initiates punishment.[23]

3. *Lay-out*

The Uppsala papers are a collection of notes, and in editing them one has to heed whatever evidence they provide for separating or uniting the material they contain. In several instances, Linnaeus' indication of a separation or division has been overlooked by Malmeström and Fredbärj, and this has had to be corrected in the translation. There are, however, any number of more complicated cases, in which it is the subject matter being dealt with rather than the manner in which it has been noted down, which provides us with reasons for re-arranging the lay-out of the text. The Swedish edition has not been followed in retaining clearly incongruous material at the end of the case-history of Cicero, although it has been heeded in including in the main text the note by Linnaeus' son at the end of the Nordenflycht entry. It has certainly not been followed in treating the note concerning *Parliament* as a case-history – not only the placing of this sheet in the Uppsala papers, but also the wealth of material from the collection of case-histories that can be subsumed under it, indicates that the heading has to be regarded as referring to one of Linnaeus' general categories.

Linnaeus was quite evidently aware that alphabetical listing has nothing to do with natural or intellectual interconnectedness. In the footnote on Man in the final edition of the *System of Nature*, he indicates how he would have

[23] Fries, T.M., 1903, II: 256; Wijkmark, O.H.V., 1913; Malmeström, E., 1959, 1964b; Linné, C. von, 1968, 60, 74, 101 (erroneous numeration); 102, 196 (identification); *Nemesis Divina*, III.i.3.1; III.iii.5.1; IV.iii.2.1; III.ii.2.4; IV.iii.2.1 (*Tobit* 4: 13).

arranged the subject matter of the *Nemesis Divina* had he prepared the work for publication. It is therefore essential, when editing the Uppsala papers, that some attention should be paid to the taxonomic principles of the general system. We have already noticed that three of the main issues raised by these principles are already apparent in the opening section of the 1735 edition of the work – by deriving the necessity of classification from the very nature of human knowledge, Linnaeus raises the question of the relationship between what is mental and what is physical, by classifying man with animals, he forces us to ask how what is mental and what is physical inter-relate within man himself, by relating the apparent fixity of the species to the idea of a Divine creation, he requires that we should consider the relevance of our scientific views to our religious convictions. In the footnote on Man in the final edition of the work, he attempts to mark out the ground for establishing answers to these questions by indicating that our ability to classify effectively is rooted in our bodily make-up and our psychological capabilities, and that it derives much of its vigour and scope from our moral and religious convictions. He justifies this general position in a thoroughly traditional manner, by invoking the time-honoured principle of the great chain of being, within which man holds a crucially intermediate position, partly confined to the world of dumb creatures, partly capable of comprehending it:

> Theologically, man is to be understood as the final purpose of the creation; placed on the globe as the masterpiece of the works of Omnipotence, contemplating the world by virtue of sapient reason, forming conclusions by means of his senses, it is in His works that man recognizes the almighty Creator, the all-knowing, immeasurable and eternal *God*, learning to live morally under His rule, convinced of the complete justice of His Nemesis.[24]

The material collected in the Uppsala papers is therefore to be regarded as filling in the detail on this conception of the higher capabilities of man. If there is to be any proper understanding of the theological and moral framework of Linnaeus' attitude to the natural world, it is therefore essential that the subject matter of the *Nemesis Divina* should be brought into harmony with the general principles pervading the rest of his scientific work.

In the present edition, the attempt has been made to subsume the case-histories under the relevant general categories, and then to range the outcome within the teleological structuring outlined in the *System of Nature*. If any of the fragments were regarded as having more than one plausible interpretation, they have been given more than one systematic placing. The result is the first properly co-ordinated presentation of the material contained in the Uppsala

---

[24] Linné, C. von, 1968, 58, 79, 139 (division); 126, 168 (incongruous); 175 (Parliament); *Nemesis Divina* II.i.1.5; II.ii.1 (Uppsala ms. 12b, 12); II.i.3.2 (Uppsala ms. 54, 1); IV.i.5.9 (Uppsala ms. 147, 16); IV.ii.2.8 (Uppsala ms. 181, 8, title accepted); III.iii.4.1 (Cicero), III.iv.3.13 (Nordenflycht, last line), IV.i.2.1 (Parliament); Lovejoy, A.O., 1936; Broberg, G., 1975; Linnaeus, C., 1735b (1766/1768[12]) I: 32.

papers. The general reader is now able to judge for himself how the *Nemesis Divina* as a whole relates to the rest of Linnaeus' work as a systematic thinker.

Linnaeus distinguishes seven main facets to the understanding of man, which he ranges in order from the most basic and limited to the most elevated and comprehensive. If we consider these in the light of the material collected in the Uppsala papers, we find that they fall into three main groups.

At the most basic level man is to be understood in predominantly biological terms, by taking into consideration his *physiological* characteristics, his *dietary* needs, and the uncertainties of his growth and development, or what Linnaeus refers to as the *pathological* aspect of his existence. In the newly-structured *Nemesis Divina*, the material most directly related to this level reflects the nature of man as such, prior to his proper involvement with others. In sorting it out, the attempt has therefore been made to establish a progression from essentially physical matters such as birth, life and death, to the awakening spirituality of man's becoming aware of his own inner life, relating intuitively to those about him, entering into conscious rapport within a social context.

At the second level, man is to be understood in predominantly social terms – by taking into consideration what Linnaeus characterizes as his *natural* state, that is to say, the individual and collective psychology of his activities in groups or associations. In the new *Nemesis Divina*, the material relating to this level is particularly extensive. It reflects the limited preoccupations of civil society, prior to their having organized themselves into self-conscious political activity. Sorting it out has therefore involved distinguishing between the various degrees of interest in which political activity is rooted. The natural point of departure here is Linnaeus' extensive survey of the unbridled sexual activity which saps human affection and disrupts family life. This leads easily on into his analysis of the more constructive relationships between parents and children. The passions which disturb and dislocate the healthy and ordered development of social life are then brought under consideration, together with the principles of the laws by means of which they are restrained and regulated. This provides the basis for the final phase of the treatment of man as a social animal – the complexities of the individual and legal decision-making involved in the distribution of wealth, the enigmatic interplay of fortune and misfortune so characteristic of civil society.

The third level constitutes the final focal-point of both the treatment of man in the *System of Nature*, and the material collected in the *Nemesis Divina*. The central theme here is an increasing degree of clarity in distinguishing between what appears initially to be fate – the evident inscrutability of fortune and misfortune – and what is eventually seen to be Providence – God's complete command and control of every aspect of His creation. In the *System of Nature* there is a succinct but well-balanced account of this crucial aspect of Linnaeus' theodicy, in which he makes the point that although the enlightened investigator may be aware of the theological implications of the laws regulating civil society, its members are more likely to see them as being restrictive and coercive. Man as a *political* animal tends to live in accordance with custom rather than reason, and if he organizes himself into churches, to persecute the truly religious for entertaining speculative opinions different from his own. Man is also capable of acting *morally*, however, of ridding himself of the wantonness, ambition, malice, greed learnt at lower levels, and living irreproachably – conducting himself chastely, unassumingly, graciously, contentedly. And it is this moral potential, and this alone, which makes man wholly acceptable in the sight of God – which opens up the possibility of dispelling the illusions of a blind fate and a capricious fortune, and entering into conscious accord with Providence. It is evident from numerous observations scattered throughout Linnaeus' writings, that he saw his scientific work as an integral part of this moral development, and of the *religious* awareness in which it finds its consummation. In 1748, for example, he concluded one of his *Academic Delights* by reminding his readers that:

> The contemplation of Nature is a foretaste of the bliss of Heaven, the soul finding it a constant source of joy, the initiation of complete solace, the culmination of human felicity; contemplating nature is like awakening from a deep sleep and wandering out into the light – the soul is forgetful of itself, and may be said to be living in a heavenly world, or an earthly heaven.[25]

Although the treatment of these matters in the *Nemesis Divina* is different on account of its involving the amassing of such a wealth of illustrative case-histories, it is exactly parallel in respect of its categorial structure. Most of the case-histories illustrating the manners and customs of political life are drawn from the nation states of Linnaeus' own time, and especially from the goings-on in the Swedish Parliament and civil service. The kings and emperors who people his gallery, ruling though they were "by the Grace of God", can hardly be regarded as more ethically acceptable than the politicians who milled about them. The theologians and churchmen he pictures for us, entrusted though they were with their pupils and their flocks, are scarcely distinguishable from their political counterparts in moral stature. It is most noticeable, moreover, that the two final levels of the *Nemesis Divina* are almost entirely admonitory in tone. Linnaeus calls attention to the absolute importance of living irreproachably, the dangers of pride, the glory of doing good gladly, the way in which fate, fortune and retribution fall into place within the all-pervading blessedness of the final judgement of God – but as is perhaps to be expected, he has scarcely a single case-history lending positive support to this crowning vision.

---

[25] *System of Nature*, 1735b (1766/1768, 12) I: 28–32: physiology, diet, pathology; state of nature; politics, morality, theology; *Nemesis Divina* (revised): man – life, self, going out, you too; society – affection, children, rights, riches; God – nations, churches, conscience, Judgement; Linnaeus, C., 1749/1769 I: 563; Ehnmark, E., 1951/1952 85; Linnaeus, C., 1748 §7; 1759a; 1905/1913 II: 263, Wikman, K.R.V., 1964b.

II. THEODICY

i. *Divine Economy*

It could be argued that Linnaeus' conception of the relationship between morality and religion is independent of any historical associations. It is evident from the way in which he mines the Old Testament for quotations, that he is interested in it mainly as a guide to ethical action, not as the history of a people with a Divine mission. By and large, if he turns to the New Testament, it is, once again, in order to find support for moral precepts, not in order to consider the historical context of Divine Revelation or to trace the way in which this revelation was universalized through the teaching and organization of the early church. Bringing the subject matter of the *Nemesis Divina* into a rational order involves paying close attention to the purely formal aspects of Linnaeus' thinking in the *System of Nature*. It therefore tends to confirm the view that in matters of morality and religion, as in so many other aspects of his scholarly and scientific work, he was an essentially ahistorical thinker.

This is, however, a view which needs to be carefully qualified. Modern assessments of the significance of his taxonomic work have often fallen short of the mark by overlooking the fact that from the very beginning of his career, he was well aware of the importance of distinguishing between artificial and natural systems of classification. He certainly realized that although a formal or artificial taxonomy is a matter of practical necessity to the working scientist, it is only a means to an end – that those making use of it should never regard it as anything but a working hypothesis. The data collected by making use of a formal taxonomic concept have to be submitted to constant analysis, in order that it may approximate ever more closely to its natural equivalent. All scientific classification has, therefore, to be regarded as a developing or historical process, and Linnaeus is perfectly prepared to acknowledge this. What is more, his more purely taxonomic work within any field of enquiry, regardless of how self-contained it is, is always set within the wider context of the living wholeness of the natural world.

To a certain extent, this living wholeness is reflected in the analytic and synthetic procedures which force the enquiring mind to conceive of it as a series of asymmetrical relationships, a hierarchy of relative complexities. Newton, through the research leading up to the publication of the *Principia*, had been led to distinguish the relatively abstract mathematics of motion as such, from the motion of bodies in resisting mediums, and the actual movements of planets, satellites and pendulums within the solar system. A similar pattern had emerged in his *Opticks*, in which the basic problems presented by the nature of light and colour as such, had led on naturally into a consideration of interference phenomena, the permanent colours of natural bodies, and what was then the enigma of diffraction. Linnaeus, in an exactly analogous manner, was led by the enquiries leading up to the publication of the *System of Nature*, to distinguish between the three great realms of the organic world, to note that: "*animals* are sustained by plants, *plants* by the soil, the *soil* by the Earth". What is more, when he considered the elements basic to all physical being – the earth, water, air and fire informing the sterility and fecundity, the generative power and animation of all living things – he came across the same hierarchical order. It is hardly surprising, therefore, that he should have felt justified in allowing the age-old concept of the great chain of being, which on account of his theocentric Platonic inheritance he tended to interpret in terms of levels of perfection, to merge with the then fashionable idea of an all-embracing "economy" of nature. What we are classifying is derived from a world of living processes:

Everything the Almighty Creator has instituted on our globe occurs in such a wonderful order, that no one thing subsists without the support of something else: the *Globe* itself, with its Stones, Ore and Gravel, is nourished and sustained by the Elements: *Plants*, Trees, Herbs, Grass and Mosses, grow out of the globe, and *Animals* eventually grow out of the plants. All of these are finally transformed back into their primary substances, the *Earth* feeding the Plant, the *Plant* the Worm, the *Worm* the Bird and often the *Bird* the Beast of prey; then finally the *Beast of prey* is consumed by the Bird of prey, the *Bird of prey* by the Worm, the *Worm* by the Herb, the *Herb* by the Earth: *Man*, indeed, who turns everything to his needs, is often consumed by the Beast, the Bird or the Fish which preys on him, by the Worm or the earth. It is thus that everything circulates.

In his *Economy of Nature* (1749) and *Polity of Nature* (1760), Linnaeus invokes the same principle with reference to what we should now call general ecological issues, showing that behind the violent competition and apparent confusion of the universal struggle for existence, the merciless, repulsive and seemingly meaningless war of all against all, the circumspect scientist can catch glimpses of the essential prerequisites of a well-ordered creation, the finely-tuned strategy of Divine Wisdom, which maintains "an exact equilibrium, nothing being redundant or useless". Consequently, although Linnaeus' scientific method and his conception of the relation between the Creator and the natural world can hardly be said to be historicist, they do involve thinking in terms of development and living processes, and they cannot, therefore, be regarded as entirely ahistorical.[26]

It is one thing to regard the natural processes of the mechanical, physical and organic world as an equilibrium sustained by Divine Wisdom. It is another thing to take the further step of thinking analogically about the world of human actions and decision-making, since this immediately raises the thorny issues of the freedom of the will, and the nature of the man-God relationship. It is important to note, therefore, that Linnaeus, like so many of his contemporaries,

[26] Malmeström, E., 1959; Larson, J.L., 1967a; Linné, C. von, 1792; Malmeström, E., 1925, 23–33; Linnaeus, C., 1761^{2}, introduction, "gradus perfectionis"; Lovejoy, A.O., 1936; Bremekamp, C.E.B., 1953; Anderson, L., 1976; Cain, A.J., 1992; Ereshefsky, M., 1994; Linnaeus, C., 1739, 3; 1749; 1760, §§4, 35; Hofsten, N. von, 1958.

did just this, and that it is in the *Nemesis Divina* that he attempts to work out the ultimate implications of the move.

1. *Löwenhielm*

Not long after the publication of Linnaeus' *Economy of Nature*, the lawyer, administrator and politician Carl Gustav Löwenhielm (1701–1768), spokesman for economic affairs for the Hat Party, addressed the Academy of Science on the subject of estate management. In his lecture he put forward what was then regarded as the standard solution for the difficulties involved in transferring a concept of law from one field to another. He distinguished three complementary levels of economic activity:

> First there is the great natural Economy or *Oeconomia Divina*, instituted between all animals, plants and minerals, in accordance with which one thing so serves and sustains not only the other but also the continued existence of each family of things, that nothing is lost and there is a general increase. Secondly there is the general economy or *oeconomia publica*, in accordance with which everything so falls under governances, that each limb and each greater or smaller community is able to subsist without there being any mutual damage, any deficiency in what is necessary, any compromising of convenience or satisfaction. – Thirdly there is the particular economy or *oeconomia privata*, in accordance with which each household or each member in the general functioning of the community, enjoys the gifts nature has provided for the sustenance of man, putting each particular thing to its proper use, heightening the effectiveness of whatever is found to be necessary, useful or enhancing.

It is worth pausing to consider Löwenhielm's argumentation, since it is typical of the age, and differs only marginally from that of Linnaeus himself. One of the most basic of his tenets is that man is free to accept or reject the fundamental premise of an all-wise Creator's having instituted the globe and endowed man with abilities, to acknowledge or not to acknowledge that there is a man-God relationship. In his view, the exercise of this basic freedom has direct practical consequences, since it determines man's ability to think comprehensively and accurately, and so formulate a realistic assessment of his material resources. It follows, therefore, that the extent to which man falls short of giving due acknowledgement to this fundamental premise is the exact measure of his being able to make proper use of his natural environment. Looking at the matter in a more modern light, one might say that Löwenhielm is concerned with the broad cultural and material factors involved in any informed view of scientific and ecological issues, that is, with the functioning of any truly healthy economy.

It is certainly worth noting, therefore, that in the course of developing his central theme, Löwenhielm is taking up issues closely interwoven with the psychological, ethical and religious aspects of the man-God relationship that so concerned Linnaeus – that is, with the central themes of the *Nemesis Divina*. There is, moreover, ample evidence scattered throughout Linnaeus' writings that he was by no means indifferent to Löwenhielm's particular preoccupations, that he too had given careful thought to the immediate social implications of the great wealth of case-studies he had collected concerning the distribution of riches. On December 11th 1759, for example, in his capacity as vice-chancellor of the University of Uppsala, he addressed the, "patricians and inhabitants of every degree, in both the Academy and the City", on the occasion of the appointment of Johan Låstbom (1732–1802) to the newly-instituted Erik Borgström chair of practical economics with special reference to agriculture, the first of its kind in the country. In this address, theology and economics are certainly not conceived of as unrelated disciplines:

> An almighty God having instituted this Earth, . . . everything under the Sun included in the three realms of nature, and nothing more than that, has been provided to meet man's requirements and serve his convenience. And since it is only through efficiency in putting each and every thing to its proper use that man can derive support and enjoyment from these resources, the various disciplines of what is called economics have emerged among mankind as a matter of necessity, all of us being involved first and foremost with satisfying the inner drive to provide for our own maintenance.[27]

2. *Object and subject*

In considering the *Nemesis Divina* as a theodicy, it is important to take note of the way in which freedom and necessity inter-relate within this general manner of thinking. Both Löwenhielm and Linnaeus regard man as free – not only to accept or reject the fundamental premise of an all-wise Creator's having instituted the natural world and endowed man with abilities, but also to act or not to act in ways which help to sustain, further or enhance life. They also maintain, moreover, that any fully effective exercise of the second kind of freedom depends upon the decision reached in respect of the basic premise, upon recognizing that decisions can only be made within the necessity of the God-given order. The effectiveness of this secondary or relative freedom is therefore determined by the systematic context of the field in which it is being exercised. God has instituted the natural world, but has provided man with no more than the potentiality for making proper use of it. The University of Uppsala has appointed Johan Låstbom to his chair, but it cannot guarantee that he will bring about the required improvements in Swedish agriculture. Although it can provide him with his chance, it can neither change the ways in which agriculture relates to resources and economics, nor endow him with abilities he does not possess.

Man's awareness of the systematic context within which his freedom has to be exercised, derives from the analytic and synthetic procedures intrinsic to all fields of enquiry. Since any systematic context can be brought to light by either of these two procedures, evaluating the taxonomic principles being employed involves taking into consideration their overall circularity. This can be illustrated by bringing out the complementarity of the very different ways in

---

[27] Löwenhielm, C.G., 1751, 9; Frängsmyr, T., 1971/1972; Taylor, H., 1760; Linnaeus, C., 1759b; Lidén, J.H., 1778, 336–338.

which Löwenhielm and Linnaeus regarded the internal structure of economics. For Löwenhielm, as we have seen, it was the great economy of the world of nature which was truly divine, the more limited economies of the community and the individual being derivative. For Linnaeus, on the other hand, although nature certainly had the mark of Divinity upon it, the clearest and most immediate evidence of Divinity was to be found in the individual's awareness of Nemesis.

ii. *Leibniz*

Since the suggestion has often been made that Linnaeus' physico-theology must be related in one way or another to eighteenth-century Leibnizianism, it may be of value to pay some attention to the thesis. It is certainly the case that in their approaches to reconciling the omnipotence and goodness of God with the existence of evil, they shared much common ground. Both saw the natural sciences as the allies rather than the enemies of morality and religion. Both were convinced that traditional theology needed to be modernized rather than abolished. Both regarded the justifying of God as a genuine intellectual task, capable of being carried out by means of evidence and reason, and not simply as something which ought to be consigned to the realm of faith or enshrined in ecclesiastical dogma. It can hardly be denied that when Leibniz distinguishes the absolute necessity of geometrical and physical laws from the moral necessity of ethical choice and the "equilibrium of sovereign indifference", he is indeed opening up perspectives on the world of Linnaeus. It is certainly the case that his corresponding distinctions between metaphysical, physical and moral evil throw a lot of light on the broader significance of the *Nemesis Divina* papers. There is no evidence, however, that Linnaeus had any direct knowledge of Leibniz' work and in several important respects their basic conceptions of theodicy are very different.[28]

1. *Metaphysical evil*
Unlike Linnaeus, Leibniz was a philosophically-minded mathematician, fully aware of the metaphysical implications of the fact that all particularized scientific research has its wholly abstract and universal presuppositions. Linnaeus saw clearly enough "that all was sustained within this empty nothingness by the incomprehensible first *Mover*", but it would never have crossed his mind to associate this insight with Leibniz' definition of metaphysical evil as "simple imperfection". Since he attached no more importance to the ontological status of mathematics than Newton did, he remained very largely indifferent to that aspect of the age-old Platonic tradition which took the discipline to be the

[28] Hammarsköld, L., 1821, 207; Malmeström, E., 1926, xii, 21–27; Frängsmyr, T., 1972, 155–156; Linné, C. von, 1981, 333; Lepenies, W., 1982, 17–18; Leibniz, G.W. von, 1710 §§349, 21, 1875/1879, VI: 321, 115.

only true science it had pleased God to confer on mankind. He therefore differed radically from Descartes, Spinoza and Leibniz, in not attaching any real significance to the closely-related ontological proof of God's existence – the wholly abstract and universal coupling of thought and being implicit in the assertion that "God is that than which nothing greater can be conceived". Had Linnaeus been required to respond to the distinction between the truths of reason and the truths of fact, to the principles of identity and sufficient reason, he would probably have taken them to be self-evident to the point of tautology. What particular advantages did one have in the furthering of the sciences, simply because one had armed oneself with the conviction that a predicate must be explicitly identical with its subject, that there must be some reason why everything is as it is, and not otherwise? One can imagine him asking what exactly Leibniz was doing, when he claimed to be resolving the apparent disparity of matter and mind into the general concept of God's pre-established harmony. What sort of explanation did he think he was giving of "the marvellous formation of animals", when he related it back to such a concept? Given such radical differences between the basic approaches of the two thinkers, it is hardly surprising that we should find very little indeed in the *Nemesis Divina* corresponding to the Leibnizian conception of simple privation, the "origin of evil in the region of eternal truths".[29]

2. *Physical evil*

The same is true of Leibniz' corresponding concept of physical evil. Although he took the essence of this to be suffering, he was mainly concerned not with exploring the ways in which it affected living beings, but with relating it back to its origins in metaphysical evil. Wisely enough, perhaps, instead of attempting to provide his readers with specific instances of justified suffering, he contented himself with supporting his thesis by means of an abstract analogy – just as in mathematics apparent irrationalities are finally found to resolve themselves, so in the physical world anomalies and suffering are eventually to be seen as necessary parts of a rational whole.

While admitting that animals do indeed suffer, Leibniz maintained that their pain is less intense than that of man. Wolff reiterated this doctrine in his *Natural Theology* (1739/1741), and his disciples at Uppsala, notably the prominent theologian Nils Wallerius (1706–1764), went out of their way to emphasize that it was only right and proper that physics should be acknowledged as based in metaphysics, and not simply subservient to subjective ends. Such a view came in for heavy criticism from many quarters after the dis-

[29] Leibniz, G.W. von, 1710, §21; Garrison, J.W., 1987; Hobbes, T., 1651, I.iv; Anselm, 1965 II–IV; Descartes, R., 1641, III, V; Spinoza, B. de, 1677a, I prop. xi; Leibniz, G.W. von, 1875/1890, IV: 358, V: 418, VI: 614, VII: 261; 1710, Disc. prélim. §2, 1 §§7, 20, préface 41; see the critique of Leibnizianism, by Linnaeus' pupil Pehr Forsskål (1730–1763), *Dubia de principiis philosophiae recentioris* (Göttingen, 1756), III.ii.2.4.

astrous Lisbon earthquake of November 1st 1755. It is, moreover, completely alien to Linnaeus' whole manner of thinking in the *Nemesis Divina*, the guiding principle of which is that all intelligible instances of physical suffering, including those associated with the Lisbon earthquake, are to be traced back in one way or another to a moral fault.[30]

3. *Moral evil*

It is, indeed, in their treatment of moral evil that the theodicies of Leibniz and Linnaeus differ most widely in tone, even if they do share common ground in respect of basic principles. If God himself is not to be regarded as directly responsible for individual moral aberrations, some status has to be ascribed to the freedom of the individual. Leibniz tackles the problem by referring once again to the basic metaphysical concept of simple privation – individual actions, though adequately determined by their God-given context, will always involve some element of imperfection. Man sins, exercizes free choice in the creation of moral evil, on account of his capacity for willing in the light of this imperfection, acting in accordance with his own limited insights.

For Linnaeus, the freedom of the individual is far less problematic. No one would want to deny that the way in which man acts is determined by a whole series of factors. It is important to note, however, that these factors are freely available for analysis, and that the individual is also capable of deciding freely and self-consciously, especially in respect of moral and religious matters:

> A man can hang himself, drown himself, cut his own throat, and can also choose not to do so. But if, for any reason, he is condemned to death by the supreme judge, he no longer has any choice in the matter; his execution is an unavoidable necessity. A man is therefore at liberty to commit or not to commit a crime, but not to avoid the issue once he has committed it and been condemned.[31]

iii. *Numen*

1. *Divine justice*

Leibniz is equivocal on the crucial issue of fate. He appreciates the universality of the concept, the importance to his central endeavour of its being recognized by Stoics, Turks and Christians alike, the effectiveness with which it draws together the various strands of his speculation. He sees it as having enabled him to demonstrate that abuse brings about its own punishment, that habitual drunkenness necessarily results in degeneration, for example, that it is through the natural course of things that good is rewarded and evil punished. In that

---

[30] Leibniz, G.W. von, 1710, §§21, 241, 10, 242, 250; Wolff, C., 1739/1741, I: 373, 549–563; Wallerius, N., 1752; Crocker, L.G., 1959; Barber, W.H., 1955; Wade, I.O., 1958, 42–48; *Nemesis Divina*, III.iii.5.1; IV.ii.1.2.

[31] Leibniz, G.W. von, 1710, §§287–328, 21, 121. 151: *Nemesis Divina*, IV.iv.1.3; Geyer, C.-F., 1992; Streminger, G., 1992; Neuhaus, G., 1993.

he identifies fate with the natural course of things, however, it is difficult to see on what grounds he can criticize Spinoza for having conceived of a nature "devoid of choice, goodness and understanding", and working by means of "a blind and wholly geometrical understanding". How does what he is doing in the *Theodicy* differ from what Spinoza is doing when he maintains that law is either a matter of natural necessity or of human will – what Hobbes is doing when he derives his "immutable and eternal laws of nature" from the individual's basic right of self-preservation?

In this particular respect, the case-histories collected in the *Nemesis Divina* establish a standpoint diametrically opposed to that of Leibniz. It is precisely the unnaturalness of what happens which guarantees its importance as evidence, and what is to be learnt from the evidence is that natural reason provides us with very little insight indeed into the ways of Divine Justice. One Christmas, a member of the parish in which Linnaeus was born refused to use his hands to save his wife from drowning. It can hardly be maintained that it was simply through the natural course of things that five years later the same hands so rotted away that the man died. Many of the Swedish nobility were reduced to homelessness through the policies of Charles XI. What, then, were the immutable and eternal laws of nature in accordance with which Stockholm castle burst into flames while the king's body was lying in state? By what natural necessity does it come about that when Nemesis strikes those who seduce another's wife, "she often dies of some disease of the uterus, while the most frequent end for him is drowning". Artedi may have so detested the Dutch, that he had no desire to go to heaven if there were any of them there, but by what act of human will did it come about that he should have been drowned in Amsterdam?[32]

2. *Authority*

The best way of getting to the heart of the difference between the two theodicies, is by comparing the two conceptions of what is being justified, the two views of the Divine Being.

The whole point of Leibniz' work was to show that this could not be regarded as a meaningless operation, that it is possible to discourse rationally concerning the nature of God. Unless good grounds can be found for questioning the validity of the ontological argument, any enlightened thinker is obliged to take omnipresence and omnipotence into consideration, and the ascertainable order and harmony of the world necessarily bring wisdom and goodness to mind. In fact despite Leibniz' preoccupation with the total abstraction of universals – the logical coherence of mathematics, the principles of identity and sufficient reason, the general concept of the pre-established harmony of matter and mind – it was what he saw as the Creation which determined his

[32] Leibniz, G.W. von, 1710, préface 8, 43–44, §§112, 74; Hobbes, T., 1651, I, xiv; Spinoza, B. de, 1670, IV, 1677b, II, 3, 4; *Nemesis Divina*, III.ii.1.5; III.iv.3.5; III.i.3.1; IV.iii.2.8.

conception of the Deity, and not vice versa. He therefore propagated a sort of deistic naturalism, maintaining that the existing order is the best of all possible worlds, that although God is indeed the true reason of all things, He also wills the natural order because it is in itself good. It follows of necessity, therefore, that God is to be regarded as "permitting" or "going along with" evil, that although He wills what is good, He is obliged to acquiesce in the best possible.[33]

Although to a certain extent Linnaeus is in agreement with Leibniz concerning the possibility of discoursing rationally on the nature of God, their basic conceptions are in fact very different. This becomes evident once we take a look at the language Linnaeus employs in this respect. Leibniz had extracted his metaphysics and his conception of the attributes of God from his response to what he saw as the Creation, the world of nature. Linnaeus quite evidently viewed the matter from the other end of the telescope. In his conception, as in Newton's, it is God Himself who has sovereign authority over all created things, and the words which come most frequently to his mind when he wants to express this are, - King, Ruler, Master, Steersman, Manager. Since authority implies power, the natural world may be said to have its Creator, Author, Founder, Originator, Begetter. Power involves wisdom, so that the order may be said to be the work of the Designer, Framer, Architect, Mechanic, Maintainer of everything. In respect of human actions, Linnaeus speaks of the authority being exercised by the Judge, Magistrate, Arbiter, Custodian, Guardian of mankind. In this respect, therefore, the whole of his vocabulary is based on the idea of an entity distinct from nature, and essentially remote from human comprehension.[34]

There is, of course, a converse side to this, and the best way of grasping the significance of it is by considering carefully the quotation from Ovid which he placed on the title-page of the *Nemesis Divina – Innocue vivito, Numen adest*, live irreproachably, God is near. God, conceived of as *Numen*, as He often is in the everyday language of Linnaeus, is the king, creator, designer and judge who is *nodding* his assent in what is happening. It is on His authority, in accordance with His sovereign will, that the sparrow falls to the ground. While ever present in every moment of our lives, and closely resembling us in that He "sees, hears, knows all", He also keeps His own counsel, judging, rewarding and punishing in the light of it. Linnaeus introduces his own version of the ten commandments by exhorting us to be, "persuaded by nature and experience" of His presence, and to acknowledge Him as the ultimate source of human law. In attempting to deal comprehensively with the nature of God's judgement, he piles up Biblical quotations, nearly all of them from the Old Testament, and many from the Apocrypha, in support of the Psalmist's sweeping rhetoric: "He that planted the ear, shall he not hear? He that formed the eye, shall he not

---

33 Leibniz, G.W. von, 1710, §§60, 7, 21, 23, 107.

34 Malmeström, E., 1926, 247–250, 1964a, 323–342.

see?" As Linnaeus illustrates by several of his most striking case-studies, it is God in this sense who supplements the rewards and punishments imposed by human authority within established legal systems, bringing them into accord with the secrets of the heart and the dictates of conscience.[35]

When dealing with Linnaeus' conception of the economy of nature, we noticed that unlike his contemporary Löwenhielm, he did not speak of it as being in itself divine. This final focal-point of his theodicy makes it evident why this was so. As he saw it, man's most immediate awareness of God derives not from contemplating nature, but from the constant demands being made upon him by this numinous awareness of Nemesis.

III. THE BACKGROUND

By bringing the Uppsala papers into line with the general principles of the *System of Nature*, we have been able to throw new light on the ways in which Linnaeus' religious views relate to his scientific work. Comparing the *Nemesis Divina* with Leibniz' *Theodicy* has enabled us to bring out the extent to which the particular manner in which he intermingles scientific interests and religious convictions reflects the mainstream preoccupations of eighteenth-century physico-theology. It remains for us, therefore, to look more closely at his personal career, as well as the more distinctly Swedish elements in his cultural background, since these too had a decisive influence on the collecting of this material.

i. *Sweden and Denmark*

1. *Politics*

The political situation in the Sweden of the time provides us with a convenient starting-point. Linnaeus was born just over two years before the battle of Poltava. As a boy of twelve or thirteen he had heard of the humiliating negotiations associated with the country's final capitulation at the end of the Great Northern War. Throughout the years of his maturity he had witnessed the sorry spectacle of a marginalized monarchy and a parliament reduced to ineffectiveness by party wrangling – in 1741 the declaration of the ill-advised war of revenge against Russia, in 1756 the remorseless crushing by the Estates of an attempt by the monarchy to regain a modicum of its basic prerogatives, in 1757 the reckless intervention in the Seven Years' War. It was not until the closing years of his life that the country's fortunes began to take a turn for the better, and then only as a result of the revolution of 1772, when the new

[35] Ovid, 1985, I: 640; Widengren, G., 1953^2, 37–53; *Matthew* 10: 29; *Nemesis Divina*, III.iii.8.1; IV.iv.1.4; *Psalm* 94: 9; *Nemesis Divina*, II.ii.3; II.iv.1.6; III.ii.3.2; III.iii.5.3; IV.iv.1.7; Cicero, 1970, II, viii; Hooker, R., 1594/1662, I, ix, 2.

king Gustavus III, aided by the populace and the military, outmanoeuvred the Estates and imposed a more effective constitution at bayonet-point.

All this contrasted sharply with the course of national events during the two generations preceding his birth. The country had emerged from the Thirty Years' War as the champion of European Protestantism and a major power in her own right. Under Charles X her natural boundaries to the south had been established at the expense of Denmark, and her control of the continental coast of the Baltic at the expense of the German Empire and Russia. Under Charles XI the monarchy had enlisted the support of the peasantry in reducing the power of the aristocracy. Under the new constitution the king was sovereign lord, responsible for his actions to God alone, bound only by the law and the statutes, requiring no intermediary between himself and his people. Charles XII had inherited a sound economy, a well-administered empire, and the finest fighting troops in Europe. When on the strength of his military successes he forced the German Emperor to grant religious toleration in Silesia, he was hailed throughout Protestant Europe as the agent of Providence.[36]

There can be no doubt that although Linnaeus in the *Nemesis Divina* makes little direct reference to this dramatic switch in his country's fortunes, the stark contrast between the imperial grandeur of the period preceding Poltava and the shameful shambles of the social and political world he saw about him, it was in fact a reversal which had a profound effect upon him. Dr. Johnson (1749) saw in the downfall of Charles XII yet another instance of the vanity of human wishes:

> On what foundation stands the warrior's pride,
> How just his hopes, let Swedish Charles decide,

and the basic inspiration for Linnaeus' collection of case-histories was almost certainly very similar.

Linnaeus came of Småland peasant stock. His father had studied theology at the University of Lund, and returned to his home area to marry the daughter of a local clergyman. Two years after Linnaeus was born he was appointed Rector of Stenbrohult, a village on Lake Möckeln some thirty miles south-west of Växjö. Growing up in a country rectory during those years, Linnaeus certainly realized at a very early age that the clergy had played no small part in the initial successes of the king's military campaigning. When the army was in the field, it was customary for one of the regimental chaplains to hold a service in camp every morning and evening. The doctrine expounded was clear and to the point – nothing was expected of the men than that they should fear God and honour the king. And was it not the case that when the troops were being landed on Sjælland and the stormy sea threatened to make the operation hazardous, the waves had been quelled by a glance from the monarch? At Narva

[36] Roberts, M., 1973, 1979, 1986; McTurk, R.W., 1974; Sandblad, H., 1942; Lundström, H., 1902; Ruuth, M., 1914.

God had sent a heavy fall of sleet to conceal the assault at precisely the right moment. The whole of the daring crossing of the Dvina had been favoured by providential good fortune. At Fraustadt snow drove into the eyes of the enemy, and ceased as soon as the Swedish battalions had made their breakthrough. It had been long predicted that the king would not be defeated until he had taken Rome. When he captured the Russian stronghold of Romni at the end of 1708, the similarity between the two names gave rise to a rumour in the army that the long-standing prediction was about to be fulfilled. According to the peculiar Swedish calendar of the time, the opening moves of the battle of Poltava were made on the second Sunday after Trinity. The cavalry were ordered to mount at midnight, and although everything went smoothly as the squadrons began to move to their allotted positions, some men noted with surprise that no prayers were held, something which had never before been omitted.[37]

It is a marked characteristic of the *Nemesis Divina* that although it contains so much material drawn from the military and political events of the time, it offers little in the way of analyzing or explaining them. An ordinary historian might attempt to account for Sweden's ascendance and decline by taking basic geographical, economic and social factors into consideration, by approaching the success or failure of the military campaigns in terms of alliances, logistics and tactics. Linnaeus simply notes, under the heading of *Retribution*, that: "Many always do badly after a certain day. Everything then began to go wrong. Charles XII's first nine years went well, then nine years of misfortune". It is interesting to note, moreover, that although in most cases he manages to trace back the individual's subsequent misfortune to an initial moral fault, in this particular instance he does no more than record that the exact date of the king's death had been accurately predicted by the man who was reputed by some to have assassinated him.[38]

If we are to judge from the case-histories included in the *Nemesis Divina*, it looks as though it may well have been the Russian War of 1741–1743 which first encouraged Linnaeus to set about extending his collection. On the international scene, this conflict might be accounted for by taking into consideration Russia's support for Maria Theresa, and France's attempt to counteract this by sending an army into Bohemia and encouraging Sweden to think of regaining the Baltic provinces. Internally, the decision to attack Russia was made possible by the demise of Count Horn's Cap administration, and the emergence of the bellicose Hat Party under the leadership of Linnaeus' patron Carl Gustav Tessin. Linnaeus therefore had good reasons for analyzing the course of events either in terms of party conflict, or with reference to the rights and wrongs of international obligations and alliances.

---

[37] Johnson, S., 1749, lines 191–192; Braw, C., 1993; *Krijgz Articlar*, 1621; Hildebrand, K.G., 1942; Normann, C.-E., 1948; Englund, P., 1992, 25, 64, 77.

[38] *Nemesis Divina*, III.ii.4.1; II.iv.3.5.

It is interesting to note, therefore, that he chose instead to concentrate exclusively, and in a curiously impartial manner, upon the fate of various individuals, to show how certain specific moral faults were eventually counterbalanced by equally specific instances of retribution. Malcom Sinclair, for instance, the Swedish national hero of the time, whose murder by Russian agents helped to create the atmosphere which made war possible, had brought his end upon himself by knifing a fellow officer when a prisoner-of-war, and observing on a certain occasion that if there were any Russians in heaven, he had no desire to go there. Burkhard Münnich, the Russian minister who had commissioned Sinclair's murder, also ordered the mistreatment of a political rival, and was eventually mistreated in precisely the same manner by the Empress. Andreas Ostermann, another German who rose to governmental prominence in "that dreadful country" and was eventually banished to Siberia by the Empress, had brought the fate upon himself many years before, when as a university student he had killed one of his fellows in a brawl.

Linnaeus does all he can to show that the same working out of Divine justice becomes apparent once we look closely enough at the corresponding course of events in Sweden. During the late summer and autumn of 1741, the web of subconscious tensions which followed the declaration of war at the end of July gave rise to forebodings and portents. The palace revolution which brought Elizabeth to power in St. Petersburg on November 25th was foreseen in a dream in Stockholm some five weeks earlier. General Buddenbrock may not have been to blame for the fall of Willmanstrand and the capitulation at Helsingfors, but he was made the scapegoat for these disasters by the government and executed in Stockholm – on precisely the same spot as that on which his father had had Otto Paykull unwarrantedly executed thirty-six years before. General Lewenhaupt was no more to blame for the final outcome of the war than Buddenbrock, but he suffered the same fate on account of the unjust manner in which he had had a political opponent committed to prison some years before.[39]

This is the pattern of Linnaeus' "experimental method" throughout the whole of the *Nemesis Divina*. It is not movements or policies, events or organizations which are investigated, but individuals thinking and acting in concrete and specific situations. The method is, therefore, the direct corollary of the final focal-point of the theodicy on which it is based – its central principle being that man's most profound understanding of man derives not from attempting to analyze or explain random aspects of the macrocosm, but by concentrating upon the moral capabilities of individuals. This is indeed a far remove from the simplistic generalizations of Descartes and Spinoza and the facile rationalism of Leibniz. Linnaeus' most prominent forerunner in this line

[39] Danielson, H., 1956; Holst, W., 1936; Odel, A., 1739, 1963; Hörnström, J., 1943; *Nemesis Divina*, IV.iii.2.8; IV.i.3.5; II.ii.4; II.iv.2.7; IV.i.4.3; II.ii.3; III.iii.4.4.

of thinking would appear to be Lord Bacon, who opens *The Advancement of Learning* by reminding his readers that:

If any man shall think by view and inquiry into these sensible and material things to attain that light, whereby he may reveal unto himself the Nature or Will of God, then indeed is he spoiled by vain philosophy: for the contemplation of God's creatures and works produceth (having regard to the works and creatures themselves) knowledge, but having regard to God, no perfect knowledge, but wonder, which is broken knowledge.

Is there any point in calling Linnaeus in question when he crowns his own magnificent contributions to the advancement of learning by admitting that the human mind knows neither fate nor the future, that if man is unable to take measure of himself, it is hardly remarkable that he should be unable to grasp God, that although we are able to understand certain effects, we know and perceive nothing of that which "moves invisible and alone", bringing "all that seems the work of human hands" under the sway of the Almighty?[40]

2. *The Church*

Given his family background, it is conceivable that Linnaeus might have attempted to combine his theodicy and his conception of individual morality with some acknowledgement of the intrinsic importance of an organized church. The case-histories concerning the clergy included in the *Nemesis Divina* are evidence enough that he was very far from doing so. He evidently thought it worth his while to suggest to his colleagues in the Faculty of Theology at Uppsala, that they ought to ignore petty heresies and concentrate instead on refuting the freethinkers on essential points of doctrine, but he was just as evidently under no illusions as to the moral capabilities of theologians.

To a certain extent, his position in respect of the established church of his country was similar to that of the pietists of the time, who found its organization and ritualism spiritually dissatisfying, and were persecuted on this account, not only by the ecclesiastical authorities but also by the state. The latitudinarian Anglicanism of the time tended to further the study of the early church in order to find grounds for justifying the Reformation. Linnaeus was perfectly content simply to adopt the view of mediaeval Christianity put forward by his Danish contemporary Ludvig Holberg (1684–1754), and dismiss papistry as little more than a means of blinding and exploiting a superstitious peasantry. In fact there are good grounds for thinking that Linnaeus regarded the enquiries he was undertaking in the *Nemesis Divina* as confirming the view that all forms of organized religion are to some extent at variance with spiritual truth. The following passage in his Uppsala lectures seems to indicate that it may well have been his visit to The Netherlands which first fixed this idea in his mind:

What strikes you when you travel abroad and see places where people have freedom in the practice of religion? In Amsterdam, on the Holy Days, I went into one place of worship

[40] Bacon, F., 1605, I, i; *Nemesis Divina*, IV.iv.4.1; II.ii.1; I.v; IV.iv.1.1.

after the other, and saw the devotion with which the congregations celebrated God, each in its own manner, – the Jews rocking back and forth and howling, the Papists crossing themselves, sprinkling water and falling to the ground, the Quakers and the Anabaptists, the Reformed and the Arminians, all having their own way of doing things, all showing reverence; all having a God, all being convinced that they were in the right.[41]

3. *The Law*

Although he was reluctant to attach much spiritual importance to the coerciveness of the church, Linnaeus accepted as a matter of course certain aspects of the established laws of his country which may well appear to us to be harsh in the extreme. We have already noticed how he moralized on the sight of the farmhand who had been broken on the wheel for committing incest with his step-daughter. Under the law of the land as it stood at that time, not only blasphemy, perjury and holding God in contempt, but also maltreating one's parents, manslaughter, usury and bearing false witness, as well as a whole series of precisely-defined sexual offences, were punishable with death. The code under which this state of affairs prevailed had first come into force in 1608, and during the period immediately following the initial legislation, had been strictly and rigorously applied. During the middle years of the seventeenth century, the period of foreign wars and imperial expansion, there was a certain reluctance to exact the death penalty for such offences, a certain re-emergence of the earliest principles of Swedish law, and indeed of the Swedish crown courts of the sixteenth century, which like those of all other Germanic societies, tended to enforce order by imposing fines. This drift into laxity did not go unnoticed, and in 1686 a Law Commission was set up to look into it. The legislation of 1608 was more strictly enforced in the courts, and in 1734 a new code was drawn up, confirming it.[42]

The key to understanding this development, which at first sight seems to be so at odds with what might be expected of early eighteenth-century legislators, lies in taking into consideration the Bible-based Lutheranism of the time. The Reformation had left Swedish society cut off from the broad western-European tradition of canon law. It was only natural, therefore, that in considering their responsibilities to society at large and to the newly independent monarchy in particular, the church leaders of the country should have turned for guidance to the Bible. The key figure in this respect was Olaus Petri (1493–1552), translator of the Bible, reformer, historian, elder brother of the first Lutheran archbishop of Uppsala, who towards the end of the 1530s drew up an extremely influential set of *Rules for Judges*. Although Petri was by no

[41] *Nemesis Divina*, IV.ii–IV.iii.1.3; Linderhielm, E., 1962; Pleijel, H., 1935; Malmeström, E., 1942a, 1962; Whiston, W., 1711; McAdoo, H.R., 1965; Holberg, L., 1748/1752, nos. 37, 47, 56, 67, 73, 97, 191, 204, 206, 352, 443, 465; Linnaeus, C., 1733, 193.

[42] *Sweriges Rijkes Landzlagh*, 1608; Posse, J.A., 1850; Östergren, P.A., 1902; Sjögren, K.J.V., 1900/1909; *Minnesskrift*, 1934.

means indifferent to the New Testament, to the injunction to judge not, to the idea that God desires not the death of a sinner but rather that he should turn from his wickedness and live, it was the Old Testament doctrine of judges' being instituted "to please not man but the Lord", which provided him with the main thrust of his argument. As he saw things, God presides in judgement over the whole of His creation, and certainly over the fate of nations – protecting and furthering the righteous, punishing and annihilating the wicked. The judge's overriding duty is therefore to further the well-being of the nation by helping it to avoid the wrath of God, to protect the people as a whole from "hunger and pestilence, insurrection and war" by seeking to extirpate sin, by seeing to it that those who break God-given laws are duly punished:

> The first thing a judge must know is that he is God's administrator, that the office he holds is not his own but God's, and that therefore the judgements prepared or pronounced by him are also God's. Since the office is held on God's behalf, it is not a human judgement which is passed, but most certainly God's ... The judge has also to remember that just as he himself is God's administrator, so too are the people he is judging God's people: the administration and the office being God's and not his own, he is judging not his own people but God's.[43]

These *Rules* helped to create a legal-theory which had difficulty in drawing effective distinctions between Divine, ecclesiastical and civil authority. Although everyone accepted the fundamental metaphysical assumptions, – the authority of the Bible, the absolute need for maintaining law and order – there were no ready means for working out the details of the legislation required. In 1602, for example, when plans for a general revision of the law were put before the Estates, the peasantry passed a motion requiring that since God's laws were the pattern for all others, the new code should be in harmony with them. The clergy naturally tended to advocate the promulgation of laws based on those communicated by God to Moses, and were therefore quite prepared to countenance the death penalty. As has already been noticed, the Crown tended to be more in favour of codifying earlier Swedish law and enforcing it by means of fines.

Even the most striking of the case-histories included in the *Nemesis Divina* can scarcely match the curious course of events by which this impasse in the Estates was resolved, and the draconian code of 1608 became law – became such a marked feature of Sweden's national culture during the period of imperial expansion.

By 1603 it was not only the Chancery but also a committee set up by the nobility which was preparing proposals for legal reform. Two years later, when Charles IX assumed the title of king, it was becoming increasingly apparent that the nobility were intent on using the proposed reform in order to curtail the power of the monarchy. This resulted in the king's paying more attention to the views of the peasants and the clergy, and looking again at Mosaic law,

[43] Petri, O., 1914/1917; Bergendorff, C.J.I., 1928; *Matthew* 7: 1, *Luke* 15: 1–32, *Ezekiel* 23: 11, 2 *Chronicles* 19: 6; Holmbäck, Å.E.V., 1928; Ståhle, C.I., 1968.

especially in so far as it had a bearing on regal authority. He and the members of his Chancery were well aware that since there was no tradition to build upon, any such fusing of Divine and civil law was bound to be fraught with difficulties. At the time, the head of the Chancery was Nils Chesnecopherus (1574–1622), a close confidant of the king, who had taught mathematics at Marburg for a number of years. In July 1606, while working on the proposed reform, he was commissioned to travel to Viborg Castle, at the head of the Gulf of Finland, to negotiate with the Russians. The secretary he employed there on this occasion was a local customs officer by the name of Henrik Jönsson (d. 1628), son of a clergyman with a living at Sysmä in Karelia. Jönsson had studied law somewhere abroad, and during the course of their working together Chesnecopherus learned that he was in the process of translating a law-book into Swedish. When he asked him about it, he was informed that it was the fourth German edition of a treatise on law and the Scriptures by François Ragueau (1540–1605), professor of civil law at the University of Bourges, first published in Latin at Frankfurt on Main in 1577.

Chesnecopherus looked the work through, and was intrigued by the fact that it appeared to provide precisely the framework required in order to resolve the difficulties they were facing in their attempt to reform the law. Ragueau's main objective was to show that the laws of Justinian are borne out by Scripture, that is, that temporal and divine authority constitute a harmonious whole. Each section of Justinian's *Digest* is illustrated and confirmed by means of extracts from the Bible. It is therefore the emperor himself, author of the codification, who emerges as the focal point of both civil and divine law. The prefaces to the German edition Jönsson was translating obviated unmanageably religious interpretations by pointing out that it was an essentially historical work, that it simply showed how God's decrees were revealed to the Jews and the Greeks at those particular periods in the past, that if Mosaic law was found to be at odds with natural law or conscience, it was not to be regarded as binding.

Chesnecopherus realized that by preparing legislation in the light of this treatise, it would be possible to reconcile and co-ordinate the objectives of all the interested parties. Those who looked to Scripture, and saw it as the central task of the judge to avoid the wrath of God, to protect the people as a whole from hunger and pestilence by enforcing Divine law rigorously and so helping to extirpate sin, could be met by making talion, the Mosaic principle of an eye for an eye and a tooth for a tooth, the central tenet of the legislation. Just as God, by means of Providence, protects and furthers the righteous, punishes and annihilates the wicked, so the judge, by means of the law, should be able to balance the scales by bringing home the harm that has been done, by seeing to it that the punishment fits the crime. Those who look to the Crown to protect the other estates from the ravages of the nobility, should be able to find in the new legislation, especially in the provisions for the punishment of high treason, further justification for regarding the monarchy as instituted to this end by Divine Providence. Those who look to the law mainly as a means for furthering secular ends, ought now to recognize that there are good reasons for thinking very carefully indeed before they consider holding it in contempt. Jönsson was therefore encouraged to finish his translation, which was published in 1607 and by the middle of the century had been republished on at least three further occasions. The legislation of 1608 was, therefore, the immediate outcome of his work, and as we have seen, it was still in force during the lifetime of Linnaeus.[44]

4. *The Kolding Society*

When Linnaeus began to transcribe the *Nemesis Divina* notes onto octavo sheets during the autumn of 1765, he had already been collecting "experimental" evidence of the divine nature of talion for over a quarter of a century. One of his reasons for attempting to reorganize his material seems to have been the publication in 1758 of a new book on the subject by the Dane Frederik Christian Friess (1722–1802). Friess was a cleric, and had gathered his materials while working as a catechist at Kolding in Jutland. In 1759, partly on the strength of his publication, he was appointed parish priest at Hjarup with Vamdrup, villages close to Kolding, in the diocese of Ribe. Linnaeus evidently read the book in the Swedish translation, which was prepared by the Stockholm schoolmaster and hymn-writer Olof Rönigk (1710–1780), and published at Stockholm in 1763.

Friess had assembled about fifty case-histories illustrating the divine nature of talion, the great majority of them similar in one way or another to those Linnaeus had been collecting. Linnaeus took over about a quarter of them – generally, but not always, taking note of his source. Friess's literary sources were somewhat different from those of Linnaeus, although he too drew the great bulk of his instances from Scandinavian, German and general European history, as well as from Latin literature. Like Linnaeus, he had a propensity for backing up his main points with quotations from Scripture. The immediate inspiration for making the collection seems to have come from pietist circles, and more especially from his relative Erik Pontoppidan (1698–1764), vice-chancellor of the University of Copenhagen, who had studied in England and the Netherlands, and had himself published works of a similar kind.[45]

Although Friess makes no mention of English sources, there is evidence that he was influenced by the work of Bacon, which he may have become acquainted with through Pontoppidan. It was generally assumed within the Baconian tradition, that the foundation of all sound knowledge lies in the es-

[44] Strömberg-Back, K., 1963: Munktell, H., 1936: Almquist, J.E., 1942; Ragueau, F., 1577, 1579, 1597, 1607; Günther, L., 1889/1895; Koch, K., 1972.

[45] Friess, F.C., 1758, 1763; Pontoppidan, E., 1758; *Nemesis Divina*, IV.iv.3.1, see the analysis of what Linnaeus took over from Friess in the note on this section.

tablishing of a reliable and comprehensive system for classifying the sciences. The particular branches of science are to be cultivated by drawing up "histories" of accumulated data, which constitute the raw material out of which the inductive method develops pragmatically effective "axioms" or general principles. Such a programme can best be carried out not simply by individuals working in the light of their particular interests, but by organizations engaged in furthering communal objectives by means of co-ordinated research projects. At the beginning of book four of the Latin version of the work in which Bacon put forward this blueprint for the reform of the sciences, he gave an account of what he thought it would entail in respect of furthering man's understanding of himself:

The doctrine concerning the Person of Man takes into consideration two subjects principally, the Miseries of the human race, and the Prerogatives or Excellencies of the same ... And certainly I think it would contribute much to the magnanimity and the honour of humanity, if a collection were made of what the schoolmen call the *ultimities*, and Pindar the *tops or summits*, of human nature, especially from true history.

As is well known, Bacon's description of Salomon's House in the *New Atlantis* bore fruit in the establishment of the Royal Society of London, and in the founding of many similar societies throughout Europe during the seventeenth and eighteenth centuries. As is perhaps less well known, his suggestion for the furthering of man's understanding of himself gave rise to a whole series of English works, similar in design and execution to Friess's treatment of talion and Linnaeus' *Nemesis Divina*. It is worth noting, moreover, that although many of Friess's case-histories were derived from literary sources, about a third of them were collected from direct experience, by members of a Baconian-style Kolding Society, "set up for the furthering of various useful undertakings", of which Friess himself was director and chairman. What is more, Friess managed to develop a sophisticated system of classification for his case-histories, distinguishing between ordinary cases, in which talion is "bound in a certain manner to certain laws and persons", extraordinary cases, in which it is evident that it "derives from the Allhighest, who is bound by neither laws nor persons", and finally putting forward the postulate of the all-pervading justice of the Almighty:

There is no need for me to give any elaborate proof when I say that to me God's right to retribution appears to be unexceptionably just, and worthy of unending praise. For dust though I am, I can say with all due respect that it flows from the very nature of God, that if He were not just, in talion as in all else, He would not be God: being just is both an actual reality and an essential property, and is to be found in all things as pertaining absolutely to the divine and most perfect Being.[46]

---

[46] Bacon, F., 1623a, IV, i; 1857/1874, IV: 374; Beard, T., 1597; Bogan, Z., 1653; Clarke, S., 1646; Wanley, N., 1678; Turner, W., 1697; Jenks, B., 1700; Friess, F.C., 1758, preface 35–36 (§5), 57 (§6), 149 (§13); Fyhn, J.J., 1848; Jensen, F.E., 1944.

The kind of argumentation Friess is employing here is precisely the same as that Descartes, Spinoza and Leibniz had made use of when countenancing the ontological argument as proof of God's existence, when maintaining that in the case of God it is being which is both an actual reality and an essential property. They found this a particularly congenial proof, because it enabled them to bring out the importance of pure mathematics by calling attention to the fact that in this realm too, we have direct insight into necessary truth. Since thought and being are inseparable aspects of the proof, it also provided a philosophical justification for the applicability of mathematics. This led them on to think of nature as a whole pervaded by necessity, and even into identifying nature with God. It is important to note, therefore, that Friess employs the argumentation with reference not to the being but to the justice of God, and that it is this that leads him to emphasize the providentiality rather than the necessity of events. The line of thinking is typical of the Baconian tradition, and as is evident from our examination of the status accorded to the natural world by Linnaeus, it is an aspect of Friess's general approach which must have appealed to him.

It was not only with the pietists and the Baconians that Linnaeus shared the view that it is not natural science but morality which provides the readiest link-up with religion. Although little was then known of Newton's theological views, it was apparent from the Leibniz–Clarke correspondence, as it was from the concluding sections of both the *Principia* and the *Opticks*, that in this respect he too stood firmly within the Baconian tradition. Roger Coates (1682–1716), in his preface to the second edition of the *Principia* (1713), saw off the Cartesians, Spinozists and Leibnizians in no uncertain manner, picturing them as "sinking into the mire of that infamous herd who dream that all things are governed by fate and not by providence, and that matter exists by the necessity of its nature always and everywhere, being infinite and eternal", and then maintaining that in the light of Newtonianism it was:

> Without all doubt that this world, so diversified with that variety of forms and motions we find in it, could arise from nothing but the perfectly free will of God directing and presiding over all. From this fountain it is that those laws, which we call the laws of Nature, have flowed, in which there appear many traces indeed of the most wise contrivance, but not the least shadow of necessity. These therefore we must not seek from uncertain conjectures, but learn them from observations and experiments.[47]

Friess's final postulate concerning the absolute nature of God's justice, like his subordinate distinction between ordinary and extraordinary cases of talion, was wholly in line with Linnaeus' own conceptions. What is more, nearly all the distinguishing characteristics of the cases which Friess regards as essential to his classification, are also to be found in the material collected by Linnaeus. Both undertakings contain instances of apposite retribution being brought about by means which cannot possibly be regarded as natural, of

[47] Plantinga, A., 1968; Petry, M.J., 1994; Newton, I., 1729 (1934) I: xxxi–xxxii; Alexander, H.G., 1956; Malmeström, E., 1954/1955.

the miscarriage of human justice being set right by super-human means, of something offensive done on the spur of the moment precipitating awesome consequences, of retribution taking place in a situation analogous to that in which the offence was committed, or after a corresponding period of time, of punishment being meted out not to the offenders themselves, but to their children or children's children. It is therefore in the taxonomic aspect rather than the content of their work, that their conceptions diverge. Friess centres the whole of his classification on the distinction between ordinary and extraordinary cases of talion. Linnaeus shows little inclination to follow him in this, probably because he regarded his seven aspects to the understanding of man as more in accord with the general principles of the *System of Nature* than Friess's more explicitly theological orientation.[48]

ii. *England and Holland*

1. *Baconianism*

Bearing in mind Linnaeus' non-Leibnizian theodicy and avoidance of theological niceties, his anthropological ethics and lack of ecclesiastical commitment, his empiricism and his penchant for classifying, it is difficult not to think of him as part of the general Baconian tradition. One almost expects to find him opening his *Fundamenta Botanica* with the quotation from that part of the *De Augmentis* in which Bacon is discussing the general classification of the sciences. The Linnean version of what Bacon wrote reads as follows:

> For well I know that a botany, a natural history, is extant, large in its bulk, pleasing in its variety, curious often in its diligence; but yet weed it of follies, antiquities, quotations, idle controversies, philology and ornaments (which are more fitted for table-talk and the lucubrations of the learned than for the instauration of philosophy), and it will shrink into a small compass. Certainly it is very different from that kind of history which I have in view.

Joshua Childrey (1623–1670), in his *Britannica Baconia*, first published in 1660, had put into practice the precepts of Lord Bacon and given an account of, "the natural rarities of England, Scotland and Wales, according as they are to be found in every shire". It does not seem very likely that there was no connection at all between this work and Linnaeus' famous series of books on the Swedish provinces.

In *The Wisdom of the Ancients*, first published in 1609, Bacon had taken Nemesis to be the personification of:

> The vicissitude of things ... a goddess venerated by all, but feared by the powerful and the fortunate ... The word Nemesis manifestly signifies revenge, or retribution; for the office of this goddess consisted in interposing ... in all courses of constant and perpetual felicity, so as

[48] *Nemesis Divina*, II.ii.5, III.ii.1.5, III.iii.2.6 (Friess, §6); II.ii.3, III.ii.3.2, III.iii.5.3 (Friess, preface); II.i.1.8, III.iii.5.2, IV.iii.2.7 (Friess, §7); II.ii.2, III.iii.1.11. III.iv.3.9 (Friess, §11); III.iv.2.4, III.iv.3.14, IV.i.2.1 (Friess, §12); III.i.3.4, III.ii.4.1, III.iv.2.3 (Friess, §12); Linnaeus, C., 1735b (1766/1768[12]) I: 28–32.

not only to chastise haughtiness, but also to repay even innocent and moderate happiness with adversity; as if it were decreed, that none of the human race should be admitted to the banquet of the gods, but for sport.

It is difficult to believe that this manner of interpreting classical mythology, which was very popular throughout the whole of the seventeenth century, should not have had an influence, in one way or another, upon Linnaeus.

In *A Compleat History of the Most Remarkable Providences*, published in 1697 by William Turner (1653–1701), vicar of Walberton in Sussex, we are informed that the author had, "read of a Man, that was haled out of doors in a violent manner by his own Son, who cried out to him: Oh! pray, no further; for just so far I dragged my Father". It is certainly interesting, therefore, to find Linnaeus in the *Nemesis Divina* telling precisely the same story, and adding that, "This happened near to where I was born, when I was a child". It looks very much as though the flourishing English seventeenth-century tradition of books such as Turner's, deeply influenced as it was by the Baconian approach to practical ethics and religion, also spread abroad, probably in the form of homiletic compilations, well-used in country vicarages such as that at Stenbrohult.[49]

So little is known about the reception of Bacon's ideas in Sweden, that it is difficult to draw any real conclusions as to how Linnaeus may have fitted into it. In the case of other major philosophical movements such as Aristotelianism, Ramism, Cartesianism and Wolffianism, the basic documentation is extensive and there has been a fair amount of scholarly enquiry. In the case of Baconianism, the outcome of the curious dearth of direct evidence is that no one has undertaken a comprehensive survey. The enormously influential career of Olaus Rudbeck (1630–1702) is a good example of the general situation: professor of medicine at Uppsala, technician and master-builder, distinguished contributor to such diverse fields of empirical enquiry as astronomy, botany and anatomy, even author of a gigantic work on the myth of the lost Atlantis which inspired a whole school of Swedish historical research: what he accomplished might well appear to be the result of having put Bacon's ideas into practice in the most effective manner imaginable. As a matter of hard fact, however, there is only one known instance of his ever having mentioned his supposed intellectual progenitor.

The lawyer Bulstrode Whitelocke (1605–1675) was despatched by Cromwell on a mission to queen Christina while she was holding court at Uppsala during the winter of 1653/1654. In his account of the embassy, he records that on February 24th 1654 he met officers and gentlemen, including the diplomat Lars Cantersten (1615–1658), who had studied at Leiden and had dealings

[49] Linnaeus, C., 1736a (iii); Bacon, F., 1623a, II, v; 1857/1874, I: 501, IV: 299; Childrey, J., 1660; Bacon, F., 1609, no. xxii; Turner, W., 1697, I, ch. 99, no. 11; *Nemesis Divina*, III.ii.3.3; Ehnmark, E., 1941, 34–35; Beard, T., 1597; Clarke, S., 1646; Bogan, Z., 1653; Wanley, N., 1678; Jenks, B., 1700.

with Charles II in Scotland, who discoursed with him concerning: "matter of learning, and particularly of English authors, as Selden, Milton, the Viscount of St. Albans, and others, whom they much admired and commended". There is no evidence that the company had any more than a superficial acquaintance with these writers, however, and even in Swedish university publications of the time, the few references there are to Bacon's works do not create the impression that they were very well understood. In a doctorate defended successfully at the University of Åbo in 1665, for example, we find Bacon being hauled over the coals on account of his anti-Aristotelianism:

> He treats physics as a practical discipline, confusing it with the manual arts which are derived from it. He thinks natural bodies have strange principles, ... that they are all living. He coins notions, concepts and new terms which have never been heard of before, mixing them up with a new method and a new manner of teaching. He treats substantial form as being nothing more than a co-ordination and deployment of corpuscles, as it is in other aggregates.

It is almost certainly a matter of some significance that several eminent Swedes had dealings with the Royal Society of London during the first forty years of its existence: Johannes Schefferus (1621–1679) on account of his work on Lapland, Georg Stiernhielm (1599–1672) as the result of his philological interests, Urban Hjärne (1641–1724) on account of his work in medicine, geology and chemistry, Olaus Rudbeck because of his pre-eminence in so many fields of enquiry. Stiernhielm and Hjärne also had a first-hand knowledge of Bacon's writings. Erik Benzelius (1675–1743), after meeting Leibniz in the August of 1697, went on to strike up a friendship with Hans Sloane (1660–1753), secretary of the Royal Society, when he visited London a couple of years later, but it was not until 1710 that he became involved in the founding of a comparable society in Sweden.[50]

Nevertheless, it looks as though it may well have been Benzelius who first introduced Linnaeus to Bacon's writings. The account given of the churchman's extraordinary private life in the *Nemesis Divina* indicates that Linnaeus must have known him well during his early years at Uppsala, and this was just the period when Benzelius was most active in promoting Bacon — notably by encouraging the hapless antiquary Nils Hufwedsson Dal (1690–1740) to translate selections from his works into Swedish. Linnaeus had every reason to respond positively to the enthusiasm, and it may well be significant that in his 1733 lectures on healthy living he should have made public acknowledgement of his indebtedness to Bacon's "history of life and death, with observations natural and experimental for prolonging life". At this stage, it looks as though he must have found Bacon a help not only in organizing but also in condensing

[50] Hammarsköld, L., 1821; Kallinen, M., 1995; Sellberg, E., 1979; Lindborg, R., 1965; Frängsmyr, T., 1972; Eriksson, G., 1984, 1994; Whitelocke, B., 1772, I: 454; 1855, I: 439–440; Thuronius, A., 1665, 25; Uggla, A.Hj., 1940; Seaton, E., 1935; Schefferus, J., 1673; Swartling, B., 1909, 112–113; Nordström, J., 1924; Wieselgren, O., 1910; Liljencrantz, A., 1939/1940.

his ideas. It is evident from the manuscript material relating to the *Fundamenta Botanica*, for example, that the original Uppsala version of the work was preceded by a somewhat rambling preface, which sometime before the summer of 1735, when the work was finally handed over to the printers in Amsterdam, was replaced by the quotation from *De Augmentis*.[51]

2. *Bayle's Dictionary*

After taking his degree in medicine at the University of Harderwijk in June 1735, Linnaeus spent some three years in the Netherlands, mainly in and around Amsterdam and Leiden. During this period he got to know many of the foremost experts in his field, and established his international reputation by publishing a whole series of major works on natural history, including the first edition of the *System of Nature*. It was only natural that when he left Sweden he should have made straight for Holland, for many of those who had advised him on his career had done the same in their early years, and after returning home again had continued to cultivate the contacts they had made there. This was true, for example, not only of Johan Stensson Rothman (1684–1763), the teacher at Växjö Grammar School who had first encouraged him to take up medicine, but also of Olof Rudbeck (1660–1740) and Lars Roberg (1664–1742), the professors teaching the subject at Uppsala while he was studying there. It was also true of Nils Rosén (1706–1773), his rival in the faculty, and Johan Moræus (1672–1742), his future father-in-law, a general practitioner at Falun in Dalecarlia. In fact ever since the beginning of the Thirty Years' War, Swedes had been following the trade routes south and getting themselves enrolled at Dutch educational establishments — especially Groningen, Franeker, Amsterdam and Leiden. The result was an extensive and intricate interweaving of the two cultures, not only in such major academic fields as theology, law and medicine, but also in philosophy, history and philology, as well as technology, the natural sciences and mathematics.[52]

It is certainly worth asking, therefore, if these Dutch contacts throw any light on the distinctive characteristics of the *Nemesis Divina*.

Although there appears to be no evidence that Linnaeus had any direct knowledge of Bayle, the central theses put forward in the famous *Dictionary* undoubtedly influenced his particular brand of theodicy, even if only through the medium of the Wolffianism he encountered among his colleagues at Uppsala. Like Bayle, he recognized the necessity of not allowing theologians alone to tackle the task of reconciling the omnipotence and goodness of God with the existence of evil. Through his own tendency to pietism, he found some common ground with Bayle's radical and uncompromising fideism, and

---

[51] *Nemesis Divina*, IV.ii.2.14; Dal, N.H., 1726, 1729, 1736; Gibson, R.W., 1950; Pérez-Ramos, A., 1988; Bacon, F., 1623b; Linnaeus, C., 1733, 20; 1736a; Fredbärj, T., 1964.

[52] Wrangel, E., 1897; Boerman, A.J., 1953, 1978.

certainly agreed with him in remaining intensely sceptical of the value of all forms of theological dogmatism and ecclesiastical organization.

On the other hand, he was bound to regard the case-histories he was collecting as an eloquent refutation of Bayle's thesis that an ethically sound society might well be founded on atheistic principles, and he agreed with Leibniz that the justifying of God is not simply a matter of faith, but also a genuinely intellectual task, involving some reference at least to the findings and accomplishments of the natural sciences.[53]

3. *Natural Theology*

When Linnaeus arrived in the Netherlands, nearly half a century had passed since Boyle's work on *Final Causes* had first appeared in Dutch translation. During this period Cartesianism and Spinozism, and to a certain extent even Leibnizianism, had become associated in the public mind not only with a tendency to atheism, but also with unreliability in the interpretation of the natural sciences. The prestige of Newton, together with the popularization of his doctrines by means of the Boyle lectures, had created an intellectual atmosphere in which mathematics and natural science combined easily with the everyday religious culture of what was in effect the established church. Experimental philosophy, as expounded by professed Newtonians, had become a normal aspect of the religious awareness of the ordinary educated citizen, and societies for the cultivation of an interest in it, like that founded by Friess at Kolding, had sprung up all over the country. Books dealing with the theological implications of the most diverse aspects of science, from astronomy and snowflakes to gardens and grasshoppers, found a ready market in all walks of life.[54]

The basic assumption in works of this kind was that nature was potentially a chaos: evidence of order and design was amassed, it was pointed out that there could be no design without a designer, and this designer was then identified as God – the Creator and Sustainer of "all that shimmers here before our eyes". Linnaeus must have found reasoning of this kind congenial enough, since he himself had already developed something very similar to it.

In its Dutch variation, natural theology of this kind had two further features which may well have made an impression upon Linnaeus. Bernard Nieuwentijt (1654–1718) had not only given its central tenets their classic formulation in his more popular work. In his abstruse but well-written *Foundations of Certitude* (1720), he had also developed a highly sophisticated critique of Spinozism, putting forward views which in many important respects anticipate those of modern logical positivism. He based it on the distinction between pure mathematics, which he conceived of as being completely abstract and tautolo-

---

[53] Bayle, P., 1695/1697; Lempp, O., 1910, 13–32.

[54] Boyle, R., 1688; Derham, W., 1714; Engelman, J., 1747; Hervey, J., 1748; Rathlef, E.L., 1748/1750; Bots, J.A.H., 1972.

gical, and applied mathematics, which he regarded as necessarily involving experimental work. He pointed out that Spinozism, especially in its pronouncements concerning God, had overlooked this distinction, and ascribed an ontological significance to what was in fact no more than an abstract concept. He illustrated the point by calling attention to Euclid's treatment of incommensurables in book ten of the *Elements*, and then bringing out the similarity between the mathematician's employment of the infinitely small, and the Spinozistic theologian's invocation of the infinitely large. Essentially, his exposition is an anticipation of Kant's treatment of the first two antinomies of pure reason:

I would ask you, therefore, to judge for yourselves: if the *Mathematicians* have to admit that their science is incapable of so revealing and demonstrating the truth of such entities as the *Geometrical Infinite* that each of us may be convinced by his own understanding: what certitude can be expected from an intuitive attempt to envisage certain objects, which unlike the simple *Geometrical Infinite*, based as it is on nothing more than extension and multiplicity, are rendered Totally Infinite by involving Incomprehensible Perfections? Any attempt to grasp this Infinitude by means of our finite understanding faces constant and inevitable frustration – simply serves to bring to light the deficiencies of the understanding, the shortcomings of all our envisionings and capabilities.

In the *Foundations of Certitude*, Nieuwentijt also makes mention in passing, of the tradition among the mathematically-minded philosophers of the time, of ascribing moral certainty to laws of nature, on account of their practical or social applicability. It was, however, Willem Jacob 's Gravesande (1688–1742), professor of philosophy at Leiden, Baconian, Newtonian, fellow of the Royal Society of London, who first popularized the concept of moral certainty within the general Dutch tradition of natural theology, by maintaining in his widely-read textbook on the *Mathematical Elements of Physics* that:

In physics, it is not the case that our senses enable us to pass immediate judgement on everything. There is another manner of reasoning, which although it is not mathematical, also has its validity: it is based on the axiom *that one has to accept as true everything which would either destroy human society or deprive people of the possibility of life, were it to be denied.* No one can call this proposition in question, and it is perfectly clear that it is the basis of the second and third of the Newtonian rules of reasoning in philosophy.[55]

Although there would appear to be no evidence that Linnaeus had any direct knowledge of Nieuwentijtls criticism of Spinoza or 's Gravesande's axiom of moral certainty, they were widely-discussed topics in the circles in which he moved, and they may, therefore, have played some part in the further development of his own views, which were certainly not at odds with attempts to modify the basic tenets of rationalist metaphysics. The axiom of moral certainty accorded well with the final focal-point of his own theodicy –

[55] Nieuwentijt, B., 1715, 1720, 179 (pt. ii, sect. 23, §19), 182; Kant, I., 1781, 426–443, 1787, 454–471; Vermeulen, B., 1987; Petry, M.J., 1979; 's Gravesande, W.J., 1720/1721, preface; Pater, C. de, 1988, 1994; Gori, G., 1972.

with the conviction that man's most immediate awareness of God derives not from contemplating nature, but from the constant demands being made upon him through the regulating procedures of Nemesis. The similarity between the infinitely small of the mathematicians and the infinitely large of the would-be metaphysicians certainly appealed to his imagination, since he returns to it time and again in his writings. The microscope and the telescope provide us with partial access to "the invisible world" of the microcosm and the macrocosm. It could well have been Nieuwentijt's criticism of Spinoza which Linnaeus had in mind, when he observed in his treatise *On Barren Barley*:

It is strange that we should have such a propensity for disregarding the tiniest of things, that we should be so ready to blame one another for splitting hairs and wrangling over details: for if we think about it, we soon realize that it is the tiniest of things which have the greatest effect, both in nature and in general.[56]

4. *Medicine*

Investigating the ways in which the Netherlands influenced the *Nemesis Divina* also involves taking into consideration Linnaeus' work as a medical practitioner. Immediately after leaving the country and returning home, he established a flourishing practice in Stockholm by specializing in the prescription of mercury ointment as a cure for venereal diseases, and quite a number of the case-histories included in part three of the present edition date from this period.

In 1740 Nils Rosén succeeded Olof Rudbeck as professor of medicine at Uppsala. Soon afterwards Linnaeus succeeded Lars Roberg as holder of the University's second medical chair – partly on the strength of the degree he had acquired at Harderwijk, but mainly, of course, as a result of the epoch-making works he had published in Leiden and Amsterdam. At first there was some uncertainty as to how the teaching and research should be apportioned. It was eventually decided that Rosén should be responsible for anatomy, physiology and practical medicine, Linnaeus for botany, dietetics and materia medica, and that they should share chemistry and pathology between them. The arrangement worked well, and under their direction the Uppsala Medical School established a European reputation for itself. Through his work in anatomy and pathology, Rosén managed to harmonize the more abstract mechanistic conceptions of the Cartesians, with the more concrete and practically-orientated approach Sydenham had developed out of the Baconian tradition. The shared appreciation of Baconianism enabled Linnaeus to highlight the systematic significance of Rosén's work in the *System of Nature* – by analyzing out the physiological, dietetic and pathological aspects of man's understanding of himself.[57]

---

[56] Linnaeus, C., 1750; Linné, C. von, 1767; Hult, O.T., 1934; Wilson, C., 1995.

[57] *Nemesis Divina*, III.i; Pehrsson, A.-L., 1965; Rosén, N., 1738; Linnaeus, C., 1735b (1766/1768[12]) I: 28–29.

Both were aware that this harmonious co-operation had been made possible by what they had learnt in the Netherlands, that is to say, by what the Dutch medical tradition had learnt in the course of ridding itself of dogmatic Cartesianism. Descartes had taken a great interest in medicine, and for a few decades after his death his mechanistic concept of the body and its vital functions had had a considerable influence on Dutch medical faculties, evidently because it appeared to open up the prospect of introducing easily understood causal explanations into medicine. It is doubtful whether it ever had much effect on practising physicians, however, and in its original form it had largely disappeared from the Dutch medical scene by the turn of the century – partly as a result of the increased use of the microscope, which had exploded most of its mechanistic hypotheses. The last Cartesian-style *Oeconomia Animalis*, compiled by Pieter Muis (1645–1721) of Rotterdam, was published at Amsterdam in 1697.

The best-documented account of the way in which late seventeenth-century medical thinking rid itself of Cartesianism, is to be found in the exchange of ideas between John Locke (1632–1704) and Thomas Sydenham (1624–1689). Faced with the virtual bankruptcy of the attempt to derive causal explanations from mechanistic concepts, they decided that the best way forward, especially in the crucial field of pathology, was the Baconian one of sorting out what was effective from what was not, by concentrating on the proper classification of individual case-histories. It was not by chance, therefore, that it was the physician and botanist Johan Frederik Gronovius (1690–1762), thoroughly versed as he was in the merits of this taxonomic approach, who first recognized the importance of the *System of Nature*, and saw the initial Leiden edition of it through the press.[58]

To a large extent, the success of the co-operation between Rosén and Linnaeus was due to the common influence of Herman Boerhaave (1669–1738), professor of medicine at Leiden. Rosén may have attended some of Boerhaave's classes during the latter part of 1730, and Linnaeus was in constant contact with him throughout the greater part of his stay in The Netherlands. Ever since his inaugural oration in 1709, Boerhaave had been calling attention to the shortcomings of Cartesianism and the merits of the Baconians:

> Since the Cartesian school deduced most things from fictitious causes, since they put their trust in mere generalities, making huge leaps to particularities, their work is so useless to the physician that medical science may safely shed this tremendous burden.

Six years later, in an oration delivered on the occasion of his relinquishing the vice-chancellorship, he bid his audience:

> Praise the good fortune of this same physical science, in that it had the enormous advantage of acquiring as its adept the great Bacon. He was the greatest by far in investigating all subjects

[58] Lindeboom, G.A., 1978; Folter, R.J. de, 1978; Dewhurst, K., 1958, 1963, 1966; Romanell, P., 1984; Nordström, J., 1954/1955.

that can be attained by human science; it is difficult to make out whether it was his advice or his example, exertions or liberality, that were most influential in restoring the disfigured science of physics.

Boerhaave replaced the simplistic reductionism of his predecessors with a comprehensive approach to anatomy, physiology and practical medicine which did justice to the findings of a whole range of subsidiary disciplines – mathematics, mechanics, hydrostatics, hydraulics and chemistry. He also gave a clear account of his methodological reasons for doing so. In an oration on chemistry delivered in 1718, he pointed out the futility of a one-dimensional approach to science:

This is why these people form so many sects, – starting from different experiments, everyone sets up a general theory which fits in with, and is closely determined by, his own observations. Consequently, it is almost impossible to find among them a single person who does not quarrel with others on any given subject ... Chemistry has groaned under this burden, but it has pulled through under its own power ... For each time the chemists brought certain bodies in contact with others, they discovered new phenomena, different reactions, and dissimilar effects, which refused to be encompassed by some universal and common law ... It has become clear that nothing is more fallacious than to explain everything from a single point of similarity, and to measure one and all by that single yardstick.

It was well known, moreover, that in matters theological Boerhaave was on the side of the angels. When Voltaire happened to arrive during dinner he was turned away without an interview, since the professor of medicine: "was not prepared to rise for someone who did not rise for God". It is almost certainly the case, therefore, that when Linnaeus extended aspects of man's self-knowledge from the basic Boerhaavian trio of physiology, dietetics and pathology, to the wider concerns of psychology, social ethics and theology, he saw himself as completing the general programme of enquiry outlined in the life-work of his mentor.[59]

iii. *Stenbrohult and Växjö*

1. *The Wisdom of the Ancients*

Given the first chapter of *Genesis* and the chronological arrangement of the books of the Bible, it is hardly surprising that natural science developed within a Christian context should involve thinking historically. Although Bayle, Leibniz and Linnaeus attempted to formulate the problem of theodicy in philosophical terms, it was in fact an issue inextricably bound up with an interpretation of the past, and in the last instance they themselves had no real interest in divorcing it from its traditional context. Like the great majority of their contemporaries, they were well aware that the philosophical significance of the Biblical narrative does not depend on its being literally true in a purely

---

[59] Boerhaave, H., 1709, VII: 12–13; 1715, XXXVI: 48–49, 1718, XV: 19–20; 1983. 132, 177, 202; Lindeboom, G.A., 1978, 310.

historical sense. What is, is not simply what ought to be: through the continuous reiteration of an errant wilfulness, mankind violates primal innocence and persists in a state of sin. The everyday implications of the history of Adam and Eve are therefore clear enough – falling from grace is a matter of free human choice, and is constantly changing our common lot for the worse. Man's actual state is therefore different from that purposed for it by the Creator – there can be no justification for denying the freedom of the will and maintaining that both good and evil are of purely Divine origin, or for ignoring the difference between creature and Creator and postulating a pantheism.

In the period preceding the full discovery of geological time, rational interpretation of Biblical historicism tended to accord readily enough with its profane counterpart. Hesiod and Ovid had distinguished the ages of gold, silver, bronze and iron. Varro had invoked a triadic chronology – from the beginning to the Deluge – of which we have no account apart from Scripture, from the Deluge to the first Olympiad, which we know of through the myths and fables that have come down to us, and from the first Olympiad to the present, which is recorded for us in the work of ordinary historians. It was, therefore, a significant move on Bacon's part, as he indicated in the opening sentence of *The Wisdom of the Ancients*, when he set himself the task of expounding the hereditary wisdom of the second age by interpreting its mythological fables in modern terms:

> The earliest antiquity lies buried in silence and oblivion, excepting the remains we have of it in sacred writ. This silence was succeeded by poetical fables, and these, at length, by the writings we now enjoy: so that the concealed and secret learning of the ancients seems separated from the history and knowledge of the following ages by a veil, or partition-wall of fables, interposing between the things that are lost and those that remain.[60]

Given this general framework, it was only natural that in the minds of those who were actually doing the research, much of the natural science of the time should have had a fairly clearly-defined historical dimension. The legendary Olaus Rudbeck, for example, had not only revolutionized Swedish historiography by amassing evidence for the view that Svea's realm was the lost Atlantis of Plato, the cradle of all civilization, the land in which the descendants of Japheth had settled after the Deluge. He had also lectured in a thoroughly professional manner on mathematics, music and astronomy, completed a vast work on botany, discovered the lymphatic gland, planned in detail a counterpart to Vesalius in the field of animal anatomy, furthered instrument-making, land-surveying and agriculture, brought about marked technological advances in the building of canals, ships and houses, as well as fortification and pyrotechnics. Newton was convinced that many of the advances he had initiated in the exact sciences were in fact re-discoveries of truths known long ago to the ancients – that Pythagoras had been familiar with

[60] *Genesis* 1: 1–31, 2: 8–3: 24; Porter, R., 1977; Hesiod, 1978 lines 106–201; Ovid, 1994, I: lines 89–150; Varro *De gente populi Romani*; Bacon, F., 1609, 1857//1874, VI: 625.

most of the actual contents of the third book of the *Principia*, that there was nothing untoward in rounding off the *Opticks* with a reference to the sons of *Noah*.

It was widely recognized, therefore, that it was perfectly possible to work within this broadly historical context in an essentially ahistorical manner. It was the historical source of Plato's *Timaeus* which had drawn Kepler's attention to the analytical transition from the four elements to the five regular solids. It was, however, in the interest of science that he applied these geometrical abstractions to the orbits of the planets, consolidating the move by also taking epistemological and theological factors into account:

Since it is the cognition of quantities native to the mind which determines the nature of the eye, it must be the mind which determines the eye and not vice versa. But why waste words? Since geometry pertains to the Divine Mind from all eternity, preceding the genesis of things, since it is therefore God himself, for what is in God must be God himself, it must have provided God with the prototypes for the creation of the world, and have passed over into man with the image of God, not been taken up into him through the eyes.

It is important to note that this conception of geometry had had a profound effect on the analytical approach to language to be found in the works of Georg Stiernhielm (1590–1672), which was well-known to Olaus Rudbeck, and may therefore have had a certain influence on Linnaeus. Evidently as a result of his contacts in the Netherlands and England, Stiernhielm had cultivated an interest in both mathematics and language-theory. He had taken up the idea, probably originating either from Johannes Goropius Becanus (1518–1572) and Simon Stevin (1548–1620), or from George Gascoigne (1525–1577) and John Dee (1527–1608), that just as geometry is to be derived ultimately from the point, so language is to be analyzed back into the monosyllable. As Stiernhielm saw it, therefore, simplicity and complexity in mathematics have their exact counterparts in language and nature.

In many fields where the primary concern was non-historical, the historical dimension was acknowledged simply as a matter of course, often as a result of controversies or polemics. Surprisingly enough, this was the case even with Descartes, who in a letter to the French Jesuit Jacques Dinet (1584–1653) came out with an unexpectedly orthodox historicist view of his own philosophy:

Everything in peripatetic philosophy, regarded as a distinctive school that is different from others, is quite new, whereas everything in my philosophy is old. For as far as principles are concerned, I only accept those which in the past have always been common ground among all philosophers without exception, and which are therefore the most ancient of all.[61]

---

[61] Eriksson, G., 1984, 1994; McGuire, J.E. and Rattansi, P.M., 1966; Kepler, J., 1619, bk. 4, ch. 1; Hultman, F.W., 1870; Swartling, B., 1909; Goropius Becanus, J., 1569; Dijksterhuis, E.J., 1970, 126–129; Gascoigne, G., 1575; Dee, J., 1570; Wal, M.J. van der, 1995, 48–52; Descartes, R., 1897/1910, VII: 580.

This readiness to acknowledge the conventional historical setting of an issue was also evident in the early eighteenth-century attitude to primitive peoples – a matter quite different from the central concerns of Kepler, Stiernhielm and Descartes, but perhaps even more important to the intellectual world of Linnaeus. When Aphra Behn (1640–1689) visited Surinam in 1663, it was not with the intention of testing the Biblical account of Eden against the new-fangled notions of Hobbes – she was simply struck by her first-hand experiences, by the contrast between the nobility and simplicity of the natives and the depravity and duplicity of the colonizers who had enslaved them. In the novel she based on her visit, however, by invoking traditional associations and characterizing the natives as "living in the first state of innocence, before men knew how to sin", she helped to change the general European conception of primitive man.

In this connection, the most intriguing anticipation of the attitude of mind we find in Linnaeus, is the significance attached to childhood in the religious writings of Henry Vaughan (1622–1695) and Thomas Traherne (1636–1674). Here once again, the basic attitude of mind is determined principally by an immediate experience, which on account of certain historical associations is then endowed with a wider significance. In Traherne's *Meditations* the childhood vision seems to have been actually experienced as God-given:

The corn was orient and immortal wheat, which never should be reaped, nor was ever sown. I thought it had stood from everlasting to everlasting. The dust and stones of the street were as precious as gold: the gates were at first the end of the world. The green trees when I first saw them through one of the gates transported and ravished me, their sweetness and unusual beauty made my heart to leap, and almost mad with ecstasy, they were such strange and wonderful things.

Linnaeus, in his lectures on *Natural Diet*, recalls the original experience primarily in order to illustrate the nature of unprejudiced insight, although he also sets it within the conventional historical context:

The world which presents itself to us through the senses is analogous to the state of innocence. I am sure there are many who go out into the countryside, and see it simply as green plus a few colours, – cloud as a shadow, the sun merely as a light. So absorbed and distracted are they by economic, political, whimsical, frivolous, avaricious, vindictive objectives, that they are blind to the goodly estate in which the Creator has placed us. I know this from my own experience, for one summer in my youth, when I fell ill of a severe fever, I did not look out of doors from the middle of March until halfway through July; and when I did go out, I saw the world as never before, everything being so exalted, so beautiful. Think of Adam and Eve, fully formed, flourishing in the very prime of health, free of prejudices, setting eyes on hills, on green valleys and running streams, in a spot where the climate was most temperate .... Did they not have occasion to admire the Creator, having only just acquired sight and for very joy not knowing which way to turn, – seeing the sun, and as they went out into the night, the moon, the stars, the heavens? When with their ears they heard the stirring of the air, the sound of animals, the song of birds, was that not evidence enough that all was well? When with their fine sense of smell, in a moment of time, they caught the scent of all plants and blossoming trees, was that not sweet and agreeable? When they tasted such wondrous fruits as the pineapple and the banana, the vine, was that not glorious? And they had not the slightest fear or care.[62]

Traherne's heart leapt on account of the transformation in his perception of everyday things. By recalling his sudden awareness of what was exalted and beautiful, Linnaeus was led to meditate on sensuous experience and prejudice. In both cases, however, it was the historical dimension, the association with the Biblical myth of the Garden, which endowed the immediate experience with a wider significance.

During the seventeenth and eighteenth centuries, the conventional conception of the myth of the Garden was such a commonplace, that there is little point in picking out instances of it and then suggesting that they might have had an effect on Linnaeus. Bacon began the most beguiling of his essays by reminding his readers that it was God Almighty who first planted a garden, and Marvell and Milton were inspired by the drama of Eden to write some of their finest poetry. Their Swedish equivalent was Haquin Spegel (1645–1714), who gets a mention or two in the *Nemesis Divina*. In his *Paradise Revealed* (1705), after dwelling on the wonders of Eden, Spegel goes on to describe the newly laid-out palace gardens around Stockholm – Drottningholm, Venngarn, Stafsund – presenting them as latter-day replicas of the primal happiness which once encompassed all mankind. It was an attractive idea, and Linnaeus' publications make it clear that he was ready enough to retail it. In one of his earliest dissertations (1731), he refers to botany as a "divine science". When dedicating *Clifford's Garden* (1736), he contrasts the pristine glory of Paradise with the degradations of later times, and declares that botany can provide us with glimpses of the divine beauty that once was. In the preface to the *Fauna of Sweden* (1746), he sings the praises of Eden as the first and finest of all gardens. It was, moreover, one of his central tenets, as he tells us in his *Library of Botany* (1736), that the perfection of Paradise involved all three of the realms of nature, and that the relative perfection of man-made gardens was to be assessed in the light of this. Evidently confirmed in his view by Stiernhielm's philosophy of language, he looked upon his taxonomic labours as the work of a second Adam, remembering well that whatsoever the first, "had called every creature, that was the name thereof".[63]

Given this commitment to the myth of the Garden, it comes as something of a surprise to find Linnaeus paying little attention to the corresponding concept of universal degeneration. For many, accepting the geographical and historical reality of Paradise led on to reading the subsequent history of the world as a

---

62 Behn, A., 1688; Vaughan, H., 1650; Hutchinson, F.E., 1947; Traherne, T., 1908, 111 §3; Salter, K.W., 1964; Linnaeus, C., 1733, §104, 168–169; Linné, C. von, 1907, no. 25, p. 48 (1718).

63 Nichols, M.A., 1966; Raynaud, C., 1976; Bacon, F., 1625, XLVI; Johnson, W.G., 1945; Spegel, H., 1705; Linnaeus, C., 1731, 1905/1913, IV, no. 10, 245; 1737b, Dedication; 1746, Preface; 1736a, 64; *Genesis* 2: 19; Broberg, G., 1978.

drifting away from man's original God-given setting. It was generally accepted that it was the task of the Church and theologians, as it was of rulers, lawyers, artists and individuals, to seek for a turning of the tide, to choose their way back to God within the realm of Providence. For Linnaeus, as we know from his oration *On the Increase of the Habitable Earth* (1744), the geographical and historical reality of Paradise must imply that the general situation has been improving, that the original sustenance provided has been on the increase. The Garden supported only one pair of each created species, whereas we now have a world sustaining individuals in their teeming millions.

It is curious that Linnaeus should have pursued this line of argument so enthusiastically, without even mentioning its equally plausible alternative, since among the Swedish peasantry of the time there was a widespread belief in a slow but steady and inexorable deterioration in the fertility of the land, in the relentless extension of the forests, in the ultimately irreversible return of pasture and arable to a stony wilderness of heather, moss and rocks. The resignation basic to this conviction was perceived by many landowners and academics as an obstacle to agricultural improvement, and various attempts were made to counteract it, notably by Linnaeus' friend and disciple Carl Fredrik Mennander (1712–1786), who published a thesis on the subject in 1748. It was, moreover, an aspect of the country-life of the time, which one might well expect to have been of particular interest to him, not only on account of its more practical consequences, but also because of the fatalism involved. It is rather strange, therefore, that there should be no mention of it in any of the accounts of his travels through the Swedish provinces.

The fact is that Linnaeus was more interested in the living processes of nature than in history. It was the here-and-now which preoccupied him, and he read the past in the light of it. It cannot be denied that he had some tendency to idealize his own childhood, but his attitude to the young in general was the severely practical one of making sure that they are not maltreated – as he saw it, a wrong in itself which could only jeopardise their health as adults. It was with this objective that in his lectures on healthy living he admonished his audience to "let the lad enjoy paradise as long as he can, for cares will drive him out of it soon enough".

As a result of having actually travelled among the Lapps, he certainly fell in with Aphra Behn's general manner of thinking, suggesting on various occasions later in life, that they and other primitive peoples might well be regarded as "living in the first state of innocence, before men knew how to sin". His reason for doing so, however, was not historical or theological but pragmatic – through the simplicity of their culture they can teach us important lessons for improving our own way of life. Precisely the same manner of thinking pervades the whole undertaking of the *Nemesis Divina*. The case-histories were collected not in order to make a theoretical point, but as a source of instruction for his son. They are centred not on movements or policies, events or organizations, but on individuals thinking and acting in concrete and specific situations. They are, therefore, as we have seen, the direct corollary of the final focal-point of the theodicy on which the whole work is based – the conviction that man's most immediate awareness of God derives not from ethical generalizations or the contemplation of nature, but from the constant demands being made upon him by his awareness of Nemesis.[64]

2. *The Garden*

There is nothing conventional or inconsistent about the place held in Linnaeus' affections by his father's garden at Stenbrohult. Throughout the whole of his later life, whenever associations turned his mind in that direction, we find him dwelling on his earliest experiences of the green world around the rectory in which he grew up. During the August of 1741, for example, when he was in the process of preparing his survey of Öland and Gotland for "the illustrious Estates of the Realm", he had occasion to pass through his home village. We find the following account in his official report on the journey:

Stenbrohult Church lies close to the shore of the great lake of Möckeln, which here opens out into a broad bay, forming a most pleasant prospect ... The garden laid out by my father Nils Linnaeus, the Rector here, contains more plants than any other garden in Småland, and it was in this garden, together with my mother's milk, that my mind was inflamed with an unquenchable love for plants.

Eight years later, by which time the rectory had been burnt down and both parents were dead, he passed through the village again. He was now engaged in the preparation of a similar government survey of the province of Scania, and we find the following in the published journal:

Whit Monday, May 15th. Travelled from Virestad to Stenbrohult Church. I found the birds all gone, the house burnt down, the younger generation scattered, so that I scarcely recognized the place where I was born. As I looked, I thought I saw *the field that once was Troy* on the spot where my late father, Rector Nils Linnaeus, laid out the garden which was once glorious with the rarest plants in Sweden, and which was destroyed by fire even before he passed away on May 12th last year. The joys of my youth, the rarest of plants, which grow wild in the area, had not yet appeared. Twenty years ago I knew everyone in the village, now I found that I scarcely knew twenty of them; those I knew in my youth as young lads, were now walking about with grey hair and white beards, worn out. A new world had come into being.

At the beginning of August 1749, while journeying back from Scania to Uppsala, he stayed overnight in the village, after which visit he never again set eyes on it.

Linnaeus' mother, Christina Brodersonia (1688–1733), had died while he was still an undergraduate at Uppsala. During his earliest years it had been her dearest wish that since he was their eldest child, he should follow in the footsteps of so many of her forebears, and also become Rector of Stenbrohult.

---

[64] Linnaeus, C., 1744; Mennander, C.F.. 1748; Wallerius, J.G., 1758; Olsson, B., 1954/1955; Linnaeus, C., 1733, §6, p. 42, §80, p. 133; 1732; 1737a, 169; Broberg, G., 1975, 254–286.

It was therefore a great disappointment to her when it became apparent that he was more suited to study medicine than theology. Their second son Samuel was born ten years after the first, and she evidently did all she could to prevent him from going the same way as his elder brother and taking such an inordinate interest in the garden. Like Carl, he was first taught at home, and during the autumn of 1725, at the age of seven, sent off with his tutor to continue his studies in Växjö. Carl left Växjö Grammar School two years later in order to read medicine at Lund, and it was evidently during that autumn that a fortune-teller visited the rectory at Stenbrohult, and so provided Linnaeus with one of the less sombre of the case-studies recorded in the *Nemesis Divina*:

My brother Samuel, quick-witted, was at Växjö school. I was regarded as dim, and had just gone off to Lund. Everyone called my brother the professor, and said that he would become one. The fortune-teller, who had seen neither of us, asked to be shown some of our clothes. She said that my brother Samuel would become a clergyman. Of me, she said: "He will become a professor, travel widely, become more celebrated than anyone else in the realm", and swore that this would be so. In order to test her, my mother showed her another piece of my clothing, telling her that it was my brother's. "No, that belongs to the one who will be a professor and live far away from here".

Some twenty years after he had last seen Stenbrohult, when his younger brother was in fact Rector there and he was long-established as Professor in Uppsala, Linnaeus wrote a New Year's letter to his relatives in the village:

Now that I am sitting here on my own over Christmas, I cannot help thinking of my beloved birthplace. "What sweetness is it which draws us to the land of our birth, and will not allow us to forget?" It is there that I have left so many of the dear ones who belong to me, living together in trust and friendship, while I have roamed all over the country ... You are able to comfort one another when in difficulty, and rejoice with one another when things are going well. There is no one I can confide in ... When the time comes and you are tired of the world, you will be gathered to your fathers and rest in their grave. That will not be the lot of my bones. You have been able to build and dwell where you were born, to live together with people brought up in the same honest and straightforward way. I have to be constantly on my guard against schemers and people with motives.

A year or so after writing this letter, Linnaeus published a short account of the natural history of Småland, in a huge, rambling and unreliable work on the general history of the province, compiled over many years and left among the posthumous papers of Samuel Rogberg (1698–1760), the subject of one of the case-histories included in the *Nemesis Divina*. Together with the ordinary presentation of facts and observations, one finds there the same tone of resignation and regret:

Unfortunately, I know more about the natural history of ... Ceylon than I do about that of my native province, which I had to leave before I had really woken up and rubbed the sleep from my eyes. I know little more of it than Stenbrohult, where I was born, and Växjö, where I first went to school, and I left them both before I was fully grown. Since then I have only viewed it as a wild-goose flying by, or a bird of passage, having passed quickly through on a couple of occasions, without stopping. That which I observed in my youth ... I now recall as a dream, which is all that I am able to communicate.

Towards the very end of his life, when he had already suffered a stroke, he noted down, in the uncertain handwriting of old age, on one of the loose scraps of paper he used for recording his passing thoughts, the heading *Nostalgia*, and then under it he scribbled the word "Stenbrohult".[65]

3. *Fate and Fortune*

In 1748 Linnaeus ran into trouble with the Uppsala theologians. He had prepared a treatise on the relationship between natural science and religion, and as was required, submitted it for consideration to Engelbert Halenius (1700–1767), dean of the Faculty of Theology. Halenius let it be known that he found nothing to object to in the work. The faculty, however, decided that he should attend its public presentation and raise a number of questions. Although he seems to have done so with tact and discretion, Linnaeus took umbrage, and is reputed to have vowed never again to trust Halenius and his like.

There was evidently more than a little misunderstanding on both sides. The theologians seem to have got hold of the idea that Linnaeus was involved in deifying nature. Given his work and his reputation this was understandable enough, but it was excessively annoying, since the whole point of his natural theology was that it should remain subordinate to ethical considerations. In response to the questioning he had had to undergo, he therefore drew up a fourteen-point critique of naturalism, in which he took the controversy into his opponents' camp, pinpointing a whole series of issues in orthodox Trinitarian theology which he suggested they might do well to concentrate upon – if they were in fact interested in the relationship between natural science and religion.[66]

Although in itself this flurry is of no great interest, it is worth looking at Linnaeus' reasons for picking out this particular field when attempting to put the theologians in their place, since it is precisely on account of its harmonizing so well with orthodox Trinitarian theology, that his religious thinking distinguishes itself most clearly from that of Newton.

If we compare the scholium Newton appended to the 1713 edition of the *Principia* with the statement concerning the Deity with which Linnaeus opens the 1758 edition of the *System of Nature*, we can hardly fail to be struck by their similarity.

For Newton, God is both transcendent and immanent, omnipresent not only virtually but also substantially, and although in no respect corporeal or human,

---

[65] Linnaeus, C., 1745, 9.8.1741; 1751, 15.5.1749, 2–3.8.1749; Ramm, A., 1907, 28, 65, 83; Tornehed, S., 1993, 17, 40; *Nemesis Divina*, II.iv.1; Linnée, C. von, 1878/1880, 11, no. 264, 104–107; Rogberg, S., 1770, ch. 4, 16–26; Tornehed, S., 1992; Levertin, O., 1906, 29; Linnaeus, C., 1729; Linné, C. von, 1888b, I: 53–105; Linné, C. von, 1951; Petersson, G., 1982.

[66] Linnaeus, C., 1748; Linné, C. von, 1905/1913, 11: 245–267; Afzelius, A., 1823, 113; Schröder, J.H., 1838, 3; *Nemesis Divina* IV.iii.1.3; Malmeström, E., 1964a, 114.

yet "all eye, ear, brain, arm, perception, understanding and action". Although God is to be admired as the perfection of life, intelligence and power, and revered on account of His universal dominion, He is to be known "only by His most wise and excellent contrivances of things, and final causes". Newton sees it as being self-evident that the diversity of natural things could only arise from "the ideas and will of a Being necessarily existing". Through our realization of this, we play some part in establishing an ethical significance, since if this Being failed to evoke our admiration and reverence, if we did not regard the diversity of things as evidence of Divine dominion, we would be incapable of conceiving of God as anything other than "Fate and Nature".

For Linnaeus too, God is both inherent and transcendent, "wholly and completely *Sense*, wholly and completely *Sight*, wholly and completely *Hearing*", and yet also "solely Himself". He is an essence, only to be seen in thought – the Lord and Master of the world, the founder and Builder of "all that shimmers here before our eyes". As we track His footsteps over the field of nature, we find "in each of them an infinite wisdom and power", and it is this which leads us to think of everything as happening in accordance with His will. We therefore revere Him as Providence, although "to say that He is *Fate* is not to be mistaken, for everything hangs upon His finger; nor is it wrong to say that He is *Nature*, for all things are born of Him".

Given this amount of common ground between them, it is important to remember that although Newton was not prepared to acknowledge it in public, he was in fact a Unitarian. In his own mind he evidently had no difficulty in reconciling transcendence with immanence. He was convinced, however, that acknowledgement of the unity of the Deity made it impossible to accept the orthodox Trinitarian doctrine of God's also being, "three Persons, of one substance, power, and eternity, the Father, the Son, and the Holy Ghost".

It is evident from Linnaeus' critique of naturalism, that on this particular point he was in much closer accord with traditional Christian thinking. He agrees with Newton that any attempt to rule out "the ideas and will of a Being necessarily existing", by asserting that the natural world is self-explanatory, is an intellectual non-starter. He also agrees with him, and, incidentally with Kant, in denying the rationality of attempting to explain morality in exclusively naturalistic terms. For him, as for them, a rational ethics necessarily acknowledges the plausibility of regarding duties as divine commandments. Unlike Newton, however, Linnaeus sees no reason why traditional Trinitarian doctrine should not be accepted as an adequate expression of immanent transcendence. In his critique of naturalism, he therefore urges his theological colleagues to concentrate on defending the orthodox view of the matter, on explaining in easily intelligible terms why it is unreasonable to confine oneself to acknowledging nothing more than the "Creator and preserver of the universe", or regarding Christ as simply a moralist, or deriving the doctrine of the Holy Ghost from the casting of a single vote at a general council.[67]

Linnaeus' polemical formulation of Trinitarian doctrine dates from 1748, the statement concerning the Deity, Fate, Nature and Providence which he inserted in the *System of Nature* from ten years later. It was precisely during this decade that he was most active in assembling the material contained in the Uppsala papers. It is certainly worth asking, therefore, what extra light can be thrown on this aspect of his thinking by the *Nemesis Divina*.

Fate, as it is conceived of by Linnaeus in the *Nemesis Divina*, would appear to be most closely associated with the first Person of the Trinity. Like the Father, it is essentially inaccessible to the human intellect. Linnaeus notes that our minds are incapable of grasping the eternal law by which everything is controlled and ordered. The decrees of fate can only be known by contemplating the ways in which they are expressed in nature. It is through nature that fate has a direct bearing upon man, and through experience that man learns to look upon fate as God's judgement and executor, from which there is no escape. Experience teaches us that our personal fate is fixed at our birth, and that it is apparent in our features. There is evidence enough that at any moment it can bear down tyrannically upon us, driving us willy nilly, annihilating all we regard as desirable, bringing home to us with a vengeance that no one is the sole author of his own fortune. Since in the face of fate all our strivings and prayers count for nothing, its decrees are only to be mastered by submitting to them, by acknowledging the absolute righteousness of God's judgement.[68]

Several of Linnaeus' points in the paper on naturalism seem to indicate that it is the second Person of the Trinity, counteracting original sin through His incarnation, working miracles within the natural order, reconciling man and God through His sacrificial death, whom he sees as having the closest affinity with fortune. Fortune is therefore the way in which fate appears when seen in the light of human passions and interests. Though its turns may seem to be fickle its pace is steady, for it is granted by the grace of God, and the whole of nature contributes to bringing it about, everyone helping along what may appear to us to be the wagon of misfortune. No one is capable of creating his own fortune, and magnanimity counts for nothing once its wheel begins to turn. By fulfilling our shortsighted passions and desires, fortune makes fools of us, encouraging our feeling of self-sufficiency and pride, luring us from the straight and narrow, preparing the way for our eventual downfall. Increasing involvement with it can only add to the overall aggravation. God will listen to no appeal against the misfortunes we bring upon ourselves, and we ourselves are incapable of understanding or rationalizing them. The only way of living

---

[67] Newton, I., 1729 (1934), II: 543–547; Linnaeus, C., 1735b (1766/1768[12]) I: 10; Bicknell, E.J., 1957[3], 22–53; Green, H., 1856; Manuel, F.E., 1974; Kant, I., 1793 , 230–231; *Nemesis Divina*, IV.iii.1.3.

[68] *Nemesis Divina*, IV.iv.4.1; II.i.3.1, no. 11; IV.iv.1.5; IV.iv.1.3; II.i.3.1, no. 1; II.iv.1.6; IV.iv.3.2, IV.iv.1.3; II.i.3.1, no. 9; IV.iv.4.1; IV.iv.1.7.

with fortune is therefore to bear its blows, rejoice in its benefits, and keep constantly in mind that in the long-run it is the meek and gentle who inherit the earth.[69]

In the language of Trinitarian theology, it is through the action of the Holy Spirit that mankind comes to terms with fate and fortune. It is very likely, given the intimate knowledge of Trinitarianism apparent in this note on naturalism, that this aspect of its doctrines played a major part in determining the most general principles of Linnaeus' thinking. In a certain sense, the whole of his treatment of nature was an attempt to overcome the vagaries of fortune, to break through to a clear comprehension of fate, to show forth the glory of God in the works of the creation. In the *System of Nature*, it is by paying proper attention to the subordinate aspects of his being, that man attains to the supreme self-knowledge of a well-founded theology. In the *Nemesis Divina*, the broader implications of this are worked out in detail. In Trinitarian terms, the whole process by which this comes about, by which apparent chaos is resolved into order, by which true science prevails, is the working of Providence, the third of the Divine Three.

Despite these differences with Newton, Linnaeus agrees with him in regarding ethics as essential to the overcoming of fate and fortune. God is not simply the supreme being, He is also ruler, creator, designer and judge, and it is as such that He plays such an essential part in the rationalizing of our lives. The order of His intelligence is reflected in that of our own. This is not simply a matter of our projecting our social awareness into our conception of Him, it is also the outcome of the awe and reverence engendered by the contemplation of nature. There are many passages in the *Nemesis Divina* celebrating the merits of withdrawing from the stream of life, letting the world go its own way, seeking peace and quiet in ourselves. There is much material taken from the Roman Stoics, extolling the virtues of mastering fate by bearing it. The broad message of what Linnaeus selects on the point is, however, much more in the line of living irreproachably because God is ever near, not falling into sin, and especially not the sin of disdaining God, of being aware that God is not to be mocked, that He knows all, that He furthers the just and punishes the wicked, that retribution is visited on those who sin, down through the generations.

As a critique of naturalism itself, Linnaeus' fourteen points are designed to call in question the validity of any philosophical standpoint which does not subordinate natural theology to ethical considerations – the views of Cartesian dualists who had opted for the supremacy of extension, of Spinozists who had deified nature, of Leibnizians who had identified fate with the natural course of things. He also notes, however, that he has drawn them up in the light of what was being asserted by freethinkers, whom he has "come across in all periods

[69] *Nemesis Divina*, IV.iii.1.3, nos. 3–7; IV.iv.2.1; IV.iii.2.3; IV.iv.2.1; IV.iii.2.1; IV.iv.3.2; IV.iv.1.1; IV.iv.2.1; IV.iii.2.1; IV.iv.2.1.

of the world", evidently the Arians who had over-emphasized the unity of the Godhead, the Gnostics who had compared Christ with the progeny of Jupiter, the Antinomians who had denied the need for observing any moral law. In fact when they are taken as a whole, the fourteen points can be seen as touching on a pretty comprehensive spectrum of highly technical theological issues – the precise inter-relatedness of the three Persons of the Holy Trinity, the existence of good and evil angels, the fall, redemption and future life of mankind.[70]

By the middle of the century, this readiness to associate naturalism with freethinking was common enough. To a certain extent, it was due to the way in which the freethinkers, who like Linnaeus were originally intent simply on removing superstition, bringing out the essential rationality of religion, had gone on to adopt so many of the tenets of the deists – to question traditional doctrine concerning the nature of the Trinity, the Creation and personal immortality. In the heat of the controversies generated by this convergence of ideas, freethinking became associated not only with naturalism, but also with libertinism, with an attitude to personal conduct very different indeed from that of Linnaeus. In Uppsala, the possibility of talking at cross-purposes when discussing such matters was increased still further when the Leibnizians or Wolffians in the Faculty of Philosophy, despite developing their own particular critique of naturalism, managed to fall foul of the orthodox Lutherans in the Faculty of Theology. In 1742, for example, Anders Knös (1721–1799) began what was to be an outstanding career in the church in the Faculty of Philosophy, by defending an elaborately Wolffian thesis on "the principles and inter-connectedness of natural and revealed religion". Although this could have been regarded as a praiseworthy initiative, it resulted in his being criticized by the theologians for attempting to prove matters of faith, and for dealing with an essentially theological issue in the wrong university context. Four years later, therefore, he attempted to set the matter right by submitting a critical analysis of naturalism in the Faculty of Theology. He was faced with such a commotion that he decided to leave academic life altogether, and take up parish work.

Linnaeus' fourteen points have therefore to be seen as part of a widespread and somewhat confused debate, in which the traditional Lutheranism of the established church of the country came into conflict with the popularized Leibnizianism then dominating Scandinavian academic philosophy. Although we know nothing for certain about his concept of freethought, it seems reasonable to assume that it was influenced in one way or another by the book Anthony Collins (1676–1729) published on the subject in 1713, since here too the movement was associated with naturalism and deism, and freethinkers

---

70 *Nemesis Divina* IV.iii.3.1; II.i.3.1; IV.iv.3.5; IV.i.1.2; IV.iv.4.1; IV.iv.3.2; IV.iii.2.2; IV.iii.2.4: IV.iv.1.4: IV.iv.2.2: IV.iv.3.2; IV.iv.3.3; IV.iii.1.3, nos. 1–6, 8 (Trinity), 11 (Angels); 7–10, 12–14 (mankind).

were identified in "all periods of the world", Socrates, Plato and Aristotle, as well as Bacon, Hobbes and Tillotson being classified as such.[71]

When considering the general significance of the fourteen points, it is important not to forget that although Linnaeus is up in arms about the way in which the Lutheran theologians he finds around him are reacting to his work, he is by no means critical of Lutheran theology as such. His main point is simply that this theology ought to be expounded more effectively, especially in respect of the relationship between natural science and religion.

The ground Linnaeus shared with his theological sparring-partners at Uppsala can best be gauged by taking into consideration his family background and early training. He had grown up in a country rectory, and although much of his schooling was designed to prepare him for a career in the church, he seems soon to have realized that he was not suited to be a clergyman. There is no evidence, however, that he ever considered abandoning the basic tenets of his religion. To some extent, his reservations concerning the career envisaged for him by his parents seem to have been due to nothing more than the contrast between his life at home, in which he found fulfilment, and his life at school, which he regarded as a chore. He was a boarder at the cathedral school in Växjö for nearly eleven years, from when he was just turning nine until he was almost twenty. The instruction provided at the school was rigorous and coercive, the main purpose of it being the training of efficient clergymen, teachers and civil servants. The day began with morning prayers and hymn-singing, either in the school itself or in the cathedral. Lessons began at six and ended at five, with a two-hour break around midday, apart from Wednesdays, when they finished at two, the rest of the afternoon being devoted to games and outdoor activities. In the senior classes there were forty-seven hours of instruction a week, – seventeen in Latin, fourteen in theology and ethics, four in Greek, two in history, one in literary style, and nine in mathematics, physics and logic. It was only during his final year that Linnaeus was able to pay particular attention to developing his interest in the natural sciences under the supervision of one of the most gifted of the teachers, Johan Rothman (1684–1763).

Religious knowledge was taught at the school partly as a matter of biblical exegesis: one of the two teachers responsible being required to "expound Greek texts from the New Testament in a straightforward and succinct manner", and partly as a matter of systematic theology, the other teacher appointed for the subject being required to give instruction "on theological topics in accordance with a publicly authorized text-book". The two instructors whose classes Linnaeus attended in the upper school were Pehr Hyllengrehn (1677–1734), who took the New Testament classes until 1725, and thereafter those in systematic theology, and Jacob Flachsenius (1683–1733), who taught him

---

[71] Clarke, S., 1706, 19–37; Thorschmid, U.G., 1765/1767; Frängsmyr, T., 1972, 156–169; Lindroth, S., 1978, 524, 549; Knös, A.O., 1742; Collins, A., 1713, 25f.; O'Higgins, J., 1970; Holberg, L., 1748/1754, no. 465.

systematic theology from 1723 until 1725. Flachsenius was by far the more able of the two, and his leaving the school in 1725 may well have strengthened Linnaeus' resolve not to go on to read theology at university. It is quite possible that he had a decisive influence on Linnaeus' intellectual development. He was the son of a professor of logic and metaphysics at the University of Åbo who had worked out a classificatory system for the natural sciences remarkably similar to that of the *System of Nature*. In 1703 he graduated at Åbo with a thesis on the sibylline oracles, and after being driven from home by the Russian raids on the city, went on to take a further degree at the University of Jena in 1711. When he first arrived at Växjö in 1713 he taught logic and physics, handing these subjects over to Rothman five years later, when he took holy orders. He left the school to take up pastoral work in the diocese of Växjö, becoming rector of Rydaholm and dean of Östbo. In September 1727, soon after Linnaeus had left Stenbrohult for Lund, he distinguished himself at a diocesan convention, gaining the approbation of the bishop, his brother-in-law, and of the cathedral chapter, by strongly defending the orthodox Lutheran doctrine of the Eucharist.[72]

As was then the case in most Swedish schools and universities, the "publicly authorized textbook" used by Flachsenius and Hyllengrehn for instruction in systematic theology was based on a work first published in Germany in 1600 – the comprehensive statement of orthodox Lutheran doctrine drawn up by Matthias Hafenreffer (1561–1619), professor of theology at Tübingen. Hafenreffer had set himself the task of combining, in one universally acceptable teaching compendium the articles of faith and the decisions concerning church government affirmed in the Augsburg Confession (1530), and the theological formulae finalized in *The Book of Concord* (1580). This had involved distinguishing the Lutheran position on such controversial topics as Divine foreknowledge and human freedom, man's depravity after the Fall, the Real Presence of Christ in the Eucharist, from that of Rome on the one hand and Geneva on the other, and it was the effectiveness with which Hafenreffer had managed to do so, which had ensured such a wide currency for his book throughout Lutheran Germany. Although the work had been rejected in Denmark, largely on account of its owing too much to *The Book of Concord*, it was readily accepted by the great majority of the Swedish clergy, evidently on account of its harmonizing so well with the current revival of Aristotelianism, and its uncompromisingly anti-Calvinist stance with respect to the ubiquity of Christ in its treatment of the Eucharist. The royal house of Sweden, not at all enthusiastic when the book was first catching on among the clergy, eventually came to appreciate its merits, not only as a magnificent exercise in concise and

[72] Ödmann, S.L., 1830; Arcadius, C.O., 1888/1889, 1921/1922; Hedlund, E., 1936; Brilioth, Y., 1943; Fredbärj, T., 1970/1971; Broberg, G., 1972/1974; Hillerdal, G., 1992; Flachsenius, J.H., 1678; Kallinen, M., 1995, 115; Scheutz, N.J.W., 1878/1880, I: 31–32; Virdestam, G., 1921/1934, 5 (1931) 285–286.

lucid theological exposition, but also as a first-rate theoretical justification for political order and monarchical supremacy. Charles XII knew the greater part of the primary-school version of it off by heart.

It was this shortened version, first published at Gothenburg in 1657 and in seven further editions before the century was out, which constituted the basis of Linnaeus' theological studies from the age of nine until he was sixteen. Once he had come under the supervision of Flachsenius, he would have been working with the fuller version of the book prepared by Petrus Kenicius (1555–1636), archbishop of Uppsala. This had first been published in 1612, and by the turn of the century had also appeared in fifteen further editions. It is certainly worth asking, therefore, what part this theological tradition might have played in the drawing up of Linnaeus' fourteen points, and in the development of the broad conception of fate, fortune and Providence pervading the *Nemesis Divina*.[73]

Hafenreffer's compendium is divided into three main sections – the first dealing with the Deity as such, the second with angels, the third with mankind. This third section is by far the most extensive, and falls into three further subsections – the first dealing with the state of innocence, the second with the state of sin, the third with the redemption of mankind through Christ, the church and the civil order. Much of Linnaeus' basic attitude to theological issues is readily recognizable here. As we pass Hafenreffer's expositions under review, it soon becomes apparent that what Linnaeus is really advocating in his fourteen points is that those teaching theology at Uppsala ought to reacquaint themselves with the basic tenets of the tradition within which they are working. In Hafenreffer, the creation and preservation of the universe are treated as one of the attributes of the Deity, and if the general drift of his exposition is carefully considered, a clear standpoint emerges in respect of the relationship between the natural sciences and religion. The subject matter being investigated by the scientist is the creation of the Almighty, and in order to understand the scientist himself we have to take into consideration not only innocence and sin but also redemption.

Trinitarian theology is dealt with in detail in the first section of Hafenreffer's book, due emphasis being laid on both the unity and the triadicity of the Godhead, and although it is admitted that in the last instance this conception of the Deity has to be regarded as a mystery, a great deal of attention is paid to bringing out the rational implications of Trinitarianism. Since Linnaeus in his points makes specific mention of "the Holy Ghost's being accepted at a certain council, at which a single vote would have decided the matter the other way", the extensive consideration Hafenreffer gives to Athanasius and the Council of Nicaea ought not to be overlooked. It seems quite likely that

[73] Hafenreffer, M., 1603 (1600); Askmark, R., 1943, 296f.; Lindroth, S., 1975, 94–106; Carlquist, G., 1918, 49; Malmeström, E., 1925, 57–61; Normann, C.-E., 1948; Hafenreffer, M., 1612, 1657.

the exposition of the doctrine of Christ's ubiquity in Hafenreffer's treatment of the Eucharist influenced Linnaeus' conception of Divine immanence. There is, however, no direct evidence for this, apart from the fact that this was a topic of particular interest to Flachsenius, and that Linnaeus does make a general point concerning those who "do not believe in the sacraments as such". Naturally enough, there is also a certain disparity between Linnaeus and Hafenreffer in respect of fate, fortune and Providence. The basic concepts employed in the *Nemesis Divina* and the *Loci Theologici* are much the same, but the language, associations and implications are somewhat different. Linnaeus' ideas take in folk-belief and the Roman Stoa he had read at school, particularly in respect of the relationship between fate and fortune. For Hafenreffer, Providence is the direct expression of God himself, it is sin and heresy which prevent man from realizing his full potential, it is faith which lights the way to salvation.

Hitherto, Hafenreffer has only emerged onto the horizon of historians of science on account of his correspondence with Kepler on the ubiquity of Christ, and his having anticipated Newton's interest in the mensural properties of Ezekiel's Temple. It should now be clear that his influence on the intellectual development of Linnaeus was by no means peripheral, and that it has to be taken into consideration by anyone seeking to understand the central concepts of the *Nemesis Divina*.

Why was it then, that sometime toward the end of his school career at Växjö, Linnaeus reacted so sharply and unexpectedly against continuing his theological studies at university? The occasion on which he did so is vividly described in a letter written by his brother Samuel Linnaeus (1718–1797) to the historian Jonas Hallenberg (1748–1834) during the spring of 1778, a few weeks after Linnaeus' death:

> Some good friends were on a visit to Stenbrohult. Father took them with him into the garden, and they gathered around a little table, sitting there and chatting while they finished off a few glasses of ale. During the conversation, father happened to say: 'Yes that's always so. If you want something, the chance will always come your way.' Charles, who was never absent when anyone went out into the garden, noticed the remark. When the company had left and father had returned to the table, Charles went up to him and asked if there were grounds for what he had said when the company was present. Father, cheerful and good-humoured as he always was, asked what he was referring to. Charles, set on getting an answer, repeated what father had said. Father said that there was certainly some truth in it, – if one wants what is good, wanting always helps. 'Then, father', said Charles, 'never ask me to become a clergyman, for that is not what I want.' O, what a thunderclap that was for father, who had always set so much store by his Charles. Hardly able to say anything, amazed and horrified, he asked him what he did want. Charles said that he wanted to read medicine and botany. Father reminded him that the family was not well off, and that taking a course such as that was extremely costly. Charles responded by taking his father at his word: 'Father, if there really are grounds for what you have said, then God will provide what is necessary. If I have a chance of getting what I want, I shall never lack the means.' Completely taken aback, and with tears in his eyes, father replied: 'May God give you the chance. It is not for me to force you from what you want.'[74]

[74] Hafenreffer, M., 1603, 24–207 (Deity), 208–227 (angels), 227–842 (man); 77–95 (Cre-

Given the demonstrably positive effect which Hafenreffer had upon Linnaeus' thinking, it would certainly be wrong to interpret this incident as simply a reaction against what was being taught at Växjö. As we have seen, dissatisfaction with school-life in general and the departure of Flachsenius may well have been decisive in hardening Linnaeus' resolve not to tie his future to the church. In fact much of the evidence available seems to indicate that this was a positive rather than a negative reaction – he had always loved plants and roaming in the countryside, and Rothman had now opened up the prospect of earning a living by continuing to indulge these passions.

It is also conceivable, however, that Linnaeus had already sensed the weaknesses and limitations of Hafenreffer's world view. Was the created world not immeasurably richer and more rewarding than one might ever have guessed from any purely theological account of the Creation? Was not the world of man more intimate and awesome than it could ever be seen to be from the study or the pulpit? Could it not be that it is by testing Providence rather than formulating theological doctrines that we come to understand God best? Hafenreffer had called the Bible to witness for each and every of his expositions, and Linnaeus was to employ the same procedure in defining the general categories of the *Nemesis Divina*. But was it not an even finer thing to submit both the world and the Bible to careful and impartial investigation, orderly and rational analysis?

4. *Providence*

Linnaeus' emotional attachment to Stenbrohult – to family life at the rectory, to his father's garden, to the countryside along the shores of Lake Möckeln, to the country people he knew in his youth – contrasts sharply with the way in which he merely assimilated intellectually what had been drilled into him at Växjö.

His father's personality and social position must have played an important part in determining this feeling for the environment into which he was born. Although he had to oppose his father's wishes in choosing his career, he must have appreciated the way in which he had dealt with the thunderclap in the garden. His father was naturally outgoing but not very articulate, and seems never to have made much of a mark as a preacher. He came of a local family which had been known in the area for generations, however, and since he also had a natural and spontaneous sympathy with people, he was effective in his pastoral work. There is record of his having given evidence of a somewhat hasty temper when dealing with officials and administrators, but among his

ation), 44–54 (Trinity), 66–74 (Athanasius); 549–662 (Sacraments), 95–137 (Providence), 262–279 (sin), 427–455 (faith); *Nemesis Divina*, IV.iii.1.3, nos. 8, 13; Kepler, J., 1938/1988, XVI, no. 586, XVII, nos. 808, 829, 835, 847; *Ezekiel* 40–48; Hafenreffer, M., 1613, 48–89, 340–344; Newton, I., 1737; Wieselgren, P.J., 1831/1843, IX (1837): 198–209; Virdestam, G., 1928, 1931.

parishioners he was known for the ease with which he was moved to tears, both by their joys and their sorrows. The parish registers he kept provide us with various glimpses of this living bond between pastor and flock, of the people among whom Linnaeus grew up. When recording a burial service, his father was in the habit of making a short comment on the departed. There were, of course, cases enough of drunkenness and evil-doing, but by and large one gets the impression of a hard-working and healthy community. One of the deceased is said to have been an "unfortunate cross-bearer", another "poor, but ever content with God's will". There were those who had "striven hard but gained little", or "lived quietly and well, and left a good name behind". Many are recorded as having been "honest, quiet and self-effacing", or "godfearing and hardworking".[75]

We know from the books that were read in the rectory at Stenbrohult, that it was there that Linnaeus first became acquainted with a powerful movement within the Lutheran church which was critical of formalities and ceremony, and which laid great emphasis on the importance of personal devotion. Although it could not be identified completely with pietism – a movement which tended to flourish in the towns rather than the countryside, and which was frowned upon both by the regular clergy and by the civil authorities – it did have certain clear affinities with it. We know, for example, that during the 1720s his father purchased a newly-published translation of *The Soul's Treasure*, a widely-read Lutheran devotional book concerned with the soul's liberating itself from tribulation by cultivating its own inwardness. Although this was a German work, written by Christian Scriver (1629–1693), court preacher at Quedlinburg to the duchess of Saxony, and had first been published half a century earlier, its message was wholly in tune with the national mood in Sweden during the period following the end of the Great Northern War. This new version of the book (there was an earlier translation published at Åbo during that "hard year of tribulation 1697") had been sponsored by Reinerus Broocman (1677–1738), formerly one of Charles XII's army chaplains. The king was now dead, the war was lost, the Russians were plundering Finland and the coastal towns. Was it not time to listen again to Scriver and make the most of inner resources?

Scriver himself had suffered personally from the ravages of the Thirty Years' War, and had been deeply influenced by an even more widely-read work, also a great favourite in the rectory at Stenbrohult – the exposition *Of True Christianity* by Johann Arndt (1555–1621) of Celle. This four-volume embodiment of reformed Lutheran doctrine, published in any number of editions throughout the whole of the seventeenth century, had for many years been the most popular book of devotion in Sweden apart from the Bible.

It is not difficult to see why the writings of Scriver and Arndt had more

---

[75] Larsson, L., 1923; Virdestam, G., 1924; 1928.

appeal for the young Linnaeus than the biblical exegesis he was forced to indulge in at Växjö. They were readable and imaginative, and brought the living Christianity he knew through life at Stenbrohult into a wider context. They highlighted the importance not only of the great tradition itself, but also of the wondrous world of nature, the immediate concerns of the individual moment. Arndt's four books deal with Scripture, the life of Christ, conscience and nature. They draw not only upon the inspiration of the mediaeval and renaissance mystics – Angela of Foligno, Johann Tauler, the *Theologia Germanica*, Thomas à Kempis, Valentin Weigel – but also upon the works of Paracelsus. Book four sets out to show "how all men are roused by creatures to love God, and convinced by their conscience that this is their highest calling". The treatment of nature is then ordered according to the work of the Creation; separate chapters dealing with the division of light from darkness, with the making of the firmament which separates the waters, the gathering together of the dry land and the seas and the engendering of grasses, herbs and trees, the placing of the Sun, Moon and Stars, the bringing forth of living creatures, the creation of man. Linnaeus, as we have seen, was never absent when anyone went out into his father's garden, and one can well imagine him dwelling upon the glorious passage in the third chapter of Arndt's fourth book:

> The plants then rise up from out of the earth, from out of their sleeping-chambers as it were. They have cast off the old body, and have taken upon themselves a new, tender, young and blossoming body, for the old has fallen away and is dead. They have shed the old garment and clad themselves anew, for the old is worn and torn, dirty and hateful: they have rid themselves of the old colour, the old form and smell. And by means of their fair and comely form, their scent and their colour, they begin to speak to us in a new tongue: Behold ye child of man, ye faithless one, we were dead and we now live; we have cast off our old body and garments, and have become a new creature ... See, we offer up to you all our powers, which serve you and not us: we blossom not for our own sake but for yours. Yea, God's goodness also blossoms in us, yielding refreshment to you through the sweetness of its scent. Is there anyone who cannot see, in the plants of the earth, so many thousands of witnesses to the love, goodness and allpowerfulness of God?[76]

The villagers and country people among whom Linnaeus grew up may not have had much idea of what the Stoics had had to say on the difference between fate and fortune, or of the theological niceties of Trinitarianism, but they knew well enough that "not even the Devil would dare to maintain that anyone hammers out his own fortune". Living, moving and having their being within the great cycle of the seasons and the agricultural year, constantly experiencing, in all around them, the equally regular rhythms of birth, procreation and death, they were under no illusions as to the supposed autonomy of individuals, congregations or nations. Prior to the tragic show-down with Russia, they had played their part as their own nation made its mark in world

---

[76] Sahlgren, J., 1922; Linderhielm, E., 1962; Pleijel, H., 1935; Lindquist, D., 1939; Scriver, C., 1675, 1697, 1723/1727; Arndt, J., 1606/1610, 1647/1648, 1695, bk. 4, ch. 3; Weber, E., 1978; Sidenvall, F., 1992.

affairs. It was the inborn fatalism of the ordinary Swedish infantryman which made the Caroline armies of the time the finest in Europe, and those who trained him and ministered to his spiritual needs were well aware of this. Recruits were discouraged from considering any evasion of enemy fire by seeking cover. They were to advance with their heads held high, never forgetting "that no musket-ball can hit a man except God will it, whether he walks straight or crooked".[77]

As we have seen, this emotional attachment to Stenbrohult and all that it symbolized for him, remained with Linnaeus to the end of his life. He may have acknowledged intellectually, in the prefaces he wrote, that Eden was the first and finest of all gardens, but what moved him most deeply was the memory of the garden in which he had grown up, the green world around the rectory in which "together with his mother's milk", his mind had first been "inflamed with an unquenchable love for plants". He certainly did nothing to avoid the public applause showered on the brilliant young research-worker in the Netherlands, the renowned professor in Uppsala, but in his quiet moments at home, as we have seen, his mind would return to his "beloved birthplace", he would ask himself what sweetness it was "which draws us to the land of our birth, and will not allow us to forget". One suspects that the same pattern of final emotional commitment permeates the whole of the *Nemesis Divina* undertaking. A theodicy had been developed which avoided the naturalism of Löwenhielm and Leibniz, which hammered home the essential point that man's most direct experience of God derives not from expounding metaphysical abstractions or contemplating nature, but from the constant demands being made upon him by his awareness of Nemesis. Traditional Trinitarianism had been invoked in order to throw light on the interrelatedness of fate, fortune and Providence, and vindicated in the face of a theology in the process of becoming alienated from its own sources.

Given the context in which Linnaeus was working, this was a very considerable accomplishment. It has to be admitted, however, that if we confine ourselves to the *Nemesis Divina* papers, it is an accomplishment which seems to lack any final focal-point. Reading through the work, one can imagine how he might have drawn the various threads of his collection together, but there is no very persuasive evidence that he actually did so.

The fact is, that if we confine ourselves to the material available, it is difficult to make much of the way in which the final judgement of God works itself out in fate, fortune, retribution and blessing. The case-histories deriving from the Stenbrohult area tend to involve intuitive rather than social or rational

[77] *Nemesis Divina*, IV.iv.1.3; IV.iv.2.1; *Krigz Articlar*, 1621, Introduction: "Since all Fortune derives from God, and every Christian people ought to worship Him, as He has revealed Himself in His Word ..."; Holthausen, C.J., *Soldatens Skyldigheter*, 1769: "Fear God, honour the king, serve faithfully"; Hildebrand, K.-G., 1942; Normann, C.-E., 1948; Englund, P., 1988, 1992, 24; Braw, C., 1993.

relationships, and they therefore throw little light on the central themes of the *Nemesis Divina*. The soldier who senses that he has met his future executioner at the inn, the fortune-teller who correctly forecasts Linnaeus' own career, the telepathic family dreams concerning death, corpses and funerals, the moral lessons to be learnt from the animal world, are quite obviously a far cry from whatever it was that Linnaeus envisaged as giving coherence and relevance to the tenets of Trinitarian theology. It should not be overlooked, however, that the motifs of nearly all the case-histories he amasses in order to illustrate the absence of natural causality in supernatural retribution, are typical enough of ordinary Swedish folk-belief, and might be multiplied indefinitely from modern folklore collections – a careless remark precipitates dire consequences, retribution occurs in the same place or at the same time as the offence committed, divine intervention sets right a miscarriage of earthly justice, the offender's descendants continue to be punished down through the generations, despite present appearances one knows that finally justice will be done, etc. In fact if it were not for the enlightening distinctions between the various aspects of man put forward in the *System of Nature*, and the possibility of ordering the material of the *Nemesis Divina* in accordance with them, there would be good grounds for arguing that Linnaeus' theodicy consists of little more than a retailing of the intuitive convictions of the folk-culture in which he had grown up.[78]

It is interesting to note, therefore, that although Linnaeus took a great interest in folk customs and superstitions, recording them diligently in his various surveys of the Swedish provinces, he seems never to have attached much intellectual significance to them. In his account of the journey through Öland and Gotland, for example, he takes note of a whole series of customs and beliefs relating to weddings, funerals and the seasons of the year, and then comments as follows:

> There is no end to this sort of twaddle. The best way of getting rid of trivialities of this kind is for forward-looking theologians to take up the study of physics and natural history, since the most effective check on superstition comes from its being held in contempt by theology. It is strange, however, how the nation has preserved these and many other such superstitions from the earliest times and from the heathen period. One encounters some of them among the poets, round about the time of Christ; in Sweden, some of them are leftovers from heathendom; some have survived from papal times, some have been artificially revived in order to replace something else. It seems to me that there is much to be said for making an extensive survey of superstitions, and indicating where each of them has originated.

[78] *Nemesis Divina*, II.iii.3.2; II.iv.1.2; II.iv.1.6; II.iii.1.3; II.iv.1.2; II.iv.3.6 (Stenbrohult): II.i.1.8; III.iii.5.2; IV.ii.2.7 (careless remark): II.ii.2; II.iii.3.6; III.iv.3.9 (same place): III.iv.2.4; III.iv.3.14; IV.i.2.1 (same time): II.ii.3; II.iv.1.6; III.ii.3.2; III.iii.5.3; IV.iv.1.7 (earthly justice): III.i.3.4; III.ii.4.1; III.iv.2.3; III.iv.3.3; IV.i.4.2; IV.ii.2.14; IV.iv.3.2; IV.iv.3.3 (descendants): II.i.3.5; II.iv.1–II.iv.3; IV.i.4.1; IV.iv.2.3 (future): II.ii.5; III.ii.1.5; III.iii.2.6 (causality); Ehnmark, E., 1941; Wikman, K.R.V., 1964a, 1968/1969, 1970.

Here we almost certainly have the key to the connection Linnaeus saw between folk-belief, theology and science: the theologians are to take up natural science and natural history in order to rid mankind of superstition. But if this is the case, it looks as though the programme of research he envisages in order to help on such a development clashes rather awkwardly with that implied by his nominal commitment to the myth of the Garden. Would it not have been more consistent of him, and would his own thinking in the final section of the *Nemesis Divina* not have been more enlightening and illuminating, if he had fallen in with the kind of historicism which usually accompanied the myth of the Garden? Should he not have followed in the footsteps of Bacon, Rudbeck, Newton and Vico, and idealized language, myths and folk-customs as a source of esoteric wisdom? He was certainly aware that this was what many of his contemporaries were doing, and confronted as we are with this apparent inconsistency in his manner of thinking, should we not ask whether he might not have done well to follow suit? He could easily have gone along with the Rudbeckians, for example, and concentrated on breathing new life into the 'Ragnarök', the twilight of the gods, the 'final phase of the powers that be'. Was it really such a trivial idea, that since fate holds sway over all, the gods themselves are subject to it and destined to pass away? According to the widespread Swedish folk-belief, which he chose to ignore, the evident deterioration in the fertility of the land, the relentless extension of the forests, the seemingly irreversible return of pasture and arable to a stony wilderness of heather, moss and rocks, were immediate and tangible evidence of a wholly universal process of degeneration. The time was approaching when the gods too would succumb to Loki and the forces of destruction. The sun would be darkened, turning the summers to winter. Men, who had struggled on through the ages of axes and swords, brother murdering brother, would have to leave their dwelling-places and pass on into the realms of the dead. The ship of death, formed from their fingernails, would sail forth. Heimdall, guardian of the citadel of the gods, seeing the giants advancing on all sides, would sound the alarm by winding his horn. Fenris the giant wolf, breaking his fetters, would rush on Woden and devour him. The sword of the giant Surt would put an end to Freyr, to all growth and fruitfulness. Thor would be mortally wounded by the Serpent of the World. The earth would sink into the sea, the stars fall from the heavens, the heat and flames of destruction reach up into the sky.

It could be that in spite of appearances, Linnaeus was not entirely indifferent to the light which such historicism was throwing on language and the meaning of words. It may be significant, for example, that when he refers to fate in the *Nemesis Divina*, he never uses the ordinary Swedish word 'skepnad', possibly because he was uneasy about its being so closely associated with initiation or creation. For the most part, he makes use of the Latin word, which has the literal meaning of 'that which has been declared' or 'the doom of the gods', concepts which fit in well with the central principle of his theodicy. He also makes use of 'öde', however, which is now the ordinary Swedish word for fate, and this too could be significant, since it was in fact a word which had been "artificially revived in order to replace something else", the noun having been extracted from a verb meaning 'to allot', evidently on account of its having been associated in heathen times with the Norns and their weavings. It is quite possible that the quality of his exposition in the final section of the *Nemesis Divina* would have been improved had he made something more of this. Although he evidently saw the point of tracing such concepts back to their origins, however, he was not prepared to treat folk-wisdom of this kind as a matter of prime importance.[79]

If we are to explain this apparent inconsistency in Linnaeus' attitude to such matters, we have to take into consideration his view of the church and church history, which was in many respects wholly conventional and straightforward. From the Växjö period on, he seems to have taken it for granted that the religious message being conveyed by the church was rooted in texts inherited from an age-old Biblical and classical tradition, which had been formalized, institutionalized and brought to Sweden by the Papal Church, and finally purified of alien elements at the Reformation. Since he seems never to have called any of these historical facts in question, since he evidently accepted them as constitutive of any well-founded view of things, it was only natural that he should have taken the traditional church view of the systems of belief it was in the process of replacing, and treated them as imperfect adumbrations of a higher and more complete truth. To a certain extent, therefore, it was on account of the long-term success of Christian missionaries such as the Englishman St. Sigfrid (d.c. 1045), patron of Växjö, during the "papal times", that Linnaeus remained resistant to current revivals of Nordic primitivism. It is interesting to note, moreover, that the martyrdom of the three nephews of St. Sigfrid does get a mention in the *Nemesis Divina*, and that February 15th, his day in the church calendar, stayed in Linnaeus' mind as that on which, in 1714, his father had first employed Johan Telander (1694–1763), brutal pedagogue though he was, to coach him for entrance to Växjö School.[80]

In one crucial respect, however, Linnaeus discovered that folk-belief and church doctrine coincided completely – both took it to be a matter of central importance that due recognition should be given to the radical duality of good

---

[79] Linnaeus, C., 1745, 311 (August 5th); Ramm, A., 1907, 25; *Nemesis Divina*, IV.iv; Lindroth, S., 1975, 297–305; 1978, 643–658; Olrik, A., 1922; Neckel, G., 1927; Rod, J., 1972; Olsson, B., 1954/1955; *Nemesis Divina*, II.i.3.1; IV.i.1.2; IV.iv.1.3 (fatum): II.iv.1.6; II.iv.3.1 (öde); Hellquist, E., 1948; *Svenska Akademiens Ordbok* 26, cols. 3769–3772; Stanley, E.G., 1975, §324; Bates, B., 1983; Mitchell, B., 1995, §§414/15.

[80] Malmeström, E., 1942a, 1962; *Nemesis Divina* III.ii.4.1; IV.iv.3.3, dating from 1765; Frondin, E., 1740; Johan Magnus, 1554, lib. 17, caps. 18–20, 560–564; Klingspor, G.A., 1932; Rydbeck, M., 1957; Larsson, L.-O., 1991, 19–28; Linné, C. von, 1957; Fries, Th.M., 1903, 1: 12; Broberg, G., 1972/1974, 12.

and evil, and the corresponding duality of good and evil angels. It is not the case that we simply decide upon and will our actions at a rational level. As we all know, to a certain extent both thoughts and deeds are determined by largely unfathomable motives. Even conscience can be played upon by the subconscious. It is perfectly understandable, therefore, that good and evil wraiths, angels and guardian spirits, should have a place in both Nordic folk-belief and the Christian tradition. Due mention is made of them in Hafenreffer's compendium, the whole of the second part of the work, that linking the consideration of the Deity to the treatment of man, being devoted to a discussion of the characteristics they all have in common and of their dual role in the God-man relationship. There is nothing in the *Nemesis Divina* papers, however, which might enable us to integrate the subjective tension between fortune and virtue into the exposition of the objective manner in which the final judgement of God works itself out in fate, fortune, retribution and blessing. The two references they contain to "the angel who protects us day and night from misfortune" and, "the shadows Virtue and Fortune", occur in contexts which clearly have more to do with our subconscious than with eschatology. It is certainly of some significance, however, that Linnaeus is even prepared to call in question an orthodox interpretation of the Scriptures in order to add credence to the doctrine of a plurality of wraiths. Disregarding Luther and the official Swedish translation of the time, he pluralizes the guardian angel celebrated in *Psalm* 34, quoting verse eight as follows: "The angels of the Lord encamp round about them that fear him, and deliver them", evidently only too pleased to exploit the slight ambiguity in the Hebrew original in order to find support for the popular doctrine.[81]

The clearest evidence that it was in fact the agreement between popular belief and church doctrine in this particular respect which provided Linnaeus with the keystone to his theodicy, is to be found not in the *Nemesis Divina*, but in certain of his other writings. The earliest documentation dates from the summer of 1741, and consists of an account which he gives in his journal of a visit to the Stenbrohult area:

Wise woman Ingeborg, of the parish of Mjärhult and Virestad, was sought after everywhere in the area as an oracle, and had created more of a reputation for herself in medicine, than many a doctor who had studied and practised during the course of a whole career ... She believed that Lucifer's followers had been cast down from heaven onto the earth, where they had found their dwellings in various places, – what are known as merrow men in the waters, those called gnomes under houses, elves and their like in reeds and under timber, wood nymphs and spirits in the forests and glades. She believed that each person had his wraith, which follows him just as the body is followed by its shadow, and that this wraith extends downwards perpendicularly, just as the person stands perpendicularly above the ground, always moving its feet as the person moves his. She found incontrovertible evidence of this in the way in which animals, woods and

---

[81] Vries, J. de, 1970, 1: 217–241, §§158–171; Hafenreffer, M., 1603, bk. 2, 208–227; Ehnmark, E., 1944; *Nemesis Divina*, II.iii.1.2: II.iv.3.1; Luther, M., 1959/1963, II: 58–66; *Biblia*, 1778, 626.

hills may be seen from many sides to be reflected in the water of a clear and still-standing lake or river, in the trees and suchlike which grow down into the water, just as they stand up above the ground. She believed that a person and his wraith are so united, *that when the person above the ground suffers, so too does his counterpart below*, and conversely, that when the person below the ground is harmed, so too is his counterpart above it.

The wise woman's scheme of things was certainly more worthy of serious consideration than the "twaddle" of superstitions Linnaeus had recorded and criticized a couple of days earlier. The realms of metaphysical and physical evil were clearly prefigured in her heaven and her earth, and she may well have added a new dimension to the concept of moral evil – if it was in fact a new thought to Linnaeus, that a person's wraith may be regarded as the counterpart to his uprightness.

In Linnaeus' mature thinking, what the wise woman conceived of in physical terms – the body being upright, the shadow passing downwards – is transformed into the spiritual polarity between ethical rightmindedness and fortune. The transformation must have taken place soon after the 1741 visit to Stenbrohult, since already toward the close of the following year, we find him contrasting the evident capriciousness of fortune with the certainty of a religiously-orientated ethics. In his obituary of his father-in-law, he observes that: "He was religious, unmoved by fortune; a word given, an agreement made was sacred, he did not waste his assets". It was not until a quarter of a century later, however, that he gave a full account of what he saw as the crucial factor in individual decision-making, in an obituary address on his fellow Smålander, Andreas Neander (1714–1765):

Anyone walking in the clarity of the sunlight has a double shade, following him through thick and thin, whichever way he turns, wherever he goes. These two shades seem to me to be his two wraiths, the Virtue and the Fortune which are in constant attendance upon him, accompanying and leading him. The darker of them is Fortune, the lighter Virtue. Regardless of how overcast the day is, they are always there, whether we are aware of it or not.

These two servants of man, say I, attend upon us whether we are walking along or travelling and *driving* on over the Theatre of this world.

In temperament and character they are quite different.

Fortune, the world's angel, is swish, flighty, rash, giddy and fickle; when we hand her the reins, she drives *recklessly* over stock and stone, and for the most part either lands us in a ditch, so that we drown, or so smashes us against a rock, that we perish miserably. Virtue, God's angel, however, is godfearing, gentle, prudent, considerate and steady; when we hand her the reins, she drives with the utmost care, avoiding even the smallest of stones, so that we are as safe on the slippery roads of the world, as we are on the floor of a room. It is in accordance with God's pleasure, and often also with our own, that the reins are handed to whichever of these servants takes our fancy. Although it is most prudent to hand them to Virtue, Fortune will avert this. The late Neander, as we shall now see from the course of his earthly life, handed them over completely to Virtue.

It is of the very essence of the Earth and of all created things, that they should praise their Maker through man, the mouth of nature. Anyone aware of the perilous world we travel through every day of our lives, knows that he can only survive under the guidance of God's own angel Virtue; it is therefore in God that he trusts, knowing that He will never allow the reins of those who keep to Him, to fall into uncertain hands.[82]

---

[82] Linnaeus, C., 1745, 312 (August 7th); Ramm, A., 1907, 26; Hyltén-Cavallius, G.O., 1864/1868, I: 355–361; Wikman, K.R.V., 1964a; Nilsson, J., 1952/1960, III: 144–150; IV: 110–116; Fredbärj, T., 1962, 106; Malmeström, E., 1925/1926, VIII: 103–104.

PART TWO

NEMESIS DIVINA

NEMESIS DIVINA

Talion is exact retribution,
the balancing of the scales;
Autopathy in Greek, – harm brought home.

Live irreproachably, God is near.

I. INTRODUCTORY

I.i

My only son

You have come into a world, which to you is unknown.
The Author is unseen though you wonder at His majesty.
What you see is confusion, raised and marked by none.
The fairest of lilies you see choked by weeds.
Yet a just God dwells here, giving to all their due.
Live irreproachably, He is near!

There was a time when I doubted if God was mindful of me.
The long years have taught me what I now pass on.
All wish to be happy, yet there are few that can be so.
If you seek happiness, know that you are always in God's sight.
Live irreproachably, He is near!

If Scripture cannot teach you, learn of experience.
I have set down the few cases known to me.
See yourself within them and take care.
Happy is he who learns from the afflictions of others.

I should have named no names, set none of them down,
Had I not to convince you of the truth.
Guard the names as your own, as you do your sight and heart.
Trust no one in the world, all can turn enemy.
Should families, kin and relatives learn of it,
You would be pursued as long as you live,
And even, perhaps, to the grave.

Keep closely that which I give you,
Do as I desire you should,
That the name and honour of none may suffer.

If you keep not my command you sin,
You steal away the good name of others.
You abuse the trust of your father,
And will certainly suffer the punishment due.

For names have been given to convince you,
When you enquire privately into the matter.
Perhaps some of the tales I have heard are untrue.
Take heed, say nothing; impugn no one's name and honour.

I.ii

Ecclesiastes 9: 2: There is one event to the righteous, and to the wicked.

I.iii

I have often wondered if the heavenly powers
Are concerned with the Earth, if a steersman is in charge,
If mortal fate is anything but uncertain chance.
Rufinus' punishment finally freed me from such doubt,
And acquitted the gods.

Claudianus

I.iv

There is nothing so sublime and above all perils, that it is not under God and subject to Him.

I.v

Thou mighty Ruler, how mighty is Thy rule,
  And how unfathomable are Thy decrees.
Though the fateful power which forms things may be seen,
  Of that which moves invisible and alone, which brings
All that seems the work of human hands under the sway
  Of Thy sceptre, we know and perceive nothing.

II. MAN

II.i WHAT IS LIFE?

II.i.1.1

It is an axiom that no body moves of its own accord. Animals move of their own accord, simply by willing.

What is it whereby the body thinks and moves itself, whereby it senses? How does the soul move the body, give rise to sensation? How does the soul think? How does it will and not will?

II.i.1.2

Dwell on what is your own.
You have arisen from a frothing drop of detestable lust.
You have emerged from a nasty hole, between excrement and urine.
You swell with the content of your bowels and void it daily: a libidinous shitbag.
Life hangs by a thread, cobweb fine, nothing more fragile.
Life is a beautiful bubble.
You are full of yourself when prospering, but a beggar in adversity. Kierman.
When dead, an abominable corpse.
Death shows what an empty bubble man is.

II.i.1.3

RUTTING

Woman. Aged fourteen, face colours, eyes radiant, breasts swell, menstrual flow, pudenda develop; she is restless, joyful, enjoys singing.

Nature. The spring: fish are tasty, birds sing and are more than usually beautiful, the hen's face reddens, the peacock acquires its ocellated plumage, the black grouse's red eyebrows swell and stand out, the thrush makes music, the nightingale warbles, the he-goat attacks, the ram butts, the velvet is shed from the stag's antlers.

Anatomy. The spermatic vessel becomes larger, dilates, in a lascivious person two or three times larger, abnormal enlargement in the most lascivious; ova swell, hence chlorosis, the green sickness, her whole body smells, the odour excites; desire makes girls beautiful, infants and old women are not so; the most learned can be the most lecherous, Tycho Brahe, Newton. Venery turns unruly youths into the best of men; memory becomes excellent; happy, bold, red-faced. The pallor of the deflowered; the woman naked, manic; ravished, depressed, needs to recover.

Flowers venereous, hence beautiful, sweet-smelling; after blossoming, lose both beauty and scent. Pubic growth strongest in the very lascivious, as with the bull; hair long, thick, coarse. The young woman loves her husband, the old woman is irritable.

II.i.1.4

Where the free child of the air speaks of his delight.

II.i.1.5

What is life? It is a little flame which lasts only as long as the oil does.
The King of Prussia.

II.i.1.6

I conceive of man as being a waxen candle.
The sun illumines the body, wisdom the soul.
The world is a palace for omnipotent wisdom.
God lights every soul with His fire.

So it is that all, in that God has formed them, shine on this stage with their own wisdom. Some He has made into a great light, others He has made into tallow. They burn as long as they exist. When they have burnt out, God sets others in their place, that there may always be candles giving light.

Man can no more say that the world has been made for his sake, than the candle can say that it is the reason for the existence of the palace. Everything within omniscience contributes to the majesty of God.

II.i.1.7

Voigtländer

Voigtländer, regimental surgeon at Uppsala, a blusterer. When drunk and annoyed, he swears he will run his sword through the first person he comes across. It happens to be his friend Cedercrona, later public prosecutor for the House of Lords, and he thrusts his sword straight through his chest. Against all expectations, he manages to cure him.

He drives like a Jehu through the streets of Uppsala, like a fury. This driving causes one wretched member of the University staff to lose an arm and a thigh, to spend the rest of his days begging for money at Kyronius' Corner.

In 1760 Voigtländer is posted to Pomerania. His horse bolts and he sustains a double fracture of the thigh. The fractures are not set properly and cause him pain. He is restless, and the bone re-fractures. When he gets home he suffers further pain and has to keep to his bed, but this does not stop him from going hunting when he is on the way to recovery. He has a stroke and dies. The crippled member of the University staff sees him borne to his grave.

II.i.1.8 Friesendorff

The two Misses Friesendorff, who lived at Hammarby before I acquired the estate, died in 1725. They were at odds with one another the whole of the time, could not get on at all; so much so that they divided the place in two. When the one died, the other was delighted, said she would mourn in scarlet. Four days later, however, she too lay on the bier. Buried on the same day, in one grave. At long last, no longer at odds.

II.i.2.1 VANITY

Wisdom of Solomon 2: 2, 3: For we are born at all adventure: and we shall be hereafter as though we had never been: for the breath of our nostrils is as smoke, and our reason as a little spark in the moving of our heart: Which being extinguished, our body shall be turned into ashes, and our spirit shall vanish as the soft air.

Ecclesiasticus 33: 12, 13: Some of them hath he blessed and exalted, and some of them hath he sanctified, and set near himself: but some of them hath he cursed and brought low, and turned out of their places. As the clay is in the potter's hand, to fashion it at his pleasure: so man is in the hand of him that made him.

Do not regard the simple life as a misfortune. To be poor and healthy is better than being a privy councillor. The poor have more to laugh and be happy about.

Man has been given a restless soul, always responding to novelties.

II.i.2.2 Rudbeck

Court physician Rudbeck, the younger, was not only a god-fearing and respectable person, but also a good man. When he married for the third time, it was with his housemaid, a virago, as hard as nails. She nags him to such an extent that he becomes weary of the bickering. It was like having a female devil in his home; day in, day out, she plays the tyrant in the kitchen; acts as if the whole world existed only for her.

Her husband dies and she moves out into the country, taking charge of the property belonging to the other children.

She eventually falls ill, and is in a state of such excruciating pain, that she has to lie prostrate for eight days, screaming as if she had been stretched out on burning coals. This could be heard all over the neighbourhood, and lasted until she finally escaped from her vile body and died.

II.i.2.3 Netherwood

Netherwood, receiver of the revenues in Växjö, when he sees that he cannot balance the books, takes his leave and so passes his position on to his son.

The son publicizes the debt. The father spends many years in Stockholm, going about like a ghost and explaining the debt, since he had lost everything on account of it.

The son does not bother himself about the father; he is well-off, purchases farms, builds, etc.

Twenty years later a far from inconsiderable robbery involving many thousands of copper dollars takes place at the revenue office. Only those who were authorized had possession of the keys, and since it looks as though the son himself may have stolen the money, he is indicted and summoned to appear before the revenue court in Stockholm. The case does not go well for him. He writes to his children, taking formal leave of them, and lies down to die in Stockholm; makes certain he takes something which kills him.

II.i.3.1 DEATH

1. Fate guides us, the length of time left to each was fixed during the hour following our birth.

2. Nothing is so deceptive, so unpredictably treacherous, as human life; by Jove, is there anyone who would have accepted it, had it not been thrust upon us in our innocence?

3. Death threatens from behind, stealing everything away.

4. The whole of life calls for tears. New problems pile up, old ones are still unsolved.

5. Death eases all sorrow, for it leads us back into the tranquillity that was ours before we were born.

6. Death is nature's finest invention, it repels calamities, terminates the weariness of old age, removes shackles, provides an escape from prison. It reserves its greatest benefits for those it visits uninvited. The blessings it bestows recompense me for the injuries of life.

7. Ageing takes place with incredible rapidity.

8. Everything human is frail and short-lived. The earth and its inhabitants can be regarded as a point, and our lifespan is not even a part of a point when

compared with the whole of time.

9. Fate goes its own way. Our prayers and strivings are of no avail. We have been walking with death since we first saw the light of day. Fate fulfils its task, taking from us the feeling of our own death, concealing itself under the name of life, that it may steal up upon us more easily. Childhood follows infancy, youth childhood, age youth. Growing is dwindling. Everything is borne down and swept away in the stream of time.

10. To reach our prime is to hasten to our end.

11. Fate possesses its own, uninfluenced by any prayer, unmoved by compassion or indulgence; it keeps unswervingly to its course, following its destined way. It is the eternal law of fate which controls the ordering of everything, its primary tenet being acquiescence in what is decreed. Death calls all alike, and pays no regard to the wrath or favour of the gods.

12. Everything is reserved for death. Every single person you see before you will suffer capital punishment. The only difference between being condemned to death and dying of our own free will is that death's delay is highly esteemed.

13. You would have willed it, had you known that everything happens in accordance with God's decree.

14. Nothing has power over us when we have power over death.

15. Nature is constantly precipitating her own destruction. In creating things, she uses her strength sparingly and apportions herself out in imperceptible increases, but she destroys suddenly and with all her violence.

16. Once the end has been reached, there is no difference between us. It is of no interest to me how large my tomb will be after I am gone, death bulks equally large everywhere.

17. Let us face this defeat with steadfastness, for it cannot be avoided or provided against.

18. There is nothing so insignificant that it cannot contribute to destruction.

19. This earth will eventually lie upon me. I can draw it over myself or it gathers of its own accord, that is the only difference. Man's life is very short. He who has contempt for it will gaze secure upon the raging of the sea, watch with calm the menacing flashes of the sky.

20. One has to die. Why ask when? Death is nature's law, the remedy for all ill.

21. It is said that none return from thence.

22. I have lived, completed the course set by fortune. Among those who are to perish, we live. Though we are born different, we die the same.

23. No animated being has come to life without fearing death.

24. All die on the day prescribed for them. You lose none of the time allotted to you, for the rest belongs to another.

25. There is no significance in the end's coming sooner or later.

26. We die daily, each hour steals away a part of our life. A large part of death has already transpired.

27. What an existence, what a world! Wars, plagues, fires, robberies, thievery, poisonings, shipwreck, sickness, injury, foul weather, tyranny, untimely grief for those most dear, slavery, universal taxation, lust, envy, enemies of all kinds. A war of all against all.

28. Ask yourself what you want of the world before you enter it and pass through. Certainly no one would have accepted life, had it not been thrust upon us in our innocence.

29. That part of infinite time filled by life is a nothing.

30. Had they died betimes, there would have been no unfortunates.

31. All who are sailing are seeking a harbour; anyone who complains once he has reached it, and is no longer being swept by the fitful winds, is out of his mind.

II.i.3.2 PASSING ON

32. The end of every one of us is fixed.

33. None is blessed before his death.

34. Fate goes its own way. Our prayers and strivings are of no avail. Each receives what has been allotted to him at birth. He has carried it through, reached the goal of his life.

35. *2 Chronicles* 34: 28: Behold, I will gather thee to thy fathers, and thou shalt be gathered to thy grave in peace, neither shall thine eyes see all the evil that I will bring upon this place.

A year before the great fire at Uppsala on April 30th 1766, a frightful number of the city's inhabitants died of Uppsala fever, the great majority of them in Kingsmeadow street; in order to blot out remembrance of the disaster, this area was subsequently cleared.

36. Nothing is permanent, few things are long-lasting.

37. What is called old age is simply the round of a few years.

> By the waters of the river of Lethe,
> They drink the soothing draught of long forgetfulness.

38. You will go where all go; it is the law and your birthright.
Cease to hope that God's will may be pierced by prayer.

39. The most ridiculous of all is he who regrets that he was not living a thousand years ago. And he who is sorry that he is not living a thousand years hence is no better.

40. You will not be, and you were not.
Fate fulfils its task, taking from us the feeling of our own death.

41. It is as foolish to fear death as to fear old age.

42. The inevitable is to be awaited, only uncertainty is to be feared. Life is granted on the condition that we shall die.

43. No man escapes paying the penalty for being born.

44. It is wrong of us to prefer a long trek to death.

45. Why fear death? We are all born to die.

46. Life is not worth getting worked up about. Is it really worth all the bother?

47. What is death? A passing-on to the warriors and their warring, to the very kingdom of God according to the Old Testament. In a single day, at a single stroke as it were, how many myriad thousands, brought up with such labour, pain and care, are being slain in battle!

48. The only thing of profit to us is the sight of the glory of God, displayed in the theatre of this glory.

Strength of mind comes from no other source than sound learning and the study of nature.

Consider how much good comes of a timely death, how many have suffered from life's having been prolonged. Death is nature's law, the remedy for all ill.

49. Once the end has been reached, there is no difference between us. Does it make any difference whether I am slain by a stone or crushed by a whole mountain? After death, there can no longer be any interest in the size of the tomb.

50. Then let us show greatness of soul in the face of this disaster, which is in any case unavoidable.

51. This I therefore regard as consolation, for it is foolish to mourn when there is no remedy. The one salvation the vanquished have is to hope for none.

52. Let us recognize that little effort is required to destroy our perishable bodies.

53. Since each of us is born to die, what could be more foolish than to fear death?

54. Somewhere, in some way, we are all bound to die.

55. What does it matter whether I draw the earth over myself or it gathers of its own accord?

II.i.3.3

Seneca. Be assured that there are no ills to be suffered after death, that that which makes the nether regions terrible to us is a mere tale. We know that no shades threaten the dead, no imprisonment, no blazing streams of fire, that in that stillness there are no judgement-seats, no accused, that one is as free of shackles as one is of tyrants. All these things have been imagined by the poets, who have terrified us with groundless fancies.

II.i.3.4

Wisdom of Solomon 2: 2: The breath of our nostrils is as smoke, and a little spark in the moving of our heart.

2: 3: Which being extinguished, our body shall be turned into ashes, and our spirit shall vanish as the soft air.

II.i.3.5 Yxkull

Johan Leijonhufvud, privy councillor to Charles IX, together with his wife Sidonia Grip, prepare a most expensive tomb for themselves in Uppsala cathedral. They are laid, together with their rings and earrings, in magnificent coffins of English pewter, embossed with their coats-of-arms and draped in velvet.

In January 1762, Baron Yxkull, the cavalry captain, son of Anna Maria Leijonhufvud, the descendant of Johan and Sidonia, prizes open the coffins, sifts their ashes through his fingers in the search for rings and gold, sells the pewter to a pewterer, and would have disposed of their very remains had he been able to make money doing so. In the vault, which he also sells off, he treats his comrades to Rhenish wine.

Posterity will witness the consequences,
Guess how it will all turn out.

II.ii THE SELF WITHIN ME

II.ii.1

What is the God who sees and hears and knows ? I see no such God.

What is it that senses within me? I do not see it. The eye is a camera obscura, it depicts objects, yet it is not through the affected nerve that I see. The nerve does not enable me to judge of anything, and although it leads to the brain, it is not there that I see.

There is, then, something which perceives and reasons, but which eludes me. Is it so strange that I should not see God, if I do not see the self within me?

What is it that motivates, operates, sustains the heart, intestines, fibres? Nothing operates of its own accord. What is it that gives rise to purging, vomiting, sweating, urinating? What excites fever, cures a wound? It is something in me, my prime constituent.

If I cannot take measure of myself, it is hardly strange that I should be unable to grasp God.

II.ii.2

Madam N.N.

When Madam N.N.'s maid is asked to carry a porcelain bowl upstairs, she stumbles and smashes it. Madam is beside herself with rage; she gives the maid a frightful thrashing, and follows this up by buying a new bowl and deducting the cost of it from her wages.

That afternoon, as Madam was going down the same stairs, she tripped and broke her leg.

Rothman

II.ii.3

Lewenhaupt

Lieutenant-colonel Dagström, a wealthy Scanian, was a member of the Holstein party opposed to king Frederick.

The way in which he spoke out in the house of lords earned him the king's displeasure and the matter was investigated.

A commission was appointed to deal with Dagström, who became vehement when defending himself. Lewenhaupt, who presided over the commission, declared him to be mentally deranged and had him imprisoned for life in Malmö castle.

As a result of the protracted imprisonment, Dagström becomes extremely obstinate.

Lewenhaupt becomes a great man, on two occasions serves with incredible distinction as speaker of the house of lords. In 1741, when the estates appoint him generalissimo in charge of the campaign against the Russians, he has doubts about his capabilities and declines. He does, however, accept their view that he has no equal.

The orders he gave were equivocal and the outcome of the war was disastrous. The vengeance of the whole country was visited upon him and upon Buddenbrock, although as was subsequently established he himself was not to blame. The outcome was that he was condemned, and beheaded at Norrmalm tollgate. Dagström was then still alive in prison.

Ihre's opinion is that it was only when Lewenhaupt had twice been speaker of the house of lords, only when he had been sent by the country to thrash the Russians, that it became apparent that he was lacking in commonsense and understanding.

In the 1738 parliament he was the keenest of all for the war against Russia. It was Buddenbrock and Lewenhaupt who strove to bring about the war.

II.ii.4 Ostermann

Ostermann, when a student at Halle, fights a duel with another student and runs him through, killing him. Makes himself scarce. Becomes prime minister, in control of the whole of that dreadful country. Always apprehensive lest the murder should out, constantly, remorseful about it. What is more, things do not go well, like Münnich he has to stretch his neck upon the block, and although he is reprieved at the last moment, he is incarcerated for life.

II.ii.5 Hamilton

General Hamilton burns down the guest-house at Rotebro, simply because it takes his fancy to do so.

It is precisely when he is lying on his death-bed that his own house catches fire; he has to be taken out of doors, and gives up the ghost with no roof over his head.

II.iii GOING OUT INTO THE DARK

II.iii.1

HAUNTING

II.iii.1.1

Haunting is mentioned in Scripture, during the earliest periods, mostly during the time of fables. In flourishing areas it eventually fails to attract any attention, although it continues to be discussed everywhere in the countryside. Most hauntings are fables. I have never seen anything.

Children are intimidated into believing in ghosts, although only when they go out into the dark, or when the shutters are to during the day. Fear of the dark remains rooted in them throughout life. Most people are afraid of churchyards, hills where there are gallows.

Scarcely one in a thousand of these tales has any truth in it.

II.iii.1.2

According to Holy Scripture, each has his angel, who protects him night and day from misfortune, perhaps also helping him by means of misfortune. Could it be that they follow the body like a shadow? When danger threatens, a hundred causes intervene in order to divert and prevent it. With another person, destined to be unfortunate, it makes no difference what preventive measures are taken. See Carleson.

II.iii.1.3

Maja Hierpe was a girl in service with me. In 1773, her mother was living im the village of Myresjö, where there was an outbreak of scarlet fever which killed off most of the inhabitants, including her mother.

The girl wants to go back to attend her mother's funeral, is earnestly entreated not to, but the next day cries and screams and wants to go home. She is infected and dies.

The night before the mother died, the girls saw a female figure in my front-room. After the girl had died, and before they knew of it, two of them heard her crying in the kitchen. Even before they knew she was ill, they were on my country estate at the time, these two heard three knocks, which were repeated. Both of them were wide awake, and made immediate mention of it.

II.iii.1.4

Caesar's soothsayer, the door bangs at night; Caesar's wife dreams he is killed the night before his murder.

II.iii.1.5

On the day when my mother died in Småland and I was in Uppsala, I was more melancholy than I have ever been, although I knew nothing of her death.

During the night, my father saw what looked like a corpse in the stove, sitting there decked in a sheet. He made mention of it to all of us. Two days later dancing-master Soberant arrived, fell ill almost immediately, and died.

One night, eight days before my wife gave birth to daughter Helen, the neighbours saw lighted candles in all the windows, and told everyone about it. My wife got to hear of it and was afraid it was a portent of her dying in childbirth, although she pulled herself together. The girl died soon after being born.

II.iii.1.6

But why do hauntings take place at night and not during the day? At two o'clock, it is night during the winter, but light during the summer. Could it be because the stars are not visible during the day?

II.iii.2 SHADES

II.iii.2.1

My room was one side of the hall, my wife's the other. My wife, together with five or more others, hears me coming into the hall, unlocking my room, going in, coming out, locking up after myself, and thinks that I have put my hat and stick, etc., away, and am about to come in to her. But since no one comes, she says: "My husband will soon be here", and half an hour later I arrive. This happened not once, but many times. It is when I have been in Stockholm, and I arrive back soon afterwards.

II.iii.2.2

The evening before my father-in-law dies, two women who are attending him hear what sounds like the nailing of coffins in the adjacent room. They come out terrified, saying that he will soon die.

While my wife is sitting with him during the night, a white cloth falls past the window, as if from the roof. She is frightened and goes into the next room, where there was someone else, sitting there equally frightened, since it had also fallen past her window. As they go in, the old man breathes his last.

II.iii.2.3

In 1728 I was lodging with court-physician Stobæus, in the top flat right under the gable, which was not to be reached by any rod in Lund. There were three blows, which were repeated, and which were so hard that they woke me up. I was thoroughly scared, and thought they were meant for me. Two days later the court-physician received the regrettable notification that a prominent person he had been attempting to cure had died. The mother was upset, caught cholera and died the day after.

This is all I heard and know for certain. It is in fact what happened.

II.iii.2.4

At twelve o'clock at night on July 22/23 1765, my wife is in our bedroom and hears something going on in my museum, the room above. Someone is walking about very heavily, on and off, for some time. She wakes me up and I also hear it. I knew there was no one there, that the doors were properly locked and that I had the key.

Some days later I received notification of the death at nine o'clock in the evening on the twenty-second of July, of my distinguished and most trusted friend inspector Carl Clerck. And truly, the step was so like his, that if I had heard it in Stockholm, I would have recognized him by means of it. But I was on my Hammarby estate, thirty-six miles away.

II.iii.2.5

Professor Dahlman's daughter, on her deathbed, longs for my daughters. They visit her; she promises to come and see them as soon as she is well again. Two days later she dies.

At twelve o'clock at night there are happenings in their room – chairs are dragged about, bedclothes pulled off, scratching under the bed; all three are wide awake. The maid wakes up, and kindles a light to see if a cat or a dog has got into the room; she finds nothing. The light is extinguished. The happenings continue again until four o'clock. They cry out and are frightened.

At six in the morning they learn that Miss Dahlman had died at twelve o'clock that night.

II.iii.3 SPELLBINDING

II.iii.3.1

It is the custom among country people, when a horse comes in sweating from a journey, to wipe the sweat off with their hands and rub it in on their own horse; the belief being that sweating is a giving off of strength, which by this means will be transferred to their own animal.

I have seen a horse, swollen with choler, lie down as if dead; various expedients proved of no avail. When an old woman pissed in her left clog and gave it to the beast to drink, it got up immediately and started eating.

II.iii.3.2

An executioner, incognito, was having a meal at an inn in Diö. A soldier broke his journey there, and while he was waiting for his horse, the hostess asked him in. He entered the room, found the unknown company uncongenial, and went back out into the rain. The hostess invited him in again, but he did not stay in the room long. When she asked him why he would not remain indoors, he said that he could not stand the other person there. She enquired of the executioner incognito why it was that the visitor had taken such exception to him. He said that he had not entered into conversation, but that when the soldier was about to leave, he had said to him: "Take care lest you become my son".

Before six months were up, the soldier had fallen under that executioner's sword.

II.iii.3.3

At Uppsala in 1767, the executioner beheaded a servant girl for infanticide. It was then thirty years since any such execution had taken place there. As he was leaving, he told the attendant clergy that he would soon be back doing the same thing again. Two months later, another servant girl in Uppsala gave birth to a child in secret. She drowned it, and suffered the same fate as her predecessor.

II.iii.3.4

An executioner in Trondheim claimed that he had healing powers. When a woman who was ill consulted him and requested medicine, his sword, which was hanging on the wall, rattled. "See to it that you are not slain by that sword", he said. Although it was not then known, she was pregnant. She smothered the child and was slain by the executioner.

II.iii.3.5

The chancellor of Denmark, held in the highest esteem by his king, roused the envy of his enemies, who managed to get him accused of high treason and condemned to death. During the night preceding the day of execution, his portrait fell onto the stone floor; although the glass fell from the frame, it did not smash. He concluded from this that he would not be beheaded, and became wholly convinced of the fact. He was reprieved from the death penalty, but imprisoned for life.

II.iii.3.6

A German had killed a young man forty years previously and been acquitted on oath. He was afflicted in numerous ways. Once he had become resigned to this, he used to say he knew the dogs would lap his blood where the young fellow was supposed to have been killed.

At the age of seventy, the old chap encountered two riders and was ridden down by one of the horses. He gave up the ghost on precisely the spot where he was said to have killed the young man.

II.iii.3.7

A country lad, Per Månsson, had come to watch an excecution. Just before the criminal appeared, the milling of the crowd made the horses restless. Månsson's horse stumbled by the block. The executioner told him to watch it, or he would stumble right on to it. He was in fact the next to be executed on the spot. He had struck his companion dead in a sudden fit of rage.

II.iv YOU TOO ARE HERE

II.iv.1 FORTUNE-TELLING

II.iv.1.1

A woman who was evidently poor and sickly, was conveyed around and presented at all the larger homesteads as a fortune-teller. She said that there was a danger of our place catching fire. This upset my mother. The woman added: "Pray to God, the fire will then be postponed until after your time". The house caught fire after my mother's death.

II.iv.1.2

My brother Samuel, quick-witted, was at Växjö School. I was regarded as dim, and had just gone off to Lund. Everyone called my brother the professor, and said that he would become one. The fortune-teller, who had seen neither of us, asked to be shown some of our clothes. She then said that my brother Samuel would become a clergyman. Of me, she said: "He will become a professor, travel widely, become more celebrated than anyone else in the realm", and swore that this would be so. In order to test her, my mother showed her another piece of my clothing, telling her that it was my brother's. "No, that belongs to the one who will be a professor and live far away from here".

II.iv.1.3

In Gävle, an officer sees a girl who is quite exceptionally inoffensive. He urges everyone to make sure that they do not become involved with her. His advice is ignored. The following year she becomes pregnant, murders the child, is burnt at the stake for it.

He is then asked if he has knowledge of his own death: "Certainly, and I am constantly depressed by it. I shall die by being burnt and drowned".

Some years later he sails from Gotland in a cargo ship carrying lime. It springs a leak, catches fire on account of the lime, and he perishes with the rest of the crew.

II.iv.1.4

The astrologer in Rome told Caracalla that when in Mesopotamia he should beware of his colonel, Macrinus. It was Macrinus who lured Martialis into killing him.

II.iv.1.5

An astrologer told Domitian that before long he would be dead.

Domitian asked the astrologer if he knew anything of his own death. He replied: "I shall be eaten by dogs".

Domitian ordered that the astrologer should be killed and burnt. An unexpected cloudburst extinguished the fire, and the unburnt corpse was devoured by hungry dogs.

II.iv.1.6

Boatman

When the Reverend Collin was studying with me in Växjö, his mother sent him a supply of food by a pack-horse, ridden by one of the servants.

A boatman up from the country also comes riding along, and accompanies the servant for a while. That evening, however, just as they are passing the gallows, the boatman draws his knife on the servant, knocks him off the horse and takes what he wants of what is being carried.

The servant revives after this assault, and reaches the town. The bailiff is sent out and captures the offender, who is condemned to death.

An aged official looks at the offender's hands, and declares that he will never die a violent death.

The offender dies while under arrest, two days before the death-warrant arrives.

If God has delineated our fate, before it comes to pass, in our hands, we ought to commiserate the unfortunates whose fate is as it is.

II.iv.1.7

Tiliander

Abiel Tiliander, the vicar of Pjätteryd, studied in Växjö. One of the teachers there had a particular propensity for the word 'finally'. On one occasion Tiliander took the chalk and wrote on the board: "Finally, despite my writing finally so very frequently, I am in fact an educated person". The teacher cursed whoever it was that had written the words.

Tiliander shot out the teacher's eye with a musket and was never found out.

Forty years later the vicar of Pjätteryd became thirsty during the night. In order not to wake the household, he went himself to fetch water from the well. He fell in and was drowned.

II.iv.2

PORTENT

II.iv.2.1

It is generally said to be the case that portents no longer occur. All are agreed that nothing is foretold by dreams. For my part, however, the day after I have dreamt of the dead, funerals, corpses, I always become angry. I have noticed this now on five occasions, and it has always been the same. Early this morning I dreamt of a funeral back home. I took this to portend an outbreak of anger, and was particularly careful to avoid any situation which might give rise to one. At about noon, however, I was provoked beyond endurance by an arrogant and insulting reply from my wife. November 15th 1765.

II.iv.2.2

Laodamia wrote as follows to her husband Protesilaos, as he was preparing to leave for Troy: "It was an ill omen when you stumbled over the threshold as you were leaving the paternal home for Troy. I groaned when I saw it". From Ovid.

II.iv.2.3

Löfling, before he left for Spain, stumbled as he was taking leave of me; he did not return. It was the same with Forsskål.

II.iv.2.4

Colonel Freidenfelt proposes to travel from Piteå to Finland over the Kvarken, during winter. As he is mounting his horse, it goes down under him twice, falling onto its knees and breathing heavily. Mrs. Solander, on seeing this, says immediately that he will never return. He was drowned on the way.

II.iv.2.5

A student, Peldan, travels from Åbo to visit Florinus in Kimito. As he is mounting his horse in Åbo, it stumbles onto its knees.

Peldan was beheaded; see the account of Florinus.

II.iv.2.6

On the very day that Lucullus gained his decisive victory over the Parthians, rumour spread among the citizens of Rome that he had completely defeated them. All enquired who had first spread it, but no one was traced. Confirmation that the victory had been gained on that very day many hundreds of miles away, came a month later.

II.iv.2.7

Princess Elizabeth's revolution in Russia was planned for the mid-January of 1742, and Lewenhaupt's army was therefore sent to Finland to open the war, everything being kept very secret.

The Russian government was forced to send its guards against ours, and since it was precisely these guards that Elizabeth was relying upon, she was obliged to undertake the coup the day before they were due to depart.

On October 21st 1741 count Magnus Stenbock, who had arrived in Stockholm from Livonia, dreams that his child's tutor had come to him during the night and told him that Elizabeth had come to the throne by means of a revolution. He tells various people about the dream.

Privy councillor Anders von Höpken, who was aware of the situation, became worried that it would become known too soon. He invites Stenbock to call on him, and asks if what he has heard about the dream is true. Stenbock confirms it. Höpken urges him, for God's sake, to stop talking in this manner, since it could cost him his head when he returns home.

Eight days later it became known that the revolution had taken place that very night, and that Elizabeth had seized power. What was the significance of this?

II.iv.2.8

Antonius Saturninus, governor in Germany, rebelled against Domitian, but Appius Maximus Norbanus was sent and slew him.

At the same time as the victory was being celebrated, news of it reached Rome, without anyone being able to find out who had brought it.

Domitian's governor Antonius Saturninus revolted, was defeated in Germany by Appius Maximus Norbanus; this was known immediately in Rome, although how this came about was never revealed.

II.iv.2.9

Dean Risell of Filipstad had a large number of children. One night his wife sees one of them come in and place a white dress on the chest belonging to their fourteen-year-old daughter.

The wife asks: "Are you asleep?" The fourteen-year-old girl is awake, and replies: "Yes, I saw the little one lay my shroud on the chest".

The next day the girl goes down to call the tutor for lunch. She says: "There is a magpie on the house and it is chattering. Shoot it".

The tutor picks up the flintlock. As he is going out the cock snaps to and the shot tears through the girl, killing her.

II.iv.3.1

Everyone is supposed to have his wraith, and is there any reason to doubt this? Everyone has his angel, which is the same thing, an angel being that which protects you on your way.

What else are these portents of death, which are so often spoken about?

What is it that manifests itself prior to a misfortune, that so often diverts misfortune, preventing it in such a variety of ways? Is it a decree?

The body always has a double shade, although it can never be seen. Is it possible that God created things like this to be in rapport with Him, and that they accompany us like shadows? I call these shades Virtue and Fortune.

Psalms 34: 8: The angels of the Lord encamp round about them that fear him, and deliver them.

II.iv.3.2

How is it that one feels anxious when evil is imminent, when misfortune is taking place far away, as I did in Uppsala when my mother died in Småland?

What is it that knocks in the walls and haunts a place when others die?

It is as ununderstandable as the soul. We have no idea of spirits, can only understand bodies.

II.iv.3.3

It is an axiom that no body moves of its own accord. Animals move of their own accord, simply by willing.

What is it whereby the body thinks and moves itself, whereby it senses? How does the soul move the body, give rise to sensation? How does the soul think? How does it will and not will?

II.iv.3.4

Caesar

Julius Caesar fought fifty pitched battles, took eighty towns, subjugated three hundred tribes, celebrated three triumphs.

He was not satisfied with becoming permanent consul, dictator in perpetuity, with the office of censor, the title of emperor, with being designated father of the country, with having his statue erected among those of the kings. He could no longer bring himself to stand before the senate: his ambition was to become king.

The soothsayer said to him: “Beware the first of the ides of March”, and did so six months before the day.

The night before the first he dreamt that he had been lifted above the clouds to Jupiter. His wife dreamt that he had been stabbed to death in her arms, and woke up terrified. At just that time of the night the door flew open, the noise of it awakening the household.

When Caesar decided to stay at home for the day, everyone was soon bruiting it abroad that Caesar was staying at home because Caesar's wife had had bad dreams. Out he went.

The soothsayer meets him. Caesar observes that the first of the ides of March has come. "But not yet gone", replies the soothsayer.

Artemidorus meets Caesar on the way to the capitol, puts into his hands an account of the whole conspiracy, pleads with him to read it immediately, but is pushed aside by the other supplicants on the way to the meeting-place.

He enters, and is stabbed twenty-three times. As he catches sight of Brutus among them, he says: "You too Brutus", sweeps his robe about him, and dies. Brutus was his own bastard, by another man's wife.

All who had conspired to murder Caesar died a violent death.

Julius Caesar was told by the soothsayer Spurina, a good six months before the day, that he should beware of the fifteenth of March. On the said ides Caesar met the soothsayer and said: "The fifteenth of March has come". The reply was: "Yes, but it has not yet gone". The same day he was murdered by the senators. When Caesar saw that Brutus was among them he said: "My son, so you too are here".

II.iv.3.5

General Carl Cronstedt gave a precise prediction of the death of Charles XII. He said he would be killed before November was out, and made the prediction known among those of the officers who were his confidants.

On the thirtieth of November one of Cronstedt's friends remarked that it was the end of the month and the king was still alive. Cronstedt replied that although that was certainly the case, the month was not yet over. That night the king was shot at Fredrikshald.

Some are of the opinion that it was Cronstedt himself who shot the king at Fredrikshald, although it is most likely to have been the Frenchman Colonel Sicre.

II.iv.3.6

Some country people were haymaking – scything the grass, piling it, pitching it into boats in order to transport it home. While this was going on, and just before the party quit, a servant-girl, unbeknown to the others, gave birth, and deposited the offspring in an out-of-the-way thicket.

As the boat was putting off, a little bitch scampered down to the shore. In its mouth it had a puppy, to which it had given birth on the island. Whimpering, it tried to swim toward the boat with the puppy in its mouth. Touched by the sight, the haymakers steered the boat back, took bitch and puppy on board, and set off home again.

Seeing this, and touched by the love which an irrational beast had shown for its offspring, the servant-girl asked the haymakers to return to the island. When they refused, she demanded, and swore that if they did not, she would throw herself into the water, drown herself. She fell onto her knees, begging them to return. Astonished and perplexed, they steered the boat back once again. The servant-girl got out, found the child, which was still alive, and took it with her.

III. MAN AND SOCIETY

III.i THE SAPPING OF AFFECTION

III.i.1.1 Joseph

A person I knew well was extremely libidinous. Although he finally got engaged and resolved to behave himself, he continued to fornicate; caught gonorrhoea which developed with cancer. After any amount of pain and worry he was eventually cured.

After a number of years he got married; continued to copulate with the girls. Admitted to having slept with more than a hundred females, most of them servant-girls.

His wife was getting on a bit, and was called away on an urgent journey for a period of three months. He fornicates, eventually shows signs of incipient gonorrhoea. Before the disease has the chance to develop, however, he undergoes a rapid cure through the administering of mercury. He has some excuse, since his wife was away longer than she had said she would be.

Finally, his wife sets out on an eight-day journey but stays away for three weeks. In order to get his own back, he sleeps with a married woman, catches gonorrhoea badly and is in danger of his life.

His conscience troubles him. He realizes that God is not only right to have punished him for all his flagrancies, but that he ought also to have brought his life to an end, that the righteousness of the Almighty demands that he should die. He finds it difficult to believe that God's grace can prevail over His righteousness. He did become well again, however, and although God in His infinite mercy spared him, he was troubled in spirit from then on.

III.i.1.2 Myhrman

Myhrman, a wealthy and respectable man in Filipstad, acquires a distinguished, virtuous and beautiful wife.

Mrs. Fernholm, who has a husband in Vadstena whom she keeps quiet about, is good-looking and frightfully promiscuous. She is tumbled by Myhrman.

Faxell, the mayor of Filipstad, had also dipped his wick with Mrs. Fernholm.

There is a court case, prosecution for seduction, and Faxell, who was a bachelor, takes exclusive responsibility on the understanding that Myhrman should allow him to fondle his wife. Faxell is constantly calling on Mrs. Myhrman, kissing her hands, etc. This depresses Myhrman.

Faxell eventually lapses into brooding on things; one day he travels to Karlstad, where he lodges with the mayor. The lord lieutenant arrives, and the mayor goes out to greet him. During the evening Faxell leaves the house; a few days later he is found drowned in the river.

III.i.1.3 ONANISTS

Farmer at Bredstedt in Holstein, afraid of having too many children, commits Onan's sin, *Genesis* 38: 9; dies of a frenzy.

Farmer, afraid of the same thing, leaves his wife and goes into service for three years. His wife becomes pregnant when he comes back, and gives birth to triplets.

Genesis 38: 9, 10: When Onan went in to his brother's wife, he spilled his seed on the ground and destroyed it, lest he should give seed unto his brother. And the thing which he did displeased the Lord: wherefore he slew him.

III.i.1.4 Brahe

Brahe, the earl, studies at Uppsala. First acts dishonourably in respect of Mrs. Rosén. After he has married the daughter of Sack, councillor of chancery, she becomes consumptive. During her illness, Brahe is said to have fornicated with a whore in front of her very eyes. His second wife was the daughter of president Piper.

In high favour with the authorities and consequently hated by the Hats. Only a few votes short of being elected speaker of the house of lords.

Non-commissioned officers told to make contact with him. They tell him of the discontent in the guards, say that if anyone started a revolution they would all be only too ready to join it. Brahe and all Caps are excluded from every deputation. A certain restlessness breaks out into mutiny. Brahe was beheaded for it, although many defended him. The truth of the matter is unknown to me, although I do know that he was said to have had trouble on the way at Whitsun. The mini-revolt did not take place until a month after that.

Puke, also beheaded for having taken part in the revolt, was reputed to have been Piper's natural son by another man's wife.

Captain Stålsvärd, engineer, was also executed for complicity. He was the most godless creature I have ever known, believed neither in God nor in religion. Never married, used the guardsmen's wives as laundresses and mistresses. When I told him this was sin, he laughed and said I was a pedant.

Marshal of the court Horn also came unstuck; a good pious chap, said to have practised sodomy.

Many were beheaded for the revolt, including:

marshal of the court Horn, reputedly a sodomist; unconcerned about his wife when she lay dying in childbirth, he was a few gunshots away in the Hat Inn at Drottningholm;

captain Stålsvärd, a veritable atheist; never married, said it was more economical to lie with a guardsman's wife and employ her as laundress; committed murder in Germany, and got away with it there;

Puke, natural son of president Piper by another man's wife; Piper therefore lost his son-in-law as well as his son.

They were executed on July 23rd 1756.

III.i.1.5 Schmidt

Munster, son of the mayor of Åbo, his wife, lieutenant Schmidt, and many others, are at a party, drinking deep into the night.

Munster has had his fill; his wife gets him to bed and then dozes off.

Schmidt comes in and gets Munster to go upstairs and drink another round; he urges him to leave the room quietly so as not to awaken his wife, and he does so.

Schmidt slips away from the group of drunken tipplers, strips off, lies down beside Munster's wife and copulates with her; she thinks he is her husband.

Munster drinks himself sober, goes to bed and wants to caress his wife. "My goodness. What again?" After she has responded thus and refused him, they realize what must have happened.

The wife is deeply disturbed and on the point of breaking down.

Schmidt finds it amusing, talks and boasts about it.

Some years later a wedding is to take place at Torneå, where Schmidt is stationed. The colonel is invited, but since he has to travel over the ice and the ice is not safe, he takes the precaution of having Schmidt accompany him in another sleigh, travelling parallel. When returning, however, the colonel decides that Schmidt can go on ahead; he says he will follow once the post has been prepared.

Schmidt himself then drives onto weak ice; his batman surfaces but is unable to help him. Schmidt is unable to avoid drowning.

Proverbs 6: 29: So he that goeth in to his neighbour's wife; whosoever toucheth her shall not be innocent.

III.i.1.6 Backman

Backman, a handsome fellow, servant to professor Asp, the girls mad about him, entered the service of alderman Borell as assistant in the linen department.

Country girls who came to buy flax from him but were short of cash could always get what they wanted, although nothing was given for nothing.

Within a year or two eight girls who had been dishonoured by him had given birth to children; he was prosecuted, got off lightly in some of the cases, swore that the rest were unfounded, and treated the whole matter as of little significance.

Early in 1774, a man travelling through Västerås goes into the toll-house in order to wake up the attendant and get the gate opened. Backman, who thinks that it is all taking too long, drives the horse and sleigh over the toll-bar and breaks one of the vehicle's shafts. The man orders him to replace it on the spot. He hurries off to a cartwright's, and simply makes off with a shaft. The result is a dispute, during which Backman knifes the cartwright in the pit of the stomach, killing him. He is arrested, condemned to be beheaded, eventually reprieved.

Exodus 20: 7: Thou shalt not take the name of the Lord thy God in vain; for the Lord will not hold him guiltless that taketh his name in vain.

III.i.2.1 INCEST

Marriage or copulation between close kin is forbidden.

The offspring derives its marrow from the mother, its body from the father.

Hereditary illnesses, which would be perpetuated by marriage between close kin, are weakened and eradicated by marriage with outsiders.

Horse-breeders are not keen on mating animals born of the same mare.

Outsiders, through the mixing, give rise to variety and diversity, of both body and soul. Nature delights in variety.

It is thus that the good gifts which would otherwise have remained confined to one family are divided among many.

It is evident from the outcome that marriage between close kin is seldom a success.

Bishop Mennander married his cousin in order to demonstrate that it was untrue that two relatives could not build up a good marriage. She died during her first confinement.

III.i.2.2 Lagerbladh

Lagerbladh, master of philosophy, from Skatelöv in Småland, is in the employ of a certain captain Eggertz. Said to have been too familiar with the captain's wife, he gets married to the captain's daughter.

Although the master of philosophy was handsome and pleasant enough, the young wife could not for the life of her put up with him. After making the acquaintance of an officer by the name of Berg, she can do nothing but think and dream of him. She turns her bottom to her husband whenever he gets into the marriage-bed.

Her parents and relations do what they can to advise, weeping and pleading with her. She promises to make a go of it, but as soon as her husband is off on a journey, she contacts Berg, writes love-letters, invites him over or goes to visit him.

Heaven and earth can do nothing about it, are quite incapable of redirecting her affections.

Her husband does not dare to petition for a divorce, since he suspects that his wife knows of the relationship with her mother.

III.i.2.3

The vow of chastity, man and wife agree to abstain, as did Eringisl Plata of Svanshals and his wife Ingeborg. He maintained to the last that he had allowed her to remain a virgin.

Egard von Kyren and his wife Catharine, the daughter of St. Bridget, lay apart on the bare floor throughout the cold of the winter, despite her being eighteen years old, and could not be induced to consummate the marriage. It was the same with privy councillor Ulf Gudmarsson, the husband of the godfearing St. Bridget, who on the advice of his wife divested himself of his offices, clad himself in sackcloth, and with a rope around his neck and a staff in his hand, went on a pilgrimage to the shrine of St. James at Compostela.

III.i.2.4 Christina Juliana Thun

Mrs. Thun, principal lady in waiting to the queen, an extremely handsome, charming, agreeable woman. She has an ordinary and impecunious husband, whom she leaves in the flower of her youth in order to take up a position at the court, where she eventually becomes first lady of the bedchamber.

A number of years after she has left her husband, she is travelling with the queen through Jönköping, which is where he lives. Despite the queen's giving her permission to spend the night with him, she does not do so.

Her husband eventually gets an appointment in Stockholm, but he never has any further occasion to rejoice in his handsome wife, remains separated from her for fourteen years in all, if not more.

Colonel Wrangel finally gains her confidence. They want to marry, and since Anders Thun is the obstacle to their doing so, they both wish him dead.

In 1767 Wrangel informs her that he has undertaken to marry someone else; disturbed by this, she contracts haemorrhage of the uterus.

In 1769 she suffers from severe suppression of the urine, and is only able to pass water by means of a catheter. A probe soon makes it apparent to all the physicians concerned that she has scirrhus of the uterus. She dies on May 7th 1769.

It had to be those parts she had denied her husband which caused her death, for she had loved another in the hope of getting rid of the man she had married.

III.i.2.5 Welin

Welin, professor at Åbo, received enough money from a young woman to enable him to travel abroad. He returned home, but since he was engaged to her, he made a second journey to Paris in order to get out of it.

He was in company in Paris; had an appointment elsewhere for a certain time. The woman he was with, however, kept him where he was. Fire broke out in the flax-storehouse on the floor below. He wanted to jump out of the window, but the woman held him back. The building collapsed, and he and six others were consumed in the flames.

He was acquainted with any number of women.

III.i.2.6 The gardening man

A worker in the university botanic garden cultivates a wealthy widow, gets her to spend all she has on him by dangling before her the prospect of marriage.

He then takes a fancy to another female and marries her.

He contracts animal scab and running cancer of the thighs, has to suffer from this for eight years before finding relief in a miserable death.

The widow fornicates and becomes a whore, a public disgrace.

III.i.3.1 ADULTERY

Adultery is copulation with another's wife.

Ecclesiasticus 23: 18, 19: A man that breaketh wedlock, saying thus in his heart, Who seeth me? I am compassed about with darkness, the walls cover me, the most High will not remember my sins. Such a man knoweth not the eyes of the Lord.

Job 31: 9, 10: If mine heart have been deceived by a woman, then let my wife grind unto another, and let others bow down upon her.

The character of a wife who is a whore:

1. she is uneasy in the house, especially when her husband is at home;
2. she looks forward to his going away, is not pleased when he returns;
3. she often feigns menstruation, to avoid his attention;
4. she lies flat during copulation, pressing down her genitals, to give less pleasure;
5. she stretches one thigh straight down while copulating, to avoid feeling;
6. she expels the penis before the climax, as if afraid of pregnancy.

2 Samuel 12: 14: David's child by Bath-sheba had to die.

2 Samuel 16: 22: Absalom went in unto David's concubine in the sight of all Israel.
Ecclesiasticus 23: 22, 24–26: A wife that leaveth her husband, and bringeth in an heir by another. Her children have to suffer for it. Her children shall not take root. She shall leave her memory to be cursed, and her reproach shall not be blotted out.
Wisdom of Solomon 3: 16: As for the children of adulterers, they shall not come to their perfection, and the seed of an unrighteous bed shall be rooted out.

Observation: A bachelor seduces a married man's wife. He subsequently enters into a happy marriage. His first wife dies in childbirth, as does also his second. His third wife is allowed to live. He pays two for one.

The contract between man and wife is breached and annulled. The true bond, the chain of love, is broken, for since weakness and circumstances result in her withdrawing from her husband in order to please the adulterer, she no longer cherishes her husband.

If there are two lovers, the one affection will sap the other.

If a pure girl, once she is married, does not involve herself with others, she will never cease to love her husband. A girl who has previously made herself generally available will never be faithful. It is therefore advisable to choose a maiden rather than a whore.

The outcome: She shares her love with the adulterer, others' children are introduced into his family, he has to manage and care for those belonging to others, to grant them part of the inheritance belonging to his true heirs, to be deprived, daily, of all they receive, to be deceived, daily, by his wife. Since it is stolen bread which tastes best, he can never be certain of his own spouse. She thinks, daily, of deceiving her husband, of pretexts for going out or for getting him to do so, and finally falls to rancouring with him.

Nemesis often strikes those who seduce another's wife. She often dies of some disease of the uterus such as haemorrhage or cancer. The most frequent end for him is drowning.

2 Samuel 12: 11–12: I will give thy wife before thine eyes, and give her unto thy neighbour, and he shall lie with thy wife in the sight of this sun. For thou didst it secretly: but I will do this thing before the sun.
2 Samuel 12: 14: David lay with Uriah's wife; his life was spared but the child had to die.

III.i.3.2

Sjöblad to crown equerry Sparfvenfeldt: "Incidentally, since we are speaking of whores, how are your sisters getting on?"

III.i.3.3 Dörnberg

Dörnberg, president in Cassel, married, becomes Frederick I's favourite in Stockholm. While in the capital, falls in love with Miss Lieven, later Mrs. Hårleman. Malicious tongues say she became pregnant by him; in any case he courted her.

Travels to Cassel. Gets a lackey to lie by his wife while she is asleep. Enters the bedroom with witnesses. Obliges the servant to quit. Divorces his wife.

Travels back to Stockholm to marry Miss Lieven. Dies on the way at Norrköping.

III.i.3.4 Meldercreutz

The father, doctor Molin, provost at Uppsala, a great schemer and money-grubber, set on scraping his shekels together. The son, a captain and a professor at Uppsala, comes into more property than anyone else in the city. He owns a works at Kalix, controls the whole of Luleå and Lappmark. It goes against the grain to persevere with the works, but he carries on stubbornly, running himself into debt when he could have lived well on very little. Ill-gotten property brings no pleasure to the third generation.

Instead of scuttling he bales out. He fornicates with the young. His wife, a good-natured woman, finds excuses for him, maintains that it is only tittle-tattle. He installs yet another beautiful maiden in his house. His wife falls ill, wastes away and dies when she becomes aware of her husband's affection for the girl. The day before she passes away, she takes her farewell of him. He replies: "Thank you my dear for having been such an honest wife. I hope you have never gone cold or hungry while you have been living with me".

The girl bruised the front of her leg and it became septic.

Meldercreutz had travelled more than anyone else in Europe, some one hundred and forty thousand miles by land. He controlled practically the whole of Luleå and Lappmark. He went hiking over Luleå fell for the pleasure of it, with Buffon's book under his arm. The most stubborn person imaginable, he persevered with his ironworks at Kalix until he was over his ears in debt.

Master Tidström said of him that he bales out everything he comes across.

He likened him, on account of his debts, to a sailor on a sinking ship.

III.i.3.5 Hökerstedt

Hökerstedt, lord lieutenant on Gotland, takes a fancy to a handsome housemaid, solicits her for some time. She eventually gives him a solemn promise that she will come to him in his room that night. He usually slept apart from his wife.

One of Hökerstedt's friends is staying with him for a few days. He catches sight of the housemaid and is struck by her beauty. The lord lieutenant tells him that he himself has been after her for some time, and that she has finally promised to come to him that very night. The friend pleads with the lord lieutenant to let him have her instead, since he will have plenty of future opportunities. The lord lieutenant agrees, and that night they exchange rooms.

During the evening, the maid goes to the wife and asks to be allowed to leave her service. She explains that the lord lieutenant has so pestered her that she has promised compliance. The wife says that she will see what she can do. That night she goes into the room instead of the maid, and thinks that she is getting into bed with her husband. She keeps quiet and says nothing in order not to give herself away. When she eventually realizes she is with someone else, it is too late. The result of the scheming was, therefore, that this gentleman became the wife's lover and the lord lieutenant's rival.

III.i.3.6 Bergqvist

Cobbler Bergqvist has an elderly and blind wife; he gets involved with another woman; his wife dies and he marries the whore. They become mortal enemies, each appealing to the magistrate against the other; for a whole year, prayers are said for them from the pulpit in Uppsala, but they have no effect.

III.i.3.7 Controller

A controller in Gothenburg sleeps with the wife of a ship's cargoman, who is away in India. When the husband returns, his wife has a child. He divorces her.

The controller wants to marry her. They both try to arrange this, but the clergy and the lawyers oppose it. The case eventually comes before parliament. Three of the four houses grant permission to marry, but not the clergy.

On the day on which this permission is granted, the controller and his whore fall out. They are prayed for in church, but prove irreconcilable.

An employee of the Gothenburg East India Company travels to the east.

While he is away, someone else sleeps with his wife. The employee returns and divorces her.

The adulterer wants to marry the woman. Faced with opposition all round, he appeals to parliament and wins the case in three of the four houses, the only opposition coming from the clergy.

On the day on which the case is decided he falls out with her. Although they marry as intended, they remain at odds, fighting and squabbling.

III.i.3.8 Canutius

Brodén, district judge on Gotland, had a good-looking wife. She was promiscuous, however, had a particular propensity for bachelors, one of whom was Canutius, the local clergyman. This so disturbs the husband that he becomes consumptive. She sweeps and dusts, gives up preparing meals for him, in order to hasten his decline. He dies.

She is pregnant and Canutius marries her, the child being called Canute after the sailor whom she considers to have been the real father. She completely abandons herself to drink.

Canutius serves two turns as a member of parliament. On both occasions he returns home to find the house in a chaos, everything has to be renewed. He beats and horsewhips her. She stews the whip and serves it up for him to eat.

III.i.3.9 Wertmüller

Wertmüller of the Lion apothecary in Stockholm was granted the title of physician-in-ordinary, and acquired the prettiest girl in Hamburg as his spouse.

The wife was fast and randy, had admirers, the main one being the commissary of a bank.

On a certain occasion she travelled to Uppsala to visit her son who was studying there, and took this admirer with her. They arranged music and dancing. “Wherever I am”, she said, “this is how it should be”. Her husband was in Stockholm.

The husband became depressed, perhaps because he was jealous, perhaps because he loved her so much, I do not know.

So it continued for many years.

Toward the end of 1769 the commissary of the bank was imprisoned for extensive debts. This shook her to the core.

Soon afterwards, in the January of 1770, the son, of whom the mother was very fond, was dancing at a party; he took a cold drink, went out to cool off, fell ill, and died a few days later.

The husband is consumptive, awaiting his fate.

Three blows, therefore, all at once.

III.i.3.10 Nietzel

Nietzel, keeper of the botanic garden at Uppsala, a German; I got him to come to the garden at Uppsala from Clifford. He lodged with a tanner, also a German. The tanner had an extremely attractive wife, who took a fancy to Nietzel, as he did to her. The tanner is aware of the situation, but cannot acquire any concrete evidence. He is not courageous enough to come out with the matter and ask Nietzel to leave. The worry brings on consumption and he dies.

Nietzel is not keen on marrying the widow, but is unable to get out of it.

They have a couple of children.

The wife falls in with certain gossips, who foresee a third marriage for her, with a professor. She begins to find Nietzel a bore, wants someone better, goes dotty, even lunatic, gives all Nietzel's things away to the gossips. On certain occasions, when her husband is listless, she has to be tied up.

Nietzel, wasting away with sorrow and vexation, begins to console himself with brandy and to cough blood. He dies of consumption.

III.i.3.11 Urlander

Urlander, commissioner and assessor in the ministry of trade and industry, lives in Norrköping and lives well.

His wife has the life-style of a queen. She is not faithful to her husband, however, and has various lovers.

Her husband dies; she is careless, squanders everything, and is put under surveillance; ends up in the home at Danviken.

III.i.3.12 Psilanderhielm

Mrs. Psilanderhielm, in Stockholm, gets lecherously involved with a courtier.

The courtly gentleman gets married. She is so vexed when she hears of this that she develops troubles in the lower abdomen, and in 1748 she dies.

When she is opened, her body is found to contain a tophus.

III.i.3.13 Jan Jansson

My farmer at Sävja, a weakly fellow by the name of Jan Jansson, marries a strong and healthy girl from Engby. Though married a full twelve years, she remains childless.

She becomes involved with a farm-hand from Uppsävja. Her husband gets to know of this; despite hostility and tears between husband and wife, the two continue to meet in secret.

Others said she became involved with two brothers.

In 1770, she is distilling brandy and the fire sets the liquor alight. When she tries to dampen it with her clothes these too catch fire; excessive burns are inflicted on her thighs, stomach and genitals. In peril of her life, she screams to high heaven for eight whole days; finally swells up, develops pains in her head, dies of this after six weeks.

She was said to have slept with two brothers. If she did so, God's retribution was not to be avoided.

III.i.3.14

A Brålanda farmer

A farmhand in Dal is in service for seven years with an old farmer who has a young wife; he violates the husband's right, which unsettles the old fellow and leads to his death.

The widow takes the farmhand to be her husband.

The young man was in the habit of fornicating with his step-daughter in the guest-house; he makes her pregnant.

The mother sees from the daughter's belly that she is pregnant; the step-father admits that he is responsible.

The mother decides, together with her mother, that a fire should be lit when the daughter has given birth, and that the child should be burnt; this is what is done.

While the daughter is lying in, she was said to be ill; when she suddenly recovers, people become suspicious; crown agents investigate; she has to admit.

On Brålanda heath in 1746, daughter, mother, grandmother and father are put to death.

III.ii THY CHILDREN SHOULD SEEK TO THEE

III.ii.1.1

O thrice and four times blest,
Whose fate it was to die before their fathers' eyes
Beneath the lofty walls of Troy.

Virgil *Aeneid* I.94

And thou, my blessed spouse,
Happy in thy death, in not having been spared for this grief!

Virgil *Aeneid* XI.158

Happy both, and laid to rest in good season!
Since they passed away before the day of my punishment.

Ovid *Tristia* IV.x.81

Ecclesiasticus 33: 19, 22, 21: Give not thy son etc. power over thee while thou livest, and give not thy goods to another: lest it repent thee, and thou intreat for the same again. In all thy works keep to thyself the pre-eminence; leave not a stain in thine honour. For better it is that thy children should seek to thee, than that thou shouldest stand to their courtesy.

III.ii.1.2

Ribbing

Captain Sten Ribbing seduces the daughter of captain Pahl in Finland after promising her marriage, and considers marrying her.

President Piper, related to Ribbing, hears of the affair and facilitates the captain's being posted to Pomerania.

In Pomerania the governor's wife takes to Ribbing, and despite the governor's being aware of this and finding it tiresome, Ribbing is accepted into their circle of acquaintances. It is, however, the governor's daughter with whom Ribbing falls in love, and since he himself finds it difficult to ask for her hand, he writes to his brother, colonel Ribbing, asking him to do so for him. The colonel complies and the governor gives a flat refusal, adding that it would have been a different matter if he had written on his own account rather than that of his brother. The colonel himself therefore asks for the girl's hand and is accepted. Captain Ribbing learns of his rejection, wastes away, coughs blood, turns consumptive and dies.

By marrying the daughter, the colonel has the prospect of coming into a very considerable inheritance; becomes a privy councillor in 1765 and is dismissed in 1769. His father-in-law the governor realizes on his property. The purchasers suddenly go bankrupt, however, and he is left with nothing.

Privy councillor Ribbing's wife is said to have had a secret affair with the French envoy.

Based on Söderberg's account.

III.ii.1.3 Solander

Daniel Solander, assistant lawyer in the court of appeal, applied for the chair of jurisprudence in succession to Reftelius. Professor Hermansson was then all-powerful in the consistory, and he saw to it that Solander was appointed. At the time, Solander was courting Hermansson's daughter, who subsequently married bishop Serenius; she had even got to the stage of sewing her wedding-dress.

The widow of apothecary Lambert was good-looking and wealthy, and as soon as Solander had been appointed he married her.

This woman eventually took to drink, spending all her own and Solander's resources on strong liquors, dying of pulmonary consumption, leaving Solander impoverished and the children neglected. He made several unsuccessful offers of marriage, became unkempt and slovenly, the children wretched.

Professor Daniel Solander, when an assistant lawyer, courts professor Hermansson's daughter, the archbishop's niece, later the wife of bishop Serenius, in order to get appointed to the chair of jurisprudence. As soon as he had been appointed, though the wedding-dress was being sewn, he married the wealthy widow of the apothecary. A few years later she takes to drink and dies of consumption. He neglects himself; on account of his economic situation he is unsuccessful in his attempts to find another wife; the children run wild.

III.ii.1.4 Lindberg

Lindberg, builder in Uppsala, had an elderly wife, took to sleeping with a handsome whore and had three children by her.

He complains that his old wife has salt rheum or venereal trouble, which only he himself could have infected her with, and so obtains a divorce.

He then marries anew with his whore.

He develops an appallingly painful sciatica, which plagues him for years, the doctors being unable to do anything about it. I was called in, diagnosed the cause as venereal and cured him; he was more of a skeleton than a human being.

While this was going on, alderman Goldhan, the belt-maker, despite his being a married man, takes to sleeping with Lindberg's new wife. He sells his own wife's wardrobe in order to buy clothes for Lindberg's wife, and eventually ruins himself financially by standing surety for Lindberg's bridge-building project.

Hates his own wife, continues to cherish Lindberg's.

Goldhan's house and property are eventually auctioned off. Lindberg's wife runs into debt. Everything goes wrong.

In 1768 the two eldest sons, which he had had by the whore while he was still married to his first wife, before he had entered into matrimony with the trollop, steal a chest containing 14,000 dollars from Wikblad, the manufacturer of twist-tobacco. Although they were caught and condemned to death, the sentence was remitted; for years now, however, they have been sitting out their time in prison.

III.ii.1.5 Såganäs

Jacob of Såganäs, in the parish of Stenbrohult, Småland, did not get on at all well with his wife. One Christmas during my youth, when she was on her way to church across the ice, she fell through it. For some time, about a quarter of an hour, she clung to the edge of the ice calling for help. It happened not far from the farm, and the husband stood there on the bank, saying that he was not going to risk himself out on the ice, that he would be glad to be rid of her. She drowned.

Five years later the fingers with which Jacob could have saved his wife began to turn putrid. They eventually rotted away on both his hands, and he died of it.

III.ii.i.6 Karl Jansson

The farmer at Kyrkebyn in the parish of Danmark was dreadful to his wife, frequently hitting her about the head. She became permanently simpleminded and took to drink.

In 1772, while on a journey to Stockholm, he went out of his mind and started raving. The whole parish had to take it in turns to keep an eye on him. Apart from this, he was a healthy, robust, industrious farmer.

A so-called village wiseman eventually opened all the patient's veins, bringing about his death.

III.ii.2.1 WIFE

1 Corinthians 7: 2, 9: To avoid fornication, let every man have his own wife, and let every woman have her own husband. If they cannot contain, let them marry.
Ecclesiaticus 25: 1: Both God and men are beautified when a man and a wife agree together.

Proverbs 31: 12: A virtuous woman will do her husband good and not evil all the days of her life.

Ecclesiasticus 26: 13: The grace of a wife delighteth her husband, and her discretion will fatten his bones.
Proverbs 19: 14: House and riches are the inheritance of fathers: but a prudent wife is from the Lord.
Ecclesiasticus 26: 14: A silent and loving woman is a gift of the Lord; and there is nothing so much worth.
Ecclesiasticus 36: 22, 23: The beauty of a woman cheereth the countenance, and a man loveth nothing better. If there be kindness, meekness, and comfort, then is not her husband like other men.
Ecclesiasticus 26: 1–4: Blessed is the man that hath a virtuous wife, for the number of his days shall be double. A virtuous woman rejoiceth her husband, and he shall fulfil the years of his life in peace. A good wife is a good portion, which shall be given in portion of them that fear the Lord. Whether a man be rich or poor, he shall at all times rejoice.
Ecclesiastes 9: 9: Live joyfully with the wife whom thou lovest all the days of the life of thy vanity, for that is thy portion in this life, and in thy labour which thou takest under the sun.

1 Corinthians 7: 3, 4, 34, 35: Let the husband render unto the wife due benevolence: and likewise also the wife unto the husband. The wife hath not power of her own body, but the husband: and likewise also the husband hath not power of his own body, but the wife.

She should take care how she may please her husband, that they may live together in love and harmony, which ought to be the heart's desire of all.
Proverbs 5: 18: Rejoice with the wife of thy youth.

Ecclesiasticus 25: 18, 19, 23: A woman in whom a man has no joy, will make him see everything as an affliction. A wicked woman abateth the courage, maketh an heavy countenance and a wounded heart. All wickedness is but little to the wickedness of a woman.
Proverbs 27: 15: A continual dropping in a very rainy day and a contentious woman are alike.
Proverbs 14: 1: Every wise woman buildeth her house: but the foolish plucketh it down with her hands.

Proverbs 6: 24: Fear God, that you may be kept from the evil woman.
Ecelesiasticus 9: 2: Give not thy soul unto a woman to set her foot upon thy substance.
Proverbs 25: 24: It is better to dwell in the corner of the housetop, than with a brawling woman and in a wide house.
Ecclesiasticus 25: 16: I had rather dwell with a lion and a dragon, than to keep house with a wicked woman.
Ecclesiasticus 25: 23: A woman in whom a man has no joy, will make him

see everything as an affliction.
Ecclesiasticus 26: 7: An evil wife is a yoke shaken to and fro: he that hath hold of her is as though he held a scorpion.

III.ii.2.2 Ihre

Ihre, Skyttean professor at Uppsala, married to the daughter of lord lieutenant Brauner. They buy Sandbro and run up a bit of a debt. His wife retires to the countryside, lives separated from her husband in order to run the estate, and does so very efficiently for a number of years, working like a slave.

When she has got the place in order and paid off the mortgage, she gives up work entirely, intends to live as other people do, to bring up their son and the three daughters.

Ihre complains of a pain in his side; she is afraid that he may be dying. They draw up a reciprocal will, according to which the spouse who survives shall possess, enjoy and manage the property and be able to dispose of it at will.

The wife dies of peripneumonia, at least that was what it was thought to be. Ihre re-marries, with Charlotta Gerner, lives handsomely, splendidly, his former wife having been so careful, having even begrudged eating. It is not for yourself that you labour.

He lives thus since the will stipulates that if the surviving spouse re-marries, the whole of the property will later pass to the children; it was for their sake that his former wife had drawn it up and arranged everything.

Ecelesiasticus 11: 18, 19: There is that waxeth rich by his wariness and pinching, and this is the portion of his reward: Whereas he saith, I have found rest, and now will eat continually of my goods; and yet he knoweth not what time shall come upon him, and that he must leave those things to others, and die.

III.ii.2.3 Von der Lieth

Colonel von der Lieth, a decent old fellow whom everyone liked, made a good job of bringing up a sailor's daughter, who turned out to be a good-looking and respectable young woman.

He married the girl, but they had no children. This was a pity, since he had an entailed estate in Germany.

Eventually, after many years, she became pregnant. People said that the child she was carrying was Dr. Wedenberg's.

The colonel and his wife were delighted, hoped that it would be a boy-child who might inherit the estate. Unfortunately, although she was delivered of a boy, he died at birth.

The colonel died three weeks later. The wife lost not only the hoped-for estate, but also her husband and his pension.

III.ii.2.4 Wallerius

Nils Wallerius, doctor of divine theology, professor at the university of Uppsala.

His second wife went out of her way to be unpleasant to her stepchildren, the servants, her husband and their child.

He eventually dies; she manipulates out of the rest of them as much of the inheritance as she can.

Shortly afterwards, early in 1766, the farm from which she was to have derived her share went up in flames.

She applies for a double year of grace, stipulating that no advantage should accrue to the stepchildren; this is granted.

In September 1766 the barn in the country in which she has just deposited the tithes due to her, catches fire. She has to relinquish the greater part of the contents to the flames. The farmer's barn next to it remains standing.

In the December of the same year the vicarage in the country to which she had moved burns down. She just about escapes, with her son wrapped up in linen; loses everything. The farmer's house next to hers remains standing.

Who can escape the retribution of the Lord? Be not cruel and hard. There is no pardon for those who show no mercy.

After the last fire, the university grants her 3,000 dollars; since there are also other contributions, she receives 18,000 dollars, which is more than she had in the first place.

Altogether she scrapes up some 21,000 dollars, which she deposits with the wealthy Julinschöld at a certain rate of interest; he goes bankrupt at the beginning of 1768.

Psalm 127: 1: Except the Lord build the house, they labour in vain that build it.

III.ii.2.5 An Uppland farmhand

The farmhand makes two girls pregnant at about the same time. The first faces public disgrace as a whore; the second does away with her child in secret.

The farmhand marries the second girl. On a certain occasion, while she is brewing, she has a fit of giddiness, falls into the brewing vat, and is boiled to death.

The first girl marries and gets on well.

III.ii.3.1

PARENTS

Ecclesiasticus 7: 27: Honour thy father with thy whole heart, and forget not the sorrows of thy mother.
Tobit 4: 4: Remember that thy mother saw many dangers for thee, when thou wast in her womb.

Ecclesiasticus 3: 16: He that forsaketh his father is as a blasphemer; and he that angereth his mother is cursed of God.
Deuteronomy 27: 16: Cursed be he that setteth light by his father or his mother. And all the people shall say, Amen.
Proverbs 19: 26: He that wasteth his father, and chaseth away his mother, is a son that causeth shame, and bringeth reproach.
Proverbs 20: 20: Whoso curseth his father or his mother, his lamp shall be put out in obscure darkness.

III.ii.3.2

A Parisienne

A certain woman in Paris has a husband who loves her faithfully and sincerely, but who is obliged to travel abroad for a year.

While he is away she lives with her butler, bearing him a boy-child which she passes to a peasant woman in the country, to be brought up incognito.

Her husband returns; ten years later he dies, having learnt nothing about this episode.

The mother eventually takes her son to be her footman. The boy is uncivil, however, and neither admonition nor chastisement brings about any improvement in him. One blunder follows upon the other, until eventually the mother is obliged to dismiss him from her service.

He returns to his foster-mother the peasant woman, and bewails his lot. She justifies having parted with him by telling him that the lady is his real mother, says she is amazed that a mother can be so hard-hearted to her son.

The lad, who now knows where he belongs, goes by night and knifes his mother to death.

There is an investigation, but no one knows who could have committed the murder. Certain flecks of blood are found on the butler's sleeves, but they are found to have originated from a bird, which had been shot, and which he had purchased and prepared the day before.

Nevertheless, the butler comes under suspicion and is put under torture, which he is unable to bear. He is condemned to be broken on the wheel and executed.

Some years later the boy commits another murder, and when he is about to be executed, admits that he also murdered the lady, his mother.

The proclamation that the butler had died innocent is drummed through all the streets of the city.

It was God who had visited retribution upon the lady and the butler.

C. Gedda

III.ii.3.3 Sånnaböke

Måns, of Sånnaböke in the parish of Stenbrohult, Småland, was a hard man, and behaved very badly toward his father.

Måns Månsson, this man's son, took his father by the forelock in order to remove him from his own cottage. When he had been pulled to the door, the father called out: "Måns, my boy, don't pull me any further. This is as far as I pulled my father". The son replied: "By the death of our Lord, man, if you dragged your father to the door, I shall drag you through it".

This happened near to where I was born, when I was a child.

III.ii.3.4 Leijel

Leijel, mayor of Stockholm, was wealthy and had no heirs; he had three sisters, however, and one of them he disinherited.

Möhlman, the son of the second sister, inherits the greater part of the property; increases it until it is worth two million dollars in gold. He wants to marry his disinherited aunt's daughter, but since the family is opposed to the match, remains a bachelor.

In his old age he becomes ill. He sends for Gerdesskőld, his cousin's son, president of the court of appeal, and asks him to draw up a will by the next day, conferring all the Möhlman property on himself. Within an hour of doing so, Möhlman dies.

It was in this way that nearly all the Möhlman money passed to the disinherited sister's offspring. Salvius and Dr. Schulzenheim, marrying into the family, come into the wealth.

Leijel's sister appealed to God when she was disinherited, and He heard her.

III.ii.4.1 RETRIBUTION

Retribution is visited upon him. Everything runs counter. No calamity comes alone.

For some, everything they undertake goes wrong, for others, regardless of their stupidity, everything goes swimmingly.

Whole families are unfortunate. The children are taught; act completely contrary. Hell bent on their own destruction. Heaven and earth can neither help nor save.

Many always do badly after a certain day. Everything then began to go wrong. Charles XII's first nine years went well, then nine years of misfortune.

One misfortune follows another. Wherever he turns, whatever he does, it goes wrong. Someone else is sitting in the lucky chair.

One dies after the other, once God begins to settle the accounts. No calamity comes alone. The house burns down, everything goes wrong. God's retribution has now been visited upon the house.

Call no one happy until his death, not even Croesus.

Gunnar Gröpe murdered St. Sigfrid's three nephews; no natural death among his descendants, the Ulfsax family, for twelve generations.

The unfortunate born of wicked parents. The piglets suffer for the piggery of the porker.

Psalms 37: 25: I have not seen the righteous forsaken, nor his seed begging bread.

Misfortune pursues him wherever he turns.

When I contemplated revenge everything went wrong for me. When I changed and put everything in the hands of God (1734), everything went well.

III.ii.4.2 Krabbe

Krabbe was a wealthy Danish nobleman in Scania, after it had become Swedish.

Sperling, the lord lieutenant in Malmö, wanted to get hold of his property. Charles XI was therefore informed that Krabbe was hostile to him, that he was in correspondence with the Danish side. The king finally sets up a commission to look into the matter.

Sperling is president of the commission; condemns Krabbe to death. The execution is due to take place on the day when the post arrives, when the king's reply or judgement will be made known. They are worried that there may be a reprieve, however, so on the morning of the day of execution, the drawbridges around Malmö go up and the gates are closed. When the post arrives, it has to wait outside for an hour. When it enters with the reprieve it is too late, Krabbe has been beheaded.

Despite Sperling's eventually coming into possession of the property, he dies in poverty.

His son falls in and is drowned while watering his horse by the moat at Malmö.

His daughter marries Horn, commandant at Narva; they were captured with their daughter by the Russians, and incarcerated in a most frightful underground dungeon.

III.ii.4.3

Appelbom

Appelbom, lieutenant in Wrede-Sparre's regiment, speaks out in 1757 against the governing party. The matter is dealt with by a committee set up by the estates.

Wrede-Sparre, Appelbom's colonel, joins the committee during the course of the investigations, maintaining that all his officers with the exception of Appelbom were well-disposed toward the party, and that he would therefore like to see him dismissed.

Appelbom is sentenced to ten years' exile. He joins the Hanoverians, becomes a major, and is in command during an action.

Wrede-Sparre's only son, a fine young fellow, goes abroad and sees active service with the French army. The contingent to which he is attached is defeated in an engagement with the Hanoverians, and he is mortally wounded. The victorious commander is Appelbom, unaware that Wrede-Sparre is one of the enemy's casualties.

III.ii.4.4

Wrangel

Wrangel, subsequently member of the privy council, while travelling in France, entices a married woman into accompanying him to Italy.

His daughter, seduced by Cederhielm, is obliged to marry a bailiff; she lives promiscuously and is disinherited by her father.

His eldest son, who is quick-witted, begins his career at Uppsala by having an affair with Braunersköld's wife. He gets involved in matters of state while in Stockholm and has to flee to Germany; dies in Hamburg.

III.ii.4.5

Morga

Jan Persson, a farmer who lived close by Alsike, had seven sons. He married them to the seven daughters of another wealthy farmer, who lived right opposite on the other side of the lake.

The last of the sons had just got married, the wedding ale had been prepared, all seven sons and daughters were crossing the water in a boat. As they came into view, the farmer walked down to the shore with drink and brandy to welcome them. As he was doing so, he saw the boat capsize in a whirlwind and all his sons and daughters drown; looking back, he saw the farm going up in flames. Hence the proverb: if Jan Persson of Morga pulled through, we can all do so.

> Come now, tell us of those who long ago
> Braved misfortune with a steady mind.

III.iii THE RIGHT COMMON TO ALL

III.iii.1.1 Seneca

Customary submission to the course of nature alleviates all unhappiness.
For those whom many fear, there are many to be afraid of.
Let them hate, as long as they are afraid.
You will see man in his audacity leaving nothing untried.
Be on your guard when prospering, never give up in adversity.
Failure changes us for the better, success for the worse.
Had you known that all is from God, you would have been justified in wishing it.
Man is a constant threat to man.
Man is ever ready to harm man,
There is no evil more frequent, persistent or insinuating.
Wild beasts are only aggressive when driven to it,
But man delights in ruining man.

III.iii.1.2

To the unfortunate, all are hostile.

III.iii.1.3

Though all can be swept by misfortune,
Fate is less unkind to the small man,
God removes the lowly with a gentler hand;
It is unobtrusive calm that preserves peace,
At home that one sleeps securely.

III.iii.1.4 The mineworker

In 1761 the mine building in Falun catches fire. A mineworker continues to ring the alarm bell in the tower until the fire is all about him. The choice is between being burnt alive and jumping to death. He chooses the latter.

As he jumps, a miner emerges from the fire carrying a table on his head. The falling man lands on the table, which gives way and so saves his life.

The miner carrying it is killed by the impact of the fall.

III.iii.1.5 A Scanian farmhand

A Swedish farmhand, eighteen years old, flees Sweden on account of having hacked off both his mother's arms. Gets work on Sjælland in the hundred of West Flakkebjerg, the parish of Gimlinge, the village of Vemmelöse. Here he is ploughing with another farmhand, loses his temper with him and strikes him dead; is broken on the wheel and beheaded for it.

III.iii.1.6

Strutz

Lance-corporal Strutz of the life guards, living at Brusarbo in the parish of Enåker, Vestmanland, stabs a tobacco-merchant's journeyman to death at Sahlberg market. Flees to Norway; returns home under a safe conduct and is eventually granted his freedom.

He buys a grey horse, which he is very fond of and keeps beautifully groomed.

As he is entering the stable the horse breaks loose; bearing Strutz out through the door, it so dashed him against a stack of timber that he died a few hours later.

III.iii.1.7

Per Adlerfelt

Condemned to be harquebussed for murder in Turkey.

Goes into hiding until the king forgets. Comes to Norway under another name, joins the king. The king recognizes him, sends him home.

Shot by the Dalecarlians.

Captain in the guards.

Came home 1709; lieutenant-colonel in the action at Gadebusch; 1714 colonel in the Nerike and Vermland regiment, 1717 in Berg's regiment; 1719 major-general, sent with propositions to Denmark; 1720 baron, envoy to Denmark when peace was signed between Sweden and Denmark; 1739 privy councillor, on the chancery council.

1743 shot outside Northgate in Stockholm during the Dalecarlian revolt.

III.iii.1.8

Planting

Second lieutenant Planting, found stabbed in his bed on Christmas night. Twenty or thirty years earlier suspected of having knifed a non-commissioned officer, but it was not possible to prove it.

There seemed to be no one who might have knifed Planting, except for a widow woman living in the house, a relative of the murdered non-commissioned officer.

III.iii.1.9

Wallrave

Wallrave, son of the treasurer royal, somewhat tipsy, going home in the dark one evening in Stockholm. A fellow in front of him is walking along with a lantern.

Wallrave walks along between the stranger and his lantern. The fellow speaks up: "Sir, this is my lantern. I'm using it to see and light my way. If you will walk to the side of me, we can both make use of it". "What was that you said, you scum?", retorted Wallrave, and he ran the fellow through with his rapier.

Wallrave flees to Norway, where he is granted a safe-conduct. He is sentenced to several years' exile, studies law and medicine, takes his doctorate at Harderwijk, returns home and is appointed professor of Roman law at Uppsala.

Wallrave enjoys his pub-life. Twenty years after the Stockholm affair he sets out late one evening in the dark from the pub run by Kähler's son Siwert, making for the stone house by the bridge. Since he knows every step of the way, he says it will not be necessary for anyone to accompany him with a lantern. He misses the bridge and is drowned. That very day he had drawn up an inventory for Rommel, the university treasurer, whom he had formerly set out to ruin. See Julinschöld.

He had killed the other fellow on account of his lantern; he now dies because he himself has no lantern.

III.iii.1.10 Tavaste-bo

Tavastehus is well out of the way on the frontier. The farmer there used to murder travellers, strangers who had gone astray.

On a certain occasion a farmer from another neighbourhood, travelling in order to purchase seed, put up at the place. He was to sleep in the same room and bed as the host's son. The son was, moreover, to sleep on the wall-side of the bed, so that if the visitor locked the door, the son would be able to open it again without this being noticed.

The son dozes off. It occurs to the visitor that he ought to go and have a look at his horse. When he returns, he finds that his host's son has turned in his sleep and is now lying up against the post of the bed. Rather than wake him, the visitor lies down on the wall-side of it.

In the middle of the night the host comes quietly into the room and brings his axe down on the skull of the person lying by the post. He then calls out to his son that the job is done, and throws down the axe.

The visitor leaps up, gets hold of the axe, and castigates the host for having done away with his own son. The host is subsequently arrested, sentenced, and broken on the wheel.

III.iii.1.11 Cronhielm

Cronhielm, son of the councillor of state, while out on a lake, encounters a farmer who drives his working sledge into the earl's sleigh. The earl strikes the fellow on the head and kills him. At the trial, he gets away with it by maintaining that the farmer must have been asleep and must have struck his head against the sleigh.

Some years later the earl is travelling the same way across the lake. A rift had formed in the ice the night before. The earl drives down into it at precisely the point at which the farmer had suffered the fatal blow. The earl's servant and the horse get out easily enough. The servant does all he can to help his master, but is unable to reach him. The earl drowns, calling out that on that spot he has witnessed the retribution of God.

III.iii.1.12 Tordenskjold

The Danish admiral Tordenskjold, the one who took Marstrand from us, shot a boy way up in the rigging, because he thought he was not getting on quickly enough with his job.

He himself was shot by a Swede in a duel in Hamburg.

III.iii.2.1 GREED

The poor lack much, the greedy everything.
The greedy man benefits none, is his own worst enemy.
Greed denies to itself what it has taken from everyone else. Seneca.

A parsimonious person takes no pleasure in his food: everything is wasted on his heirs; they become dissipated, though they could have become diligent.

Lohe, one of the nobility, inherited wealth from his father; lying sick in Stockholm, he snipped off half the doctor's prescription to reduce the cost of it. His only son died while studying at Uppsala, and was left unburied for a whole year on account of the costs. A woman living in the same house used to put out milk for her puppies every morning; it was he who drank it, the puppies languished and died.

He who has much, has much to worry about; he tells himself he will never pull through, is plagued by having acquired what he wanted.

III.iii.2.2 Broberg

Stole an *Adonis capensis*, which he took to Stockholm; after a couple of years spirits away a young plant.

Gave in his notice when he was not allowed to travel for my bulbs.

When on his own recommendation I take on a young fellow for training as an apprentice, he takes on someone else without informing me, and gets rid of the one I have appointed.

When I give him a pot to look after for me, he appropriates some of Hasselquist's material.

When I plant *Bulbos capenses* in my pot, he lets me have one, and destroys or steals the rest.

When I tell him not to remove a *Valerianam tetrandrum*, from its pot, he says that he has already done so, and shows me a different plant. The potted one has disappeared.

When I tell him not to plant out my *Antholyza cepacea*, he says that he has already done so, and when I ask to see it, he says that it has withered away.

When I told him not to transplant box trees, he does so when I am away, and sells half of them.

Geranium foliis peltatis is filched.

What happened to the *Bocconia*?

Kalmiæ, *Gardenia*, *Magnolia*.

III.iii.2.3 Kyronius

Kyronius, alderman of Uppsala, became titular mayor; though quick-witted, humorous, widely read, presentable, he was a slippery person to deal with in business matters.

The reverend Gåse's widow borrows 3,000 dollars from him, and when she eventually repays them, asks for the receipt. He looks for it for some time, saying that it is a matter of no importance, and eventually finds a piece of paper which he quickly rips up, telling her that the receipt has now been disposed of. Two years later he then presents her with the receipt, demanding a further 3,000 dollars.

Ecclesiasticus 35: 15: Do not the tears run down the widow's cheeks? And is not her cry against him that causeth them to fall? He who eats the widow's bread shall drink her tears.

His proving ungrateful to his father led his father to say that he would come to no good.

He had many mistresses, seven illegitimate children, his wife being aware of it.

In 1741 he prosecuted Herkepaeus, the mayor, for what he had said about the estates.

In 1746 he beat his opponent to become member of parliament for Uppsala, although a lot of strings had to be pulled and there were not many who voted for him. He became so powerful in the house of commons that he controlled all promotion, became known as the secretary of state. Everyone was obliged to bribe him; took 6,000 dollars from someone and still failed to help. Develops an excessively high opinion of himself; despite being a married man, lays siege to Broms' only daughter, writing her raving love-letters. He is voted the title of mayor and the reversion of the office, but not the freedom of the city; he expunges this reservation from the document. A commission of impecunious assistant justices of the court of appeal is therefore appointed to deal with him; he is obliged to fork out to them certain of his ill-gotten gains. Among them is Voltemat, whose father had been frequently fleeced by him. He returns home, opens his mouth about the estates in the council chamber, and gets lost in the same labyrinth as Herkepaeus, whom he had denounced for this in 1741. The voice of the people is the voice of God: "Crucify, crucify"

Back home again, he says he is unable to sleep and asks me to help him. Since he is afraid of opium, I am unable to do so. "No", he replies, "when I was in Stockholm I took opium like a Turk, but it had hardly any effect". I prescribe the usual dose and he administers it in large quantities to his ailing wife, in the hope of getting rid of her and being able to cultivate Miss Broms. Suspicion is aroused when his wife suffers from vomiting; then the whole matter comes to light.

His son-in-law, the son of Olof Celsius, gives him twenty-four hours, either to get out of the country, or to submit himself for arrest at the castle. In a matter of hours he has collected his belongings and set off for Copenhagen.

While he was down there he said things detrimental to the Swedish government and was therefore extradited back to Stockholm. He escapes to Germany, where for a number of years he searches around for a safe-conduct. He is finally granted not only a safe-conduct, but also a free return home without any indictment. A few days before it was granted he got married, however, so that he has never been able to return from Germany. He has quitted his all, like a sparrow quitting an ear of corn.

III.iii.2.4 Hauswolff

A young fellow by the name of Hauswolff applies for the post of assistant justice of the court of appeal in Åbo, and is accepted as a candidate. An old penurious district judge, with a large family, who had often applied for promotion and was generally down and out, is also in the running.

The king was sympathetic toward the old fellow, who had no friends, knew no one on the appointment committee.

Hauswolff has plenty of friends, who speak up for him, putting it about that the district judge is a firm supporter of the opposition party, which is quite untrue. They so manipulate the situation that Hauswolff gets the job.

The district judge meets Hauswolff, admits that he ought to congratulate him, but appeals to God instead.

That winter Hauswolff travels from Stockholm to Åbo. When the vessel runs into pack-ice, everyone thinks the end has come. Hauswolff, together with someone else, jumps into the sloop, cuts the rope, and drifts away from the vessel for five or six days. He ate leather and anything he had, including his letter of appointment, which was found in his mouth when his body was washed ashore on Gotland.

The vessel got through and was rescued together with its passengers.

III.iii.2.5 Odhelius

Odhelius, district judge, buys a farm from a widow who is in debt, on the understanding that she can continue to inhabit one of the farm cottages as long as she lives. He finds the arrangement irksome, however, gets one of the rangers to knock down her chimney during the depths of winter, is generally ruthless in his dealings with her.

His wife goes out of her mind; he eventually lands up in a debtor's prison and dies there.

III.iii.2.6 Schleicher

Schleicher of the life guards, in love with widow Von Bysing, presents her with an estate. One of his in-laws is so incensed by this that one night he fires through the window with three buck-shot, which pass straight through Schleicher's stomach and kill him.

Some years later the in-law develops stomach cancer; there are three cavities and he dies a dreadful death.

III.iii.3.1 ENVY

Envy: no one of us must excel.

III.iii.3.2 Bréant

In 1765 Bréant was secretary to the parliamentary body in charge of hospitals and children's homes. He was threatened with an indictment, since his notary Olander had gossiped and Serenius had accused him of misdemeanour; he escapes by shooting himself in his room.

Olander had banked on getting Bréant's job, and did so. He proceeded to pocket letters addressed to the management and to misappropriate money; he was arrested, got away, and went underground in Stockholm. When the court of chancery got to know of his whereabouts and instigated his arrest, he cut his own throat in desperation – just two years and a day after Bréant had done himself in.

III.iii.3.3 Westrin

In 1743 university bailiff Holmberg is fiddled out of his job by Westrin, who takes it over from him. As university beadle, Westrin had been egged on by professor Reftelius into arranging for agents to be sent to the estates managed by Holmberg, in order to find out if the farmers had any complaints against him. It was thus that he had acquired grounds for accusing him of maladministration. Holmberg vows to be avenged; eventually dies.

Olof Celsius the younger reports Westrin on a matter of no real importance. In 1751 Celsius's brother-in-law public prosecutor Grizell takes up and conducts the case against Westrin, in the hope of getting his job. Julinschöld, the university treasurer, helps him, since he has taken a dislike to Westrin. It was thus that Westrin was deprived of his good name and dismissed from his job.

III.iii.3.4 Krabbe

Krabbe was a wealthy Danish nobleman in Scania, after it had become Swedish.

Sperling, the lord lieutenant in Malmö, wanted to get hold of his property. Charles XI was therefore informed that Krabbe was hostile to him, that he was in correspondence with the Danish side. The king finally sets up a commission to look into the matter.

Sperling is president of the commission; condemns Krabbe to death. The execution is due to take place on the day when the post arrives, when the king's reply or judgement will be made known. They are worried that there may be a reprieve, however, so on the morning of the day of execution, the drawbridges around Malmö go up and the gates are closed. When the post arrives, it has to wait outside for an hour. When it enters with the reprieve it is too late, Krabbe has been beheaded.

Despite Sperling's eventually coming into possession of the property, he dies in poverty.

His son falls in and is drowned while watering his horse by the moat at Malmö.

His daughter marries Horn, commandant at Narva; they were captured with their daughter by the Russians, and incarcerated in a most frightful underground dungeon.

III.iii.4.1 INGRATITUDE

Cicero advised the Roman senate to make Augustus consul and then send him as vice prætor to Antonius.

When Augustus, Antonius and Lepidus made up the triumvirate and promised to help one another in doing away with the defenders of freedom:
Antonius let his uncle suffer the vengeance of Lepidus;
Lepidus let his brother suffer the vengeance of Augustus;
Augustus let Cicero, whom he was in the habit of calling his father, suffer the vengeance of Antonius.

"Everything is allowed to a prince", replied Julia, step-mother to emperor Bassianus, when he remarked, on happening to see her bosom, "If it were allowed, I would desire what I have seen". Julia was the real mother of Geta, whom Bassianus had murdered in her arms. Bassianus loved her for her beauty right up until his death.

Proverbs 17: 13: Whoso rewardeth evil for good, evil shall not depart from his house.

There is the tale of the man who saves a thief from the gallows. This man is then captured by enemies, who are about to hang him but cannot find a rope; the thief comes along and supplies one.

Murdered by means of his own weapon.

III.iii.4.2 Cicero

Popilius is accused of the inhuman act of murdering his father. Cicero delivers orations defending him, speaks out for an unjust cause.

Antonius becomes a triumvir and commissions his own brother to get Cicero outlawed. When he wants him beheaded he sends Popilius, for since he was capable of murdering his own father and since he owed his life to Cicero, he was just the sort of person who would be capable of executing the sentence of death upon the orator.

It was therefore through this extremity of ingratitude, at the hand of someone who was under every moral obligation to him, that Cicero was paid out for furthering an unjust cause.

III.iii.4.3 Hallman

Physician-in-ordinary Hallman is consulted by a cavalier concerning a girl who is pregnant; paid handsomely, he prescribes a means of abortion which proves successful.

The same cavalier subsequently provides the physician-in-ordinary with a wife, an extremely beautiful gentlewoman.

This wife was the so-called girl who had undergone the abortion; thus is deception deceived.

III.iii.4.4 Engberg

Engberg was the best cutler in Stockholm. The day before general Lewenhaupt was to have been beheaded, he gets him out of prison and succeeds in spiriting him away onto an island in the archipelago. He is well rewarded.

A reward of six thousand copper dollars is offered for his recapture. Carl Tersmeden promises Engberg half the proceeds if he will tell him where he is. It was thus that he found out the day and the time when the general was to have made his escape by being transported onto the English ship. Lewenhaupt is recaptured by Tersmeden.

During the next parliament Engberg is charged with the propagation of revolutionary ideas and insurrection. He just escapes the death-penalty, which is commuted to life-imprisonment in Bohus fortress.

III.iii.5.1 MALICE

Ecclesiasticus 27: 26: Whoso diggeth a pit shall fall therein.
Ecclesiasticus 4: 30: Be not as a lion in thy house, nor frantick among thy servants.

Job 31: 39, 40: If I have eaten the fruits thereof without money, or have caused the owners thereof to lose their life: let thistles grow instead of wheat, and cockle instead of barley.
Ecclesiasticus 34: 21: The bread of the needy is their life: he that defraudeth him thereof is a man of blood.
Exodus 22: 22–24: Ye shall not afflict any widow, or fatherless child. If thou afflict them in any wise, and they cry at all unto me, I will surely hear their cry. And my wrath shall wax hot; and your wives shall be widows and your children fatherless.

He who eats the widow's bread, shall drink her tears.

Ecclesiasticus 35: 17: The prayer of the humble pierceth the clouds: and till it come nigh, he will not be comforted; and will not depart, till the most High shall behold.

See to it that you do no great injury to those who are both powerful and miserable; though plundered, they still have their arms.
Proverbs 17: 5: Whoso mocketh the poor reproacheth his Maker.
Ecclesiasticus 7: 11: Laugh no man to scorn in the bitterness of his soul: for there is one which humbleth and exalteth.
Ecclesiastes 5: 8: If thou seest the oppression of the poor, and violent pervert-

ing of judgment and justice in a province, marvel not at the matter: for he that is higher than the highest regardeth; and there be higher than they.

The voice of Abel's blood and the voice of oppressed Sodom have cried unto heaven, the wages of the labourer have been withheld.

In Lisbon every All Hallows day a festive bonfire is lighted, to which the papal inquisitors, in the name of religion, lead the unfortunate sinners who are to die by burning. There is no more horrendous or cruel crime on earth than this parading of the devil's work in the name of God. It was on All Hallows day in 1755 that the earthquake, flooding and conflagration struck Lisbon, that the full force of God's punishment was visited upon these obdurate sinners. Half the earth trembled: God revealed that He could sense, hear and feel pity for the unfortunates, despite their being heretics.

Although the law allows anything in dealing with a slave, in dealing with a human being there is an extreme which the right common to all refuses to allow. Witness Pollio Vedius.

III.iii.5.2 Mrs. Dahlman

1764: the sudden death of doctor Nils Wallerius, professor of theology, whose wife, Anna Boy, was obsessed with clothing and dresses.

Mrs. Dahlman, wife of the professor of moral philosophy, ornament of her sex, was visiting Mr. Söderberg, together with me and my wife, when she heard that Wallerius had died, leaving his wife very badly off financially. On receiving the news, she expressed herself in the following words: "I wonder what dress Mrs. Wallerius will be wearing today, and whether she will be changing it tomorrow".

At the time, professor Dahlman was on his country estate. He fell ill that very hour, and eight days later was dead.

God sees and hears everything. Take care, lest Nemesis should hear.

Proverbs 17: 5: He that is glad at calamities shall not be unpunished.
Job 31: 29: Have I rejoiced at the destruction of him that hated me, or lifted up myself when evil found him?
Ecclesiasticus 7: 11: Laugh no man to scorn in the bitterness of his soul: for there is one which humbleth and exalteth.

III.iii.5.3 The farmhand at Jönköping

A farmer at Jönköping thrashes one of his workers. The fellow swears that one day he will get his own back.

The farmer keeps a knife in a rack; only he is allowed to use or touch it.

Two years after the thrashing, the farmer is rounding off the hay-harvest in the customary manner. The farmhand in question turns up at noon, encounters the farmer's wife, and tells her that since the harvest ale is now being provided, he has come for a good mug of it. She goes down into the cellar to draw it for him. He seizes the knife, follows her down, murders her in a frightful manner, and slips off.

It is almost three o'clock, and the farmer is expecting his wife to call them in for lunch. Angered by the delay, he declares that he will teach her to have the meal ready on time, that they have no intention of working themselves to death.

He soon discovers that she is not in the farmhouse. Seeing the cellar door open, he goes down and finds her lying there, picks her up and gets covered in blood, stands there thunderstruck.

Soon afterwards the guests arrive; they find him in the cellar, speechless, forlorn, blood all over him. His wife is still warm and there is the sacred knife, dripping with gore.

He is arrested and denies that he is guilty, but the circumstantial evidence is so overwhelming that he is condemned and executed.

A few years later the farmhand falls foul of someone else, and kills him by striking him on the head with a stake. When under sentence of death, he confesses that it was he who had murdered the farmer's wife.

The assistant justices of the court of appeal who had passed judgement upon the farmer had six months of their salaries docked.

III.iii.5.4 Erik Grubbe

Erik Grubbe, a wealthy gentleman, takes a fancy to the daughter of Martin Lælius, the bishop of Aarhus. He makes advances, gets the brush-off, and she marries a clergyman. Grubbe arranges for the husband to be accused of having fornicated with both the mother and the daughter.

Skonning, the sexton in Aarhus, is bribed by him into circularizing a satirical work *The Cat's Court-case against the Dogs*, and into maintaining that he had watched the alleged fornication from the lucarne windows in the church tower.

This so disturbed the bishop that he died; his wife's reputation is ruined.

Grubbe never prospered thereafter, and died thoroughly despised.

One daughter was divorced from her husband Gyldenløve and then from Palle Dyre, on both occasions for adultery. Obliged to go begging and harp-playing.

A second daughter was beheaded for having poisoned someone in Copenhagen.

A third daughter died in miserable circumstances.

Skonning fell to his death from the very lucarne window about which he had told a lie when he said he had watched the fornication from it.

Pontoppidan, *Annals*.

III.iii.5.5 Dahlberg

Lieutenant Dahlberg of Dalecarlia, a rough and powerful chap living at Hedemora, while travelling in the province, comes across two soldiers playing skittles. They pay no attention to him and simply get on with the game. This apparent lack of respect so incenses him that he swears at them. Not having intended any disrespect, they ask him if he wants to join in. After exchanging words the lieutenant completely loses his temper with one of them, beats him up so violently that he can do little more than crawl home, where he dies three days later.

The matter is investigated, and the lieutenant just about gets off.

Some years later, during the winter, the lieutenant visits his neighbours, and since there is no room for his horse in the stable, it is given its hay in a barn. He is on the point of leaving, his host and hostess are standing at the door, he is settling himself into his sleigh, when the horse bolts. The vehicle crashes into a loaded sledge and overturns, dashing his head so violently against a stone that he dies instantly.

III.iii.5.6 King Christian II

King of Denmark, Norway, Sweden, brother-in-law to the emperor. Could not be freed from the retribution of God, but had to sit out his time in prison after he had been deposed.

His bitterness was such that after he had had Sture's bones dug from the grave, he gnawed them in order to demonstrate how furious he was with the dead.

III.iii.6.1 FRIENDSHIP

So long as you are successful, there are friends enough; but they all disappear once your luck changes.

Now when that my barrel was running with cheer,
The men and the women came visiting here;
But now that my barrel is empty and dry,
The men and the women they all pass me by.

The rich give treats, attract friends; the guests drink, wipe their mouths. Then they leave and forget about it. They all disappear once your luck changes. Kierman lost all his friends once fortune's wheel had turned.

Ecclesiasticus 6: 8: For some man is a friend for his own occasion, and will not abide the day of thy trouble.
Ecclesiasticus 37: 4: There is a companion, which rejoiceth in the prosperity of a friend, but in the time of trouble will be against him.
All are hostile to the unfortunate, as dogs are to the one bitten.

III.iii.6.2

There is the tale of the man who saves a thief from the gallows. This man is then captured by enemies who are about to hang him but cannot find a rope; the thief comes along and supplies one.

Murdered by means of his own weapon.

III.iii.6.3

Sedlin

Mr. Sedlin, curate in Danmark, marries in Gävle. His mother-in-law had had no children of her own, had adopted a poor child and brought her up.

The old woman provides the curate with the girl, and with a house she intends to share with them for the rest of her days.

The adopted daughter gets on well with the foster-mother in the curate's home; she gives birth to a child and dies.

The curate then marries his child's governess, who is related to his former wife and whom he had made pregnant; they have several children.

The new wife cannot stand the old woman; she is shown the door and only allowed to take with her the things she had brought; she returns to Gävle, where she had sold her little farmstead.

From then on things go wrong.

The young wife gives birth to two children, both of whom die. During her third pregnancy she gives birth in the seventh month and dies soon afterwards.

Curate Sedlin is the friend of the bailiff in charge of the croft at Fårbro. Since he wants to get his servingman installed on this smallholding, he persuades the bailiff to give notice to the crofter, who has tended the place well for a number of years.

A few days after this the curate falls violently ill, and eventually has a stroke which paralyzes him down one side.

Ecclesiasticus 33: 12, 13: Some of them hath he blessed and exalted, and some of them hath he sanctified, and set near himself: but some of them hath he cursed and brought low, and turned out of their places.

As the clay is in the potter's hand, to fashion it at his pleasure: so man is in the hand of him that made him, to render to them as liketh him best.

III.iii.6.4 Uggla

Böstling, citizen of Stockholm, acquires a very pretty wife.

Uggla, assistant justice of the court of appeal, gets involved with Böstling's wife; in other respects he is a fine person, thoroughly pleasant.

Uggla marries and has a backward daughter; his wife dies and he marries the widow of Herkepaeus, renowned for her good manners and civility.

The revenue court writes to the lord lieutenant at Vasa, ordering him to arrest the receiver of the revenues there for the embezzlement of 70,000 dollars.

Uggla, assessor at the revenue court, delays the letter by one post day and lets the receiver of the revenues know what he is in for, so that he has time to make himself scarce.

This bit of juggling comes to light, and Uggla loses his job.

III.iii.6.5 A Swedish captain

Certain Russian soldiers are lying asleep in a barn; a Swedish captain surrounds it and has it set on fire. A Russian captain rushes out, throws himself on the ground in front of cavalry officer Johan Gyllenborg's horse, and begs for any other death than being burnt alive. Gyllenborg arranges for him to be spared.

Gyllenborg falls ill during the engagement on the Dnieper; is taken prisoner while lying incapacitated outside his tent. A lieutenant-colonel asks if there is a Johan Gyllenborg among the prisoners; he takes him home, puts him up, looks after him and gives him a considerable sum of money – for having saved him from being burnt to death.

III.iii.7.1 DILIGENCE

Ecclesiastes 9: 11: I saw that favour is not to men of skill; but time and chance happeneth to them all.

Psalms 127: 2; *Ecclesiastes* 3: 9: It is vain for you to rise up early etc.

1 Corinthians 3: 7: So then neither is he that planteth anything, neither he that watereth.

Matthew 20: 9: And when they came that were hired about the eleventh hour, they received every man a penny.

Fulfil your office as well as you can, remain on good terms with him who is set above you, and let the world go its own way.

Matthew 5: 5: The meek shall inherit the earth.

Many of my colleagues skipped classes or arrived when half the time had gone, like Frondin they were often on a double salary and made a regular habit of socializing.

I never relaxed, day or night, never took a break from reading, writing, examining.

Have I done any better? I have made a name, but so what? Others have busied themselves blotting it out. I look on while what I have accomplished is appropriated by plagiarists.

Titles and hot-air, ennobled, knighted, court-physician, but so what?

III.iii.7.2 Maja Hierpe

Maja Hierpe – the best of girls, the most virtuous, good-natured, good-looking, obliging.

In 1773 her mother and siblings died of petechial fever, which is highly contagious. She was in my service, and asked for permission to go home and attend her mother's funeral. Everyone, the whole household, pleaded with her with tears in their eyes not to run the risk of catching the fatal disease.

Heaven and earth could not have held her back; she said that not to go would be to do what was not right. She went and came back infected; in order that the household should not be infected, we moved to the country; she was sent to the hospital on the understanding that if she died she would not be anatomized.

Professor Sidrén, the head of the hospital, gave me a solemn promise to this effect. She died there and Sidrén anatomized her head. He had had many petechial fever patients at the hospital without ever having been infected, but in this case he caught the disease and became dangerously ill.

III.iii.7.3 Grizell

Grizell, crown district commissioner in Uppland, was formerly a tax-collector. He had done his job very thoroughly indeed, confiscating large numbers of stills when brandy was illegal and seeing to it that those who sold the liquor were fined. The story went that even when a woman was lying in childbed, he was there with his rapier poking about in the straw mattress looking for brandy. He raised many a sigh and made no end of people miserable.

In 1770 he was reported by his successor district tax-collector Blomberg, together with inland revenue clerk Hamrén, for not having sent in the tax-accounts, duly bound and sealed, on the day appointed. There was a financial penalty for every day overdue, and Grizell was ordered to pay six thousand seven hundred copper dollars. The matter was left to the informant's discretion, and Grizell therefore settled with Blomberg for three thousand five hundred dollars.

He had banked thirty thousand with Julinschöld, who went bankrupt; he was paid out half his deposit and lost the rest.

III.iii.8.1

Law

1. Be persuaded by nature and experience of a God who made, maintains and governs all; who sees, hears, knows all; in whose sight you have your being.

2. Never take God to witness in an unjust cause. Commit no perjury.

3. Look to God's purpose in the creation. Have faith that God guides and preserves you day by day, that all evil and all good are from His hand.

4. Be not ungrateful, that you may have long life.

5. Take care not to kill. The sin is not absolved unless the offence is made good. And who can make amends for death?

6. Put no woman to shame and steal no man's heart.

7. Never take to yourself any ill-gotten money.

8. Be honest, a man of old-fashioned virtue and trust, for you will then be loved by all.

9. Scheme not to overthrow others, that you fall not into your own trap.

10. Enter into no intrigues.

III.iii.8.2.

The more ardent the moralist, the more gall there will be;
the more stupid the priest, the more heretics he finds;
the blunter the razor, the more it tugs.

III.iv DISTRIBUTING WEALTH

III.iv.1.1

Fortune stands accused by mankind before Jupiter. Jupiter becomes angry and calls upon Fortune to give an account of herself. She says that Jupiter for the most part has created good-for-nothings, and that she will now demonstrate this. In the forum she lays out a beautiful cloth, onto which she throws gold, silver, pearls, gems, crowns, tiaras, arms. All rush forward with a mad impetuosity. One grabs a crown and puts it on his head; others trample him down and snatch it away. Another seizes a tiara, others wrench it from him, tearing it to pieces. Someone goes for gold and is forced by others to relinquish it. Trampling on one another, they all pile into a heap.

A timorous boy creeps stealthily up on one side, picks up some of the gold, then some of the silver, then something else, until he has as much as he can safely conceal, and make off without anything being suspected.

Then Fortune speaks: "This boy is my darling; decide if I did well in permitting what he did; it is he who is unassuming, the others who are evil".

"You did indeed", said Jupiter, "Good-for-nothings, get you gone".

III.iv.1.2

Plutus, accused before Jupiter of having made a bad job of distributing wealth and the gifts of God, excused himself by saying that he had become blind with age, unable to see who should have benefited, and that he had therefore distributed the gifts at random.

III.iv.1.3

The coronations of kings bring out the incurable audacity of the population. Wine flows, all mill around, the pushers get nothing. An ox is slaughtered: what is the outcome of a lawless free-for-all? Money is scattered: those who rush get very little.

III.iv.2.1

POVERTY

The poor farmer toils the whole year round and has scarcely enough straw to lie on, a minimal return from his labour. This is not working for oneself; seed has to be sold at half price in the autumn and bought back at a disadvantage in the spring. His food-basket is hung up high.

The nobleman extracts the last farthing. The farmer's children have to starve, his wife has to work without meals at the manorhouse. He drives his wagon to Stockholm; his livestock, his only support, is skin and bone. Why has not God made the lord slave and him the lord? With his wife and his children he is finally driven out of house and home.

Is it any wonder that God should finally destroy the lord's seed and reduce them to poverty?

Remember the poor slave who labours for you while you sleep. He ploughs the field, you reap. You say it is your estate, that you can do with it what you will. I say it is not, that it is on loan to you from God.

See to it that you do no great injury to those who are both powerful and miserable; though plundered they still have their arms. They cry unto God.

III.iv.2.2

The voice of Abel's blood and the voice of oppressed Sodom have cried unto heaven, the wages of the labourer have been withheld.

III.iv.2.3

Sohlberg

Sohlberg, inspector of the pumps at Falun, was wealthy. He had made his money by playing the skinflint with the poor miners.

There are five sons, quick-witted but with no sense of money, all of them hard-up and in the red.

There is no joy for the third generation from ill-gotten gains.

III.iv.2.4

Funck

Lord lieutenant Funck of Uppsala sells the local constabulary to Sylvander of Danmark for three thousand dollars. Sylvander dies a few years later, leaving four young children. Although a settlement could have been made for the widow, Funck sells the office for six thousand dollars. Despite her being in debt, he refuses her the prescribed period for moving.

Funck dies suddenly three months later. His wife is allowed to draw his salary for three months, but is refused the prescribed period for moving.

What you would not another should do to you, do not to him.

The house in which Funck died was auctioned to pay his debts.

III.iv.2.5

Do not regard the simple life as a misfortune. To be poor and healthy is better than being a privy councillor. The poor have more to laugh and be happy about.

III.iv.3.1

RICHES

Ecclesiastes 2: 22, 23: Man, for all his labour, hath grief and sorrow and taketh not rest in the night.

Ecclesiastes 5: 12: The abundance of the rich will not suffer him to sleep. He is plagued by having acquired what he wants.
Ecclesiastes 6: 5, 6: He hath no pleasure of the sun nor any rest here or there. Yea, though he live an hundred years, yet hath he seen no good.
Ecclesiastes 5: 13: Riches kept for the owners thereof to their hurt.
Ecclesiastes 5: 10: He that loveth silver shall not be satisfied with silver; nor he that loveth abundance with increase.

Ecclesiastes 6: 2: A man to whom God hath given riches, hath not received power to eat thereof, but a stranger eateth it.
Ecclesiastes 2: 21: To a man that hath not laboured therein shall he leave it for his portion.
Ecclesiastes 2: 18, 19: Yea, I hated all my labour, because I should leave it unto the man that shall be after me. And who knoweth whether he shall be a wise man or a fool?

Ecclesiastes 5: 14–16: But those riches perish by evil travail: and he begetteth a son, and there is nothing in his hand. Naked shall he return to go as he came, and shall take nothing of his labour. What profit hath he that hath laboured for the wind?
Ecclesiastes 4: 8: There is one that hath neither child nor brother: yet is there no end of all his labour; neither is his eye satisfied with riches. For whom doth he labour, and bereave his soul of good?
Ecclesiasticus 14: 4–6: He that gathereth by defrauding his own soul gathereth for others, that shall spend his goods riotously. He that is evil to himself, to whom will he be good? He shall take no pleasure in his goods. There is none worse than he that envieth himself; and this is a recompence of his wickedness.
Ecclesiasticus 14: 9: A covetous man's eye is not satisfied with his portion; and the iniquity of the wicked drieth up his soul.

All we have is lent to us by God. We have brought nothing with us and we take nothing away.

When God takes things away from us through His executor fate, we bewail the loss of what we possessed. Since it was lent, however, it was never ours.

God does not hold money in estimation. Money is of the earth, and is given to us to amuse ourselves with. Since it enables us to acquire comforts, we regard it as the means to pleasure and happiness. But it can never give rise to the happiness of the penniless farmer having a good laugh.

The cobbler always used to sing in the mornings, but since he has made his money he has been silent.

III.iv.3.2

Anders Jöranson

In 1732 Anders Jöranson was the richest mining magnate in Falun, certainly worth 200,000 dollars in gold. He had begun as a poor pitboy and worked his way up; he sucked the marrow from the bones of the poor mineworkers, lived extravagantly.

After his death his wife lived modestly, scarcely enough for a decent burial.

His son-in-law has to flee to Norway on account of his debts. The rest of his children are in extremely straitened circumstances, not a penny to spare.

III.iv.3.3

Håkansson

Olof Håkansson, a farmer from Blekinge, was speaker of his house in every parliament but one between 1739 and 1770. He took bribes and acquired an unbelievable amount of money, many hundreds of thousands of dollars in gold. Became the richest farmer in Sweden; died during the 1769 parliament.

Six months after his death, in February 1770, his younger son and then his daughter died as a result of a fever epidemic. His son, living in Stockholm, does not seem to be very robust.

Why hanker after and strive for money? Who will get it?

III.iv.3.4

Broman

Broman, marshal of the court under king Frederick: for a number of years after 1746 he received all the royal ducats for the year on the Hessian budget, 80,000 of them at sixteen dollars a piece; they were worth between eighteen and twenty-four dollars.

He reduced or abolished all royal gifts and pensions.

He was constantly acquiring and purchasing various estates, which he invariably sold at a hundred per cent profit.

He drew a huge salary, received substantial gifts from the king.

Since he was the favourite who had to be bought over and persuaded before any position or title was granted, he acquired immense sums of money.

Queen Louise Ulrica used to say that she had never seen such a fortunate fellow, since all he had to do was sit in his chair and think up what he wanted.

No Swede has ever owned so much money, had such wealth at his disposal.

It was some years after he died that his heirs finally settled with the creditors; all lost on him and not a penny went to his heirs. There was just about enough to pay for his funeral and his wife passed away in misery.

The immediate heir has no joy from ill-gotten gains.

III.iv.3.5 Charles XI

The reductions of Charles XI rendered many of the nobility homeless. Stockholm castle burst into flames when he died and it was with difficulty that his corpse was saved.

III.iv.3.6 Meldercreutz

The father, doctor Molin, provost at Uppsala, a great schemer and money-grubber, set on scraping his shekels together. The son, a captain and a professor at Uppsala, comes into more property than anyone else in the city. He owns a works at Kalix, controls the whole of Luleå and Lappmark. It goes against the grain to persevere with the works, but he carries on stubbornly, running himself into debt when he could have lived well on very little. Ill-gotten property brings no pleasure to the third generation.

Instead of scuttling he bales out. He fornicates with the young. His wife, a good-natured woman, finds excuses for him, maintains that it is only tittle-tattle. He installs yet another beautiful maiden in his house. His wife falls ill, wastes away and dies when she becomes aware of her husband's affection for the girl. The day before she passes away, she takes her farewell of him. He replies: "Thank you my dear for having been such an honest wife. I hope you have never gone cold or hungry while you have been living with me".

The girl bruised the front of her leg and it became septic.

Meldercreutz had travelled more than anyone else in Europe, some one hundred and forty thousand miles by land. He controlled practically the whole of Luleå and Lappmark. He went hiking over Luleå fell for the pleasure of it, with Buffon's book under his arm. The most stubborn person imaginable, he persevered with his ironworks at Kalix until he was over his ears in debt.

Master Tidström said of him that he bales out everything he comes across. He likened him, on account of his debts, to a sailor on a sinking ship.

III.iv.3.7 Julinschöld

In 1740 archbishop Jöns Steuch of Uppsala is energetically involved in furthering his party interest.

In the first instance he is opposed by professor Wallrave, who then joins him. Wallrave was a rake and in debt; he wanted university treasurer Rommel to advance him money. Since Rommel did not think this advisable, Wallrave attacks his management of affairs in the consistory.

Julinschöld, nephew to the archbishop and one of Rommel's clerks, is appointed to stand in for Rommel while he is suspended from office, and is eventually allocated half his salary. Rommel is ousted from his job at the university and dies destitute.

No one has any ground for thinking that Rommel had ever embezzled so much as a farthing. His book-keeping was slapdash, however, and no time was lost in relieving him of his ledgers and papers, so that he should have no means of defending himself.

Julinschöld turns out to be a splendid treasurer; he builds up the whole academy and does no end to enhance its reputation. He inherits a works from his father, enjoys a professor's salary together with every seventh sack from the mill, and through his family and supporters rules supreme in the consistory. His wealth in works and property eventually outstrips that of everyone else. No one was financially sounder, all drew interest by depositing their money with him.

Malice would have it that he shared Mrs. Rosenstein with her husband, since he was unmarried and used to eat there.

All entrusted their money to him with bonus, his notes passing throughout the realm in preference to those of the bank. He controlled everything, then all at once he crashed.

At the beginning of 1768 he went bankrupt to the tune of 1,150,000 dollars in gold. Everything he had was put under seal and taken from him, just as in the case of Rommel. During these twenty-eight years God had not fogotten what had happened; the treasurership passed to professor Berch, then living in Rommel's old house.

Tax-collector Grizell, subsequently crown district commissioner, was helped by Julinschöld, first to his job as a clerk, then in the well-remunerated case against Westrin, then to the position of sheriff, then to the 30,000 dollars he came into possession of. He did this mainly by putting in a good word for him at the archbishop's place. Grizell deposited the 30,000 dollars with Jülinschöld; when his affairs began to look a bit shaky, however, he demanded back the mortgage and then prosecuted him for it. It was this that precipitated the bankruptcy.

Julinschöld had got rid of Rommel and backed Grizell, whom he used in order to visit God's retribution upon Westrin. He then helps Grizell to acquire the 30,000 dollars which enable Grizell to avenge Rommel.

On February 14th 1768, shortly after he had petitioned to be declared bankrupt and everything had been put under seal, Julinschöld died of exhaustion and grief, just as he was getting over a severe bout of fever. There was such a general hostility to him, that he was in any case in danger of his life.

Job 16: 12: I was at ease, but he hath broken me asunder: he hath also taken me by my neck, and shaken me to pieces, and set me up for his mark.

When everything had been taken from Rommel and he was freezing during the winter, he begged a few loads of wood from the consistory. Julinschöld, after his bankruptcy, was obliged to beg wood from the consistory through Georgii.

III.iv.3.8 Hultman

De Geer, lord of Lövstad, is wealthy and unmarried; he always has a considerable amount of money in his pocket, likes to have it to hand should he need it.

He takes into his service a young barber-surgeon by the name of Hultman. It so happens that he suffers a sudden stroke and dies on the spot, the only person in the room with him being Hultman. No money is found in his pocket.

Hultman bribes his way into becoming barber-surgeon under Rudbeck and Roberg at the academy in Uppsala. Once the De Geer business has been forgotten he purchases property, – a house, land, etc.

Titular professor Ziervogel marries Hultman's daughter. Hultman is delighted with the match and willing to risk anything for his son-in-law.

Hultman spends 100,000 copper dollars on the construction of a paper-mill, which he hands over to his son-in-law; he enters into security for him to the tune of a further one hundred thousand dollars.

Ziervogel dies in 1765, leaving a debt of 260,000 dollars. This comes as a complete surprise to Hultman, who is completely ruined by the loss of the mill and having to pay out the 100,000 dollar security.

Wealth gained by evil means never passes to the third generation.
There is no joy for the third generation from ill-gotten gains.

III.iv.3.9 A Danish sheriff

A Danish mounted sheriff buys two horses from a poor farmer without settling the account. It is rumoured that the sheriff has been caught embezzling and is about to be suspended. The farmer goes at night and recovers his horses. The sheriff is let off, has the horses looked for while the farmer is away, finds them at the farmer's place. The farmer is summoned, and then hanged on account of his own horses.

A few years later the sheriff buys two horses from a stranger, a horse-rustler who was selling at reasonable prices. There is a search, and the horses are found in the sheriff's possession. The sheriff is summoned, and since he cannot name the vendor, condemned for theft and hanged on the same gallows as the farmer.

III.iv.3.10 Norrelius

Norrelius, librarian at Uppsala, an ill-natured parsimonious person, scarcely capable of getting himself to eat. Miserably married to Benzelstierna, whom he subsequently divorced.

He was already sixty when he married the penniless but good-natured daughter of baron Friesendorff. He treats her as if she were under a ban, is never anything but morose, runs the household himself, refusing to entrust her with anything. After a confinement no one is allowed to visit her and she dies from the sorrow of it.

When he is on his deathbed, he boards out his only daughter with Rosén, where she eventually dies of the small-pox.

The father-in-law old baron Friesendorff, whom Norrelius could not bear the sight of, is living in a croft without a penny to his name. His granddaughter's unexpected death as the result of the small-pox inoculation means that he inherits some twenty thousand dollars. He is delighted at the windfall, since he had believed himself to be forsaken by God, and here was something to gladden his old age. He buys himself a farm, makes a mess of it in less than a year, and ends up as poor as he was in the first place.

It was thus that Friesendorff acquired from Norrelius all that his son-in-law possessed when he used to begrudge his father-in-law a glass of water.

Plutus, who inadvertently distributes money to those who are unworthy of it, was represented by the ancients as being blind.

III.iv.3.11 Brauner

Secretary Brauner, who is quick-witted and educated, sets his sights on making a coup by marriage.

He finally picks out a widow with capital assets worth 70,000 dollars. He enters into a pre-marriage settlement with her – if she dies first he gets everything, her own child nothing.

After a few years she begins to regret this, breaks into his office while he is away and burns the document.

Hence all the tears – quarrelling and no making it up, scrapping and so on.

The final outcome is a complete separation, whereby he keeps the one daughter they have had together.

The wife eventually dies.

The following year the secretary has to pay an urgent visit to Uppsala. He had invested 30,000 dollars at interest with Julinschöld, and 10,000 with chancery officer Ihre.

Herkepaeus, a corporal in the life guards, takes his daughter.

He loses all his money when Julinschöld goes bankrupt, and legal wrangling absorbs the Ihre investment.

III.iv.3.12 Gyllenkrok

Crown forester Gyllenkrok marries the daughter of the wealthy marshal of the court Sven Cederström. He wants an estate from his father-in-law and resents it when it is not given.

He treats his wife badly during her first confinement but keeps the son. During her second confinement he packs her off to a farm, and although there was no question of her ever having been unfaithful, refuses to allow her back to home and husband.

He lives on his estate at Frötuna. His wife is not even allowed to visit her son while he is studying at Uppsala.

Gyllenkrok rebuilds Frötuna in stone, including the cottages; does indeed manage to run his finances without the help of his wife.

He has a stroke and dies soon afterwards.

The son inherits everything, and though not required to do so by the will, shares the estate equably with his mother. Before very long, however, he dies of the small-pox.

The widow re-marries with the impecunious gentleman of the bedchamber Gregory von Kothen: he begins to give himself airs, lends money without security, squanders the lot. In 1773 the bank auctions off Frötuna lock, stock and barrel.

What was the outcome of all Jan Gyllenkrok's trouble, all his building and economizing? He abandoned his wife and fornicated with his housekeeper.

III.iv.3.13 Nordenflycht

Nordenflycht, a nobleman, young, penniless, quick-witted, speculates on making a coup.

The wealthy old Lohe dies, leaving behind some sons and a daughter, all of whom are stingy and self-centred.

Nordenflycht approaches Miss Lohe during evensong, and after introducing himself informs her that he has a secret to communicate. He tells her that her brothers are plotting against her life in order to acquire her property. She believes him and breaks into tears. Nordenflycht advises her to get married in secret, says it will be all up with her if she delays at all, and offers himself as a prospective husband. She accepts, and after secret banns they are married.

Mrs. Nordenflycht becomes sickly, has no children. Nordenflycht advises her to put on an artificial stomach. He gets a guardsman's wife to agree to be wet-nurse, takes her child in order to look after it for her, and arranges for her to be sent for as soon as his wife has given birth. The guardsman's wife is relieved of her child and has no idea of its whereabouts. The child is announced as being Nordenflycht's.

The only unmarried Lohe lives in the Nordenflycht house and is frequently entertained by them. Whenever he leaves their dining-room he steals a few pieces of wood, until all his rooms are decorated with what he has acquired. Nordenflycht pretends not to notice.

Although Nordenflycht supplies Lohe with free food, he asks him to sign a chit for each meal and to pay three halfpence.

Once Nordenflycht has got things settled in this way, he gets Lohe to travel with him, a free trip, a change of air, will do him good. They travel to Gothenburg, Copenhagen, Germany, and Lohe passes away: everyone is of the opinion that Nordenflycht brought this about.

Nordenflycht was ambushed and murdered by certain Jews he had diddled.

The other Lohe dies without heirs. The guardsman's son is regarded as Nordenflycht's and comes into the whole inheritance, but he is dissolute and squanders it.

When this Lohe was ill, the doctor made out a long prescription for him, half of which Lohe snipped off. When it comes to the apothecary he is therefore unable to make it up without first consulting the doctor concerning the formula. When the doctor asks Lohe why he has done this, he says he thought the whole of it might have been too expensive for him.

Lohe's only son dies at Uppsala and remains unburied for a whole year, the father being unable to bring himself to pay for the funeral.

Died as a student at Uppsala in 1759.

III.iv.3.14 Råfelt

Lord lieutenant Råfelt, when a young man, courted Miss Stjerncrona, who was extremely wealthy and lived with her guardian, assistant governor Galle.

This Råfelt had inherited a very tidy estate from his parents, but it was not quite so tidy as he would have liked it to be.

He had a brother, and since this brother had a propensity for brandy, Råfelt made certain that there was always plenty of this liquor available, the result being that the brother eventually became a confirmed drunkard. In order to make life as pleasant as possible for his brother, he arranges for him to live on a suburban estate to the north of Stockholm, replete with a plentiful supply of his favourite drink. He also arranges for him to be constantly attended by a couple of whores. By these means he so enervates him that within two years the brother is dead, leaving Råfelt in possession of the inheritance.

Råfelt then makes up to Galle and showers expensive presents upon Miss Stjerncrona, the result being that within a year or two he has disposed of all he had inherited from his brother. What is more, he fails to win the young lady. He is very put out by this, although he does eventually reconcile himself to remaining lord lieutenant. He takes to drink, however, and acquires a bluish-reddish nose.

III.iv.3.15 Cederhielm

Lord lieutenant Germund Cederhielm passes on his position to his son. Since the son is unable to pay for the staff and the maintenance, he gives up the post and takes employment elsewhere. The renovation of the place absorbs all Germund's savings. The old man sighs at the way in which his son has behaved.

The son becomes president of the Göta court of appeal. He has two bright sons, whom he sends to Paris. They run up incredible debts. When they are about to be arrested for them, the one takes a pistol and shoots himself through the head. The other is imprisoned in Châtelet, where he remains for the rest of his life, the father being unable to secure his release.

III.iv.3.16 Wallerius

Nils Wallerius, doctor of divine theology, professor at the university of Uppsala.

His second wife went out of her way to be unpleasant to her stepchildren, the servants, her husband and their child.

He eventually dies; she manipulates out of the rest of them as much of the inheritance as she can.

Shortly afterwards, early in 1766, the farm from which she was to have derived her share went up in flames.

She applies for a double year of grace, stipulating that no advantage should accrue to the stepchildren; this is granted.

In September 1766 the barn in the country in which she has just deposited the tithes due to her, catches fire. She has to relinquish the greater part of the contents to the flames. The farmer's barn next to it remains standing.

In the December of the same year the vicarage in the country to which she had moved burns down. She just about escapes, with her son wrapped up in linen; loses everything. The farmer's house next to hers remains standing.

Who can escape the retribution of the Lord? Be not cruel and hard. There is no pardon for those who show no mercy.

After the last fire, the university grants her 3,000 dollars; since there are also other contributions, she receives 18,000 dollars, which is more than she had in the first place.

Altogether she scrapes up some 21,000 dollars, which she deposits with the wealthy Julinschöld at a certain rate of interest; he goes bankrupt at the beginning of 1768.

Psalms 127: 1: Except the Lord build the house, they labour in vain that build it.

III.iv.3.17 Stade

A certain public prosecutor in Stade, on account of his wealth and the severity of the legal proceedings by which he had acquired it, had made himself extremely unpopular. He had one child, a daughter.

A young surgeon settles in Stade. The girl takes to him and makes it known that they wish to get married. When she asks her father's permission, however, it is refused on account of the young man's impecunity.

She lets her maidservant know how distressed she is, says she wishes her father were dead.

The maidservant says she will see what she can do to help.

The daughter's handwriting is exactly the same as her father's. The maidservant suggests that she should write out an order in her father's name and hand for rat-poison, arsenic, and send an unknown maidservant in to Stade to fetch it. Familiar as they are with the public prosecutor's hand, they supply the girl with the poison.

It was a day on which her father was away and on which her mother had been invited out in Stade. The daughter stayed at home, ostensibly in order to be there when her father returned, actually in order to be there when the maidservant came back with the arsenic.

The daughter prepares pease-porridge, a dish her father was fond of, and laces it with the poison. Her father returns and asks why she has not gone out with her mother. She says she has been waiting for him, and asks him if he feels like something to eat, since she has been preparing some pease-porridge. He makes a good meal of it, feels queasy in the stomach, has to be off again and vomits up most of what he has eaten. When he returns home he takes tea, vomits again, and feels well enough. He gets peckish during the evening, asks for more of the porridge, begins to writhe in agony. The mother is sent for; he dies.

The daughter persuades her mother to allow her to marry the surgeon. A fine wedding is arranged, all the best people invited. During the festivity the rumour goes round that the bride has poisoned her father. No one can say who has started it. All who are able to, slip away from the wedding.

Scarcely a year later the barber-surgeon also hears the rumour, and finds it intolerable that this should be reported of his wife. He requests a judicial enquiry, so that the person who has spread the rumour may be punished and the good name of the family vindicated.

The corpse is exhumed, the stomach found to be eaten away by poison. The apothecary supplies the evidence of the note ordering the arsenic. The maidservant who made the purchase is identified. The daughter is arrested and grimly executed together with her accomplice.

The father had extracted much from the poor in order that his daughter might inherit. Although it was the daughter's lot to avenge this, she had to bear the punishment for the misery of the poor.

III.iv.3.18

Blackwell

Jonas Alströmer applies to England for a good economist. Blackwell, a medical doctor, is sent over, a self-assertive ignoramus, an atheist. Alströmer receives him as his son.

Blackwell makes a routine of sending Alströmer's letters over to England together with his own.

One day Alströmer opens a letter and finds a viper in his bosom. Blackwell is quite evidently of the opinion that if Alströmer and Tessin were removed the whole manufacturing capacity of Sweden might be stifled, that if the privy council were done away with the king would probably acquiesce in the appointment of an English prince as his successor. Alströmer is horrified and shows the letter to Tessin. Tessin passes it on to the privy council, which requires that Blackwell should be executed.

Blackwell is now visited by someone who says that he has come directly from the English prime minister, who orders him to approach king Frederick and solicit his support for the scheme by offering a large sum of money. If things work out as planned he will become a person of great importance.

The king is warned that Blackwell is about to make a rash move.

Blackwell has an audience with the king in order to communicate letters that have arrived from England; he promises him sovereignty. The king passes him on to the marshal of the royal household. He was arrested by Löwenhielm, tried, and beheaded in Stockholm on July 29th 1747.

The courier who delivered the letter to Blackwell was never traced; the English categorically deny ever having sent one.

When Tessin's house was being renovated a corpse was discovered in a wall. Could this have been the courier? It is certainly the case that from that day on neither lady Tessin nor his excellency ever held open house. I heard this from Mr. Söderberg, who had it from Rudbeck. To me, however, it seems unlikely that the pious Tessin would have committed such an impious crime, despite such things' being routine among the rulers of the country, and Blackwell's certainly having deserved his lot.

Blackwell lodged with a broker while he was in Stockholm. Although he had a wife in England, a botanist, he was inordinately intimate with this broker's spouse, travelling about the country with her and introducing her as his cousin. She was with him for months at a time on the crown estate put at his disposal.

One evening the broker fell ill of a colic. Medical doctor Blackwell administered something supposed to cure it. The next morning the broker was lying there dead. Blackwell was generally considered to have done away with him. The widow went into mourning for a decent period but clearly had high hopes of Blackwell.

Drake, president of the ministry of trade and industry, very keen on fostering manufacturing, fell ill and died while being treated by Blackwell. Everyone said that Blackwell had taken his life, and many were of the opinion that he had done so under orders from England.

III.iv.3.19

Only the good fortune which comes slowly lasts with us to the end.
Few of us enjoy great blessings which are also long-lasting.

IV. MAN AND GOD

IV.i THE NATIONS

IV.i.1.1

Job 12: 23: He enlargeth the nations, and straiteneth them again.

IV.i.1.2

It is to this that discord has brought the unfortunate citizenry.

A raging populace cannot be calmed by reason, nor can it be influenced by pleading.

Prayer has no effect on the decrees of fate.

IV.i.1.3

A traveller

A traveller arrives in Paris and finds lodging in a basement, where many others are lodging.

One of the others had just collected some money which had been sent to him. A certain rogue, who had discovered this, went to the basement at ten o'clock in the evening and took the money after murdering the owner of it. As he made off he ripped up with his bloody fingers the paper in which it had been wrapped, and threw it down on the stairs.

The new arrival had spent the evening visiting friends. When he arrives back in the basement at eleven o'clock, he sees the paper lying on the stairs and puts it in his pocket, thinking that it may come in useful.

The killing is discovered the next morning; a search takes place, the new arrival included; the blood-marked paper in which the money had been wrapped and which had been picked up from the stairs is found on him. The newcomer is arrested; since he cannot be brought to confess the crime, however, he is condemned to the galleys, shackled for eight years.

The killer is eventually arrested for another murder, and while on the block confesses to having committed the former crime.

The condemned man is released from the galleys at Marseilles amid the jubilation of the whole population.

As recounted by president Carleson.

IV.i.1.4

Stobée

Colonel Stobée, the later lord lieutenant, had a wife, whom he took with him when he went campaigning with Charles XII. He was captured at Poltava.

Kock von Gyllenstein was his closest friend. Stobée persuades him to get away with his wife and tells him how to do so, promising to follow on as soon as he can. He has Gyllenstein's word of honour that he will take no advantage of his charge.

On reaching the border Gyllenstein stabs to death the last of the guides.

Instead of leaving Stobée's wife in Stockholm as has been agreed, he takes her with him to West Gothland as his mistress.

Stobée arrives home, and searches in vain in Stockholm for his wife. When he hears what has become of her, he travels down to the king in Lund and writes to her. Gyllenstein follows her down to Lund, where she sickens and dies.

Stobée encounters Gyllenstein on the street in Lund. He thanks him for his word of honour, draws his sword and challenges him to a duel. When Gyllenstein declines, Stobée breaks the scabbard over his head. Gyllenstein has to be cashiered, since the other officers are no longer prepared to serve under him.

The king dies. Stobée is a staunch supporter of the Holstein party, opposed to king Frederick. Gyllenstein approaches the king and proposes a monetary reward for dealing with the Holsteiners.

One evening in the Great Cellar in Stockholm, captain Gyllenroth enters dressed to look like Stobée, and starts saying frightful things about the king. The landlord tells him that talk of this kind is not to be tolerated in public. He replies that the consequences are a matter of indifference to him. When asked his name, he says he is colonel Stobée. There are witnesses enough.

The next day an officer arrives with the guard and arrests Stobée, who has no idea what is happening. A commission is appointed, a solicitor engaged, witnesses summoned, all of them hostile to Stobée.

Stobée is summoned; as far as he is concerned all the world can bear witness. He cannot say where he was at the time since he was in a whorehouse. He cannot help laughing when all the witnesses called testify to his guilt. The judges inform him that only two of them regard him as guilty of lese-majesty, as deserving death.

The last two witnesses are a couple of young lads, who say much the same as the rest. Stobée asks them if they would recognize the colonel if they saw him, if he himself bears any resemblance to the accused. When they deny any likeness, the truth of the matter becomes obvious.

Gyllenroth was beheaded, two of the other conspirators whipped and branded.

Gyllenstein gets away to Hamburg in the first instance, where he wins the confidence of a Jew whom he subsequently throttles and robs. He then makes his way to Saxony where he courts a young gentlewoman and is rejected. After she has given him the brush-off he maintains that she is his daughter, breaks into her room at night and murders both her and her mother. He flees and is captured; terrified of dying, he is bound to a chair, broken on a wheel and beheaded.

It is generally thought to be the case that he had in fact fathered the young lady when he was in Saxony with the army, the mother having then been the wife of someone else.

IV.i.2.1 PARLIAMENT

In 1761 baron Cederhielm, councillor of the board of domains, by means of a memorandum, brings about Pechlin's being voted out of two parliaments.

In 1765 Estenberg, page to his majesty, initiates Cederhielm's being excluded from two parliaments.

In 1765, during the same parliament, Bergenstråhle, justice of the court of appeal, declares that Estenberg should be excluded, and he was in fact voted out in perpetuity.

In 1771 baron Essen and director Frietzcky, together with baron Grönhagen, declared that Bergenstråhle should be excluded, which he was, from two parliaments.

Daniel Tilas.

IV.i.2.2 Herkepaeus

Herkepaeus, mayor of Uppsala, had a foul mouth. When he returned from the 1741 parliament, Kyronius of all people accused him of having expressed himself in the council chamber in a manner detrimental to the estates of the realm.

Cederhielm, the privy councillor's son, is given the job of investigating the offence and setting it right.

Herkepaeus approaches Cederhielm, asks him to be considerate, not to forget the occasion in St. Petersburg when he dragged his father the privy councillor from the fire, saved him and his belongings from going up in flames.

Cederhielm answers him as follows: "You may well have been decent enough then, but you are now a scoundrel". He imposes a heavy penalty on him.

Ten years later Cederhielm saw his house and all the belongings he had inherited from his father, the goods the privy councillor had acquired in Russia, burnt to ashes.

IV.i.2.3 Skeckta

Captain Skeckta, during one of the parliaments of the 1740s, accused Hedman, a member of the Cap party, of offences against the state. Since Skeckta could not provide witnesses, he would have been in a difficult position had the judge not treated him leniently.

Hedman's case dragged on and left him impoverished.

During the parliament of 1758 a similar charge concerning remarks prejudicial to the interest of the state was brought against Skeckta, and he was imprisoned in Marstrand. Secret evidence against him was provided by his wife.

IV.i.2.4 Springer

Springer, a businessman in Stockholm.

Lagersparre, royal treasurer, wealthy, lives in style, travels in a coach and six. When in company he speaks disparagingly of privy councillor Lagerberg, who rides to the privy council on a white horse. Lagerberg gets to hear of this and loses no time in setting up a commission to enquire into Lagersparre's affairs. This was in 1732, Springer, then a young man, being a clerk in Lagersparre's office. Springer does what he can to help Lagerberg bring Lagersparre to his knees. Lagersparre might have got himself acquitted had he been given the time, but he was condemned to Marstrand.

Springer, who has become a well-to-do businessman and is a staunch member of the Cap party, falls out with Plomgren the mayor of Stockholm. The Hats were then in office, and a commission is set up to enquire into certain of Springer's remarks which were regarded as detrimental to the party. Lillienberg, president of the court, condemns Springer to Marstrand.

In 1747, therefore, Lagersparre is released and Springer is imprisoned in precisely that cell in Marstrand in which Lagersparre had sat out his time.

Springer eventually escapes to Russia; in 1766 he is granted the freedom to return but makes no use of it.

IV.i.2.5 Brahe

Brahe, the earl, studies at Uppsala. First acts dishonourably in respect of Mrs. Rosén. After he has married the daughter of Sack, councillor of chancery, she becomes consumptive. During her illness, Brahe is said to have fornicated with a whore in front of her very eyes. His second wife was the daughter of president Piper.

In high favour with the authorities and consequently hated by the Hats. Only a few votes short of being elected speaker of the house of lords.

Non-commissioned officers told to make contact with him. They tell him of the discontent in the guards, say that if anyone started a revolution they would be only too ready to join it. Brahe and all Caps are excluded from every deputation. A certain restlessness breaks out into mutiny. Brahe was beheaded for it, although many defended him. The truth of the matter is unknown to me, although I do know that he was said to have had trouble on the way at Whitsun. The mini-revolt did not take place until a month after that.

Puke, also beheaded for having taken part in the revolt, was reputed to have been Piper's natural son by another man's wife.

Captain Stålsvärd, engineer, was also executed for complicity. He was the most godless creature I have ever known, believed neither in God nor in religion. Never married, used the guardsmen's wives as laundresses and mistresses. When I told him this was sin, he laughed and said I was a pedant.

Marshal of the court Horn also came unstuck; a good pious chap, said to have practised sodomy.

Many were beheaded for the revolt, including:

marshal of the court Horn, reputedly a sodomist; unconcerned about his wife when she lay dying in childbirth, he was a few gunshots away in the Hat Inn at Drottningholm;

captain Stålsvärd, a veritable atheist; never married, said it was more economical to lie with a guardsman's wife and employ her as laundress; committed murder in Germany, and got away with it there;

Puke, natural son of president Piper by another man's wife; Piper therefore lost his son-in-law as well as his son.

They were executed on July 23rd 1756.

IV.i.2.6 Rogberg

Rogberg, provost in Småland, member of parliament, zealous as an assistant justice of the court of appeal during the investigations into the Brahe case; wants to get Brahe beheaded.

He returns from parliament and has abuse heaped upon him in the parish chambers by some hoodlum, who calls him a shark, a knave, a guttersnipe, etc.

This so upsets him that he becomes ill and never recovers his health; he goes to take the waters at Växjö, and dies there while in a coach.

IV.i.2.7 Renhorn

Renhorn, mayor of Arboga, an aggressive chap. In 1756 he is appointed by the estates to conduct the case against the conspirators, which he does with extreme rigour under the slogan of "life, honour, property".

The estates grant him 6,000 dollars for the trouble he has gone to.

Renhorn becomes sickly, and over the next eight years takes the waters at the Djurgård. On one such occasion he invites all the guests at the spa to a get-together, with the exception of a captain in the merchant navy by the name of Ahlström – known on account of his foul mouth as the Ottoman porte.

The next day Ahlström accuses Renhorn of having violated the bye-laws of the spa by excluding him. Recalling the way in which he had dealt with Brahe and the other conspirators, he pitches into him with a gusto under the slogan of "life, honour, property".

Renhorn is completely put out and leaves for home; a few days later he goes down with a fever and dies.

IV.i.2.8 Pechlin

Colonel Pechlin – an outstanding person, brisk, winning, forceful. He took the lead in the parliamentary debates and formed his own faction, with the result that whichever party he supported, Hats or Caps, tended to predominate.

During the 1756 parliament he advocated having the queen confined to Gripsholm, although the motion was defeated.

He took the lead in urging the execution of Brahe and Horn.

In 1772, when king Gustavus III made the wise move of changing the form of government during parliaments, Pechlin fled south to Småland. He was arrested and held in Gripsholm, the very place he had prescribed for the queen mother. Little by little, however, the charges against him were dropped.

IV.i.2.9 Gylling

In 1760 Gylling is appointed lecturer in Karlstad. During the following parliament the Vermlander Antonsson, a powerful member of the estates, requires that his brother should be appointed to this lectureship, that Gylling should therefore be removed from the post. Antonsson's brother is appointed, but a few days later he breaks his leg.

IV.i.2.10 Antonsson

Antonsson is a wealthy fellow from Vermland, a district judge and immensely influential in the parliaments. Everything he sets himself in for has to go through. The lord lieutenant and the bishop would never have dared to oppose him. He deals in property and farms, and if he sets his mind on a farm he will have to get it, or there will be legal proceedings.

Stockenström, surveyor of mines, uses bribery and does all he can to fulfil Antonsson's wishes; is eventually unable to carry out a complete absurdity, and is determined to demonstrate that Antonsson has treated him unjustly; Antonsson frightens the life out of him.

Finally, during the parliament of 1765, he sets his sights on becoming speaker of the burgesses and comes unstuck.

Everyone rushes on past him, each grabbing his own; he simply dwindles away, becoming nothing more than an ordinary chap, although he does eventually pick up a bit.

IV.i.2.11 General Rudbeck

In 1740, when he was a poor bachelor, he applied for a job at Uppsala. In 1743, during the war against Russia, he distinguishes himself in the galleys and is made a lieutenant. He marries into money at Jönköping. A large, powerful, square-built chap, well-informed in matters of politics and finance, and a strong supporter of the Caps.

In 1762 he is recommended for the privy council together with Duwall the son-in-law of Kierman, speaker of the burgesses. In order to further Duwall's interest, Kierman gets Rudbeck's recommendation quashed. Although Rudbeck was regarded as the infinitely better candidate, he pretends not to be aware that anything is amiss.

In 1765 Rudbeck becomes speaker of the house of lords, and gets his own back on the wealthy merchant and banker by seeing to it that all Kierman's possessions are confiscated and that he is imprisoned in Marstrand. He then realizes that although he had done this with his own strong hand, the axe had been wielded by an unseen power.

In 1772 king Gustavus III makes the wise move of assuming almost complete sovereignty; although Rudbeck is governor-general, he acts on his own initiative in mustering the citizenry and several companies of the Uppland regiment. He is arrested for this and then pardoned.

IV.i.3.1 Slagheck

Didrik Slagheck, a barber-surgeon, was promoted by Christian II king of Sweden to be his councillor, prime minister, archbishop of Lund.

It was Slagheck who advised Christian to undertake the Stockholm bloodbath.

In 1522 Christian summoned Slagheck from Lund to Copenhagen and had him beheaded and burnt.

IV.i.3.2 Jöran Persson

Jöran Persson of Sahlberg was the son of a priest. Eric XIV king of Sweden makes him his favourite, and is unable to do or leave anything without consulting him.

Persson incites the king into murdering Nils Sture at Uppsala.

When the king was deposed in 1568, Persson was condemned to loss of life, honour and goods; after his ears had been cut off, he was strung up, taken down, broken on the wheel and beheaded.

IV.i.3.3 Griffenfeldt

Griffenfeldt, the king of Denmark's first minister, is accused of having been disloyal to his sovereign lord Christian V.

Since no competent prosecutor can be found in Denmark, Mauritius is brought up from Hamburg.

Griffenfeldt is condemned to death.

Within a year all the judges have died a miserable death.

Two years after the event Mauritius is accused of disloyalty to his country and dies in prison.

IV.i.3.4 Görtz

Baron Görtz was king Charles XII's favourite. Since the king was hedged by divinity, the whole of the king's war was blamed on Görtz. He was said to have designed the coinage, which was Polhem's work.

When the king falls the whole wrath of the people is vented upon Görtz, someone has to die.

A commission for the trial of Görtz, with various assessors and Fehman as prosecutor, is set up under the presidency of Ribbing.

Fehman accuses Görtz of having prejudiced the good name of the king's faithful servants in the face of the crown, of insinuating that a lord lieutenant had been tardy in the execution of his duty. He produces the letter but does not read it.

Görtz asks Fehman to read the letter; in it he complains that the lord lieutenant is behind with the recruiting, suggests that he should be given another job and replaced by Fehman, who is alert and conscientious. As he reads the words, it grieves Fehman that he should be prosecuting his friend.

Rightly or wrongly, Görtz is beheaded.

The president of the commission dies during the same parliament; Fehman's face is permanently distorted as the result of a stroke; within the course of a year all the assessors have suffered misfortune, one of the most cheerful of them dying of melancholy.

IV.i.3.5 Münnich

Münnich, the Russian prime minister, has Biron transported to Siberia and imprisoned there permanently. He has upright tree-trunks set around the house in which Biron is to be confined, so that the sun should not be able to peep in on him and brighten his solitude.

After a few years Biron is removed from this prison and Münnich is confined there.

It was Russian prime minister Münnich who commissioned the Russian captain Küttler to murder Sinclair on his journey back from the Turks.

IV.i.3.6 Ekeblad

Ekeblad – privy councillor, prime minister, married: in love with the wife of the Saxon diplomat Von der Osten-Sacken. In 1757 the expedition against Prussia is decided upon, Ekeblad's being the deciding vote, and being given, so it is said, for the sake of his lover. She subsequently dies in childbirth.

In 1765 Ekeblad is removed for being member of the French party, like Mrs. Von der Osten-Sacken.

IV.i.3.7 Horn

Earl Horn, a colonel and subsequently a privy councillor and lord chamberlain, inherits an enormous amount of property from his father the privy councillor and from his father-in-law Meijerfeldt.

He soon gets rid of the greater part of it.

Travels to Russia during the 1756 parliament; is afraid to return on account of the Brahe affair and his close involvement with the Caps. His father, the great privy councillor Arvid Horn, is believed to have derived most of his wealth from Russia.

His wife becomes insane.

Though deeply in debt he eventually becomes privy councillor; dismissed in 1769.

IV.i.4.1 Marlborough

Marlborough, the English general during the reign of queen Anne, had such a series of overwhelming victories against the French during the war of Spanish Succession, that everyone was afraid of him. He was at the height of his fortune and favour with Anne when the Tory party was at its zenith; the queen made his wife mistress of the court.

Her majesty orders a pair of gloves for a certain day. The mistress of the court takes a fancy to them and keeps them for herself, ordering the glovemaker to set to work immediately and make a new pair for the queen.

The queen enquires of the glovemaker concerning her order and decides to punish him for not having carried it out. He is obliged to confess that the gloves have been delivered to the mistress of the court.

The mistress of the court falls from favour, the general is recalled, everything in the army goes into reverse. The Tory party falls from power on account of a pair of paltry gloves.

Someone sinned prior to all this, I know not who.

IV.i.4.2 Byng

The English admiral Byng takes Minorca and rampages in a frightful manner over the groaning population.

His son becomes admiral by special favour, and in 1756 is in command of transferring about a thousand troops to Minorca, where the French have landed. When he arrives, he realizes that if he lands the men they will simply be slaughtered. A council is called and all agree on the impossibility of the situation, the only exception being a captain who insists that any order has to be obeyed, be it to send the men to hell. The landing does not take place and the French take Minorca. The populace in England are outraged. When Byng returns home he is court-martialled and executed. After a while the popular cry is that he was innocent.

But the piglet has to suffer for the piggery of the porker.

IV.i.4.3 Buddenbrock

Buddenbrock, the general in the 1741 war against the Russians, had to face the full fury of the whole nation when it went wrong. When he returned he was put on trial and in 1743 beheaded at Norrmalm toll-gate – on the very spot where Paykull, condemned by Buddenbrock's father in Charles XII's time, had been beheaded in 1709. Everyone said that Paykull was innocent.

Buddenbrock and Lewenhaupt did so much to get the war against Russia under way, and it turned out badly for both of them.

IV.i.4.4 Klingspor

When the Dalecarlians were defeated at Norrmalmstorg in 1743, colonel Klingspor was a captain in the guards. The insurgants had laid down their arms and were begging for mercy when he rode at them like a madman, slashing down and massacring the poor souls, including one who was there on his knees, reaching out his hands, begging for mercy. Even his comrades found it objectionable.

In Pomerania in 1761 he was so frightfully cut about the head and face by the Prussians, that he remained an unsightly spectacle for the rest of his life, suffering infinite hardship and pain.

IV.i.5.1

The coronations of kings bring out the incurable audacity of the population. Wine flows, all mill around, the pushers get nothing. An ox is slaughtered: what is the outcome of a lawless free-for-all? Money is scattered: those who rush get very little.

IV.i.5.2

Judges 1: 6, 7: Judah defeated Adoni-bezek, cut off his thumbs and his great toes. Adoni-bezek said, Threescore and ten kings, having their thumbs and their great toes cut off, gathered up under my table: as I have done, so God hath requited me.

IV.i.5.3

Judges 9: 52–57: Abimelech, Jerubbaal's son, after the death of his father, slew seventy of his father's sons, persuaded the people of Shechem to make him king.

God sent an evil spirit between Abimelech and the men of Shechem. He fought and slew them.

A woman in a tower cast a piece of a millstone upon Abimelech so that he died. Thus did Abimelech pay God for the evil he did his father in slaying seventy of his brothers, as well as for the evil done by the men of Shechem.

IV.i.5.4

EMPERORS

Tiberius does away with both his wife's son Posthumus and his nephew Germanicus. He falls ill, and Caligula smothers him with a pillow.

After reigning four years, Caligula and his consort are assassinated during a revolt.

Claudius tyrannizes for fourteen years; is done to death with toadstools by his sixth wife.

Nero, the tyrant, in order to please his concubine Poppaea, does away with Britannicus and his wife Octavia. He also does away with his own mother, who had persuaded Claudius to accept Nero as his heir in place of his own son. He commits suicide after having reigned for fourteen years.

Many are killed by Vitellius until a revolt kills him; he reigns ten years.

Commodus is a sodomite, malicious. His concubine Marcia administers poison to him and Narcissus finishes him off. He reigns eleven years.

Severus Septimius, tyrant, father of Caracalla; his sons conspire and poison him; he reigns eighteen years.

Caracalla murders his brother in the arms of his stepmother; stabbed to death by Macrinus; reigns six years.

Macrinus is struck dead; reigns a year.

Maximinus, cruel, assassinated during a rebellion; reigns three years.

Philip rebels and kills Gordian; is himself killed in a revolt; reigns four years.

Trebonianus Gallus permits the assassination of his predecessor Decius; reigns two months.

Aemilianus assassinated during a revolt; reigns a few months.

Gallienus does not support his father, whom Shapur the king of the Persians uses as a footstool when mounting his horse; is murdered while in flight after reigning thirty-four years.

The best emperors die a natural death:

Augustus	lives to the age of	75	reigns 56 years
Vespasian		81	9
Titus the son		41	2
Nerva		71	6 months
Trajan		73	19 years
Hadrian		72	21
Antoninus Pius		70	23
Marcus Aurelius		58	19
Claudius Marcus Aurelius			10
Diocletian			20

Domitian's governor Antonius Saturninus revolted, was defeated in Germany by Appius Maximus Norbanus; this was known immediately in Rome, although how this came about was never revealed.

Caracalla, on seeing his widowed stepmother's bosom, exclaims that he would that what he sees were his. She: "All that a prince will is given freely".

Aurelius, when Avidius lost: "He could never have won, for he neglected the worship of the gods".

IV.i.5.5 Caesar

Julius Caesar fought fifty pitched battles, took eighty towns, subjugated three hundred tribes, celebrated three triumphs.

He was not satisfied with becoming permanent consul, dictator in perpetuity, with the office of censor, the title of emperor, with being designated father of the country, with having his statue erected among those of the kings. He could no longer bring himself to stand before the senate: his ambition was to become king.

The soothsayer said to him: "Beware the first of the ides of March", and did so six months before the day.

The night before the first he dreamt that he had been lifted above the clouds to Jupiter. His wife dreamt that he had been stabbed to death in her arms, and woke up terrified. At just that time of the night the door flew open, the noise of it awakening the household.

When Caesar decided to stay at home for the day, everyone was soon bruiting it abroad that Caesar was staying at home because Caesar's wife had had bad dreams. Out he went.

The soothsayer meets him. Caesar observes that the first of the ides of March has come. "But not yet gone", replies the soothsayer.

Artemidorus meets Caesar on the way to the capitol, puts into his hands an account of the whole conspiracy, pleads with him to read it immediately, but is pushed aside by the other supplicants on the way to the meeting place.

He enters, and is stabbed twenty-three times. As he catches sight of Brutus among them, he says: "You too Brutus", sweeps his robe about him, and dies. Brutus was his own bastard, by another man's wife.

All who had conspired to murder Caesar died a violent death.

Julius Caesar was told by the soothsayer Spurina, a good six months before the day, that he should beware of the fifteenth of March. On the said ides Caesar met the soothsayer and said: "The fifteenth of March has come". The reply was: "Yes, but it has not yet gone". The same day he was murdered by the senators. When Caesar saw that Brutus was among them he said: "My son, so you too are here".

IV.i.5.6

The emperor Justinian, son of Constantine, captured by Leontius, has his nose cut off and is exiled to Cherson.

Three years later Tiberius captures Leontius, mutilates his nose and casts him into prison.

Tiberius exiles young Philippicus on account of a dream, and he takes refuge in Turkey.

By marrying a daughter of the ruler of the Turks, Philippicus regains the power to drive out Tiberius.

IV.i.5.7

Birger

In 1276 king Valdemar was playing chess with his consort Sophia at Ramundeboda. The delay caused by the game led to his losing the battle of Hova, to his falling into the hands of his enemies, to his having to relinquish the realm of the Goths.

It was when Birger king of Sweden handed over his most faithful lord chamberlain to the monks and to the revenge of his brothers, when he allowed him to be beheaded, that his troubles really began.

By bending the rules of the game at Håtuna, his brothers Eric and Valdemar gain control over him, imprison him in Nyköping castle, oblige him to appease them by handing over two-thirds of the realm.

Eleven years later the king gets his brothers to visit him at Nyköping, locks them up, throws the keys into the lake, and so obliges them to die of starvation. This resulted in his having to flee and so relinquish his kingdom.

Birger handed over his most faithful lord chamberlain to the revenge of his brothers; he left him to the mercy of the monks and his brothers, who had him beheaded; it was then that his troubles really began.

By bending the rules of the game at Håtuna, his brothers Eric and Valdemar gain control over him and imprison him in Nyköping castle; he is eventually obliged to take the sacrament and hand over to them two-thirds of the realm.

Eleven years later he finally gets his brothers to visit him at Nyköping; he locks them up, throws the keys into the lake, forces them to die of starvation. This resulted in his having to flee and so relinquish his kingdom.

IV.i.5.8 King Gustavus

Gustavus I, by destroying monasteries and sacred objects, is able to extend Örby castle.

Eric XIV is deposed, and for many years held under arrest in Örby castle, where he is finally forced to commit suicide by taking poison.

John III and Charles IX, under an oak in East Gothland, enter into a solemn agreement that when they have taken the realm from their brother Eric XIV, they will reign as joint sovereigns.

John acquires the realm and breaks the contract; Charles gets nothing.

Sigismund, John's son and legitimate successor, is driven out, never takes possession of his father's realm.

Charles IX and his descendants possess the realm.

IV.i.5.9

Daniel 5: 1, 2, 5, 25–30: King Belshazzar uses for the feast the vessels consecrated to the worship of God which his father had taken from the temple at Jerusalem; it then seems as if the fingers of a man's hand write on the whiteness of the wall: "God hath numbered thy kingdom, and finished it. Thou art weighed in the balances, and art found wanting. Thy kingdom is divided". And that same night Belshazzar is slain.

IV.i.5.10 Adolf Frederick, king of Sweden

1757: when Brahe was beheaded by order of parliament,
when Palmstierna's commission was in action,
when informers were used to incriminate members of the public,
when the council and the Hats really got the wind up,
when the king and queen were so spoken and written about, that the crown itself was almost toppled.

1761: the next parliament, the services of Höpken and Palmstierna are no longer available.

1765: Ekeblad, Scheffer, ... precisely those most opposed to the king, quit the council, and for completely different reasons.

IV.i.5.11 Ziegler

The mayor of Vasa had a wife, Ziegler, who was handsome but also ill-natured and randy. When he dies she returns to Stockholm, where she was born.

In 1755 she speaks ill of the queen and is reported by her lover, who had taken control of her money but declined to marry her.

When she is sentenced to be put on bread and water and confined to a house of correction for the rest of her life, the queen intervenes on her behalf.

Some years later she is accused of a similar offence and does in fact end up in a house of correction.

IV.i.5.12 The Stuarts

Mary queen of Scots is imprisoned in England by queen Elizabeth.

While in prison Mary instigates insurrection: on the first occasion she is pardoned, on the second occasion she is condemned to be beheaded.

Mary's son James had already succeeded her as king of Scotland; when Mary is to be beheaded, Elizabeth persuades James not to intervene on his mother's behalf by proposing to recognize him as her heir, as the future king of both countries. He accepts the proposal, does not intervene, and his mother is beheaded.

James' son Charles, who succeeds to both kingdoms on the death of his father, is beheaded.

James I's favourite in England was Buckingham, who together with James' son was sent on an unsuccessful courting expedition to Spain, and while there fell in love with the prime minister's daughter. When James was about to withdraw his favour from Buckingham he fell ill, and it was the medicines Buckingham prescribed for him during this illness which killed him. It is very likely that Buckingham prescribed them in consultation with James' son, since he remained the royal favourite when the son came to the throne.

If Charles I did play a part in bringing about his father's death, he paid for it by losing his head on the block.

IV.i.5.13 Louis XIV, the great

At the beginning of the century Holland and France were at war. An audacious Dutchman proposed being given twelve of his army companions, capturing Louis XIV, and exhibiting him alive in Amsterdam. He gets what he wants, knows his route well, requisitions all his horses by force.

Every evening Louis travelled in to the theatre in Paris, but he never stayed in the city overnight.

There is a bridge between Paris and Versailles, and the Dutch reached it at nine o'clock in the evening. Miraculously enough, that night the king did stay in Paris. When one of the great nobles came along at eleven o'clock, he was taken for the king and gagged. The next morning, after having travelled back to Holland by another route, they discover that they have made a mistake. One of the group was Huydecoper, whom I knew when I was in Holland.

It was not God's will that the king should have been captured.

IV.i.5.14 Peter I, emperor of the Russians

He had those who revolted against him beheaded, and while the executions were taking place, not only carried out some of the decapitating himself, but also encouraged his ministers to do so.

During the decapitating, there was one of a batch for whom no room could be found on the block, and who therefore lined himself up with the others by lying down on the ground.

This procedure pleased the czar: he gave him a kick, told him to get up, and pardoned him.

It was the descendant of this man who strangled Peter III, the last scion of the czar.

IV.i.5.15 Elizabeth

Armand Lestocq is princess Elizabeth's physician. Elizabeth is contemplating seizing power in Russia, and someone enquires of Anne the regent what she knows of this. Lestocq gets to hear of this enquiry.

The physician goes to Elizabeth and tells her that if she does not revolt immediately he will be obliged to inform the regent of what is afoot.

Forced by him in this way, Elizabeth goes down to the guards that very evening. They proclaim her the true descendant of Peter the great, march on the palace and seize the regent.

In 1748 the physician who has helped her to the throne is banished, has all his goods confiscated and is imprisoned for life.

The revolt was planned, and the expedition we sent against the Russians in 1741 was part of the run up. The attack should have come in February 1742, when everything would have been ready, but as soon as our men arrived in Finland the St. Petersburg guards were ordered to march against them. This was a further reason for the physician's being obliged to set the revolt going, since these were the only troops Elizabeth could rely upon. It all took place too soon.

Stenbock, a Russian subject who was staying in Stockholm at the time, dreams that his child's tutor in Livonia has come to him and told him that Elizabeth is about to ascend the throne.

The next morning he tells everyone about the dream. This comes to the ear of privy councillor Höpken, who sends for him and puts questions. Stenbock tells him what he has been telling everyone else.

Höpken knows of the preparations being made for February, is afraid that this may bring them out into the open, and therefore advises Stenbock to keep quiet about his dream, lest it lands him in trouble. Stenbock agrees to say nothing more about it.

Höpken notes the date in his almanac; eight days later he learns that the coup had taken place that very night.

IV.i.5.16 Russians

Although I have been told what follows, I do not know whether it is true or not.

Although the empress Elizabeth appoints Peter III her successor and marries him to a lecherous German princess, he fathers no children on her.

The empress sees that this may have dire consequences for the country, and advises the princess to increase her chances of having an heir by taking a lover.

She selects a dashing and intelligent Polish nobleman, count Poniatowski, who was travelling in Russia at the time. It was he who was said to have engendered the prince to whom she subsequently gave birth.

Peter III gets to know of this: when he comes to the throne he takes a Miss Vorontsova as his mistress and sets about getting his empress packed off to a nunnery.

The empress revolts, captures her husband, doses him with poison, and when this fails to get rid of him quickly enough, has him strangled. He was in any case incapable of ruling.

The empress is likely to have been the reason for his imperial highness's being done away with when under arrest, and she may well have been the instigator.

The throne of Poland falls vacant and the empress sends an army into the country, ostensibly in order to keep the people under control while a free election is taking place, actually in order to ensure that her Poniatowski becomes king and to remove to Siberia those bishops who are opposed to him.

King Poniatowski is a good-natured, reasonable, learned person, a worthy ruler. He swears he will get the country straight before the Russians leave, and wants to grant the protestants the same freedom of religious practice as that enjoyed by the catholics. The Jesuits and the monks are opposed to this and form factions to promote the cause of God by frustrating the king.

The French see Russia as too powerful in respect of Sweden and Prussia, since in the last war Russia alone had proved capable of containing the Prussians. They therefore encourage the Turk to declare war on the Russians, on those who had robbed the Poles of their freedom: in order to persuade him to do so they enlist the help of a holy hermit, someone who declares that he has spoken with Mahomet and been informed by him that if the Turk now goes to war, he will bring about an incredible extension of his power. The Turk then launches the great war on Russia.

O, how many thousands of people have to bite the dust for such a trifling cause.

IV.i.5.17

The king of Prussia took Silesia from the empress while she was involved in the war with France.

Some years later she declares war on Prussia with the object of retaking Silesia, and attempts to persuade France and the empress of Russia to help her.

The king of Prussia complains that he has been set upon by three whores, the empress of Russia, the one in Vienna and mademoiselle Pompadour, the mistress of the king of France.

He himself was forced to marry by his father and never slept with his queen; if he swore anything he kept to it.

The empress of Russia swore that as long as she had one rouble left she would spend it on humiliating the king of Prussia, and the Russians did all they could to make mincemeat of him. The king of Prussia, while at table, merely observed: "The empress of Russia is getting on, and sooner or later she will be dead. If we can only hang on for another two or three months, she will have snuffed it".

Scepin, who disputed against me, the most impudent and audacious fellow that ever was, is said to have been the one who did away with the empress, so enabling so many to avoid biting the dust. A Russian monk passed on this information to a Russian envoy.

Peter III, Elizabeth's successor, immediately makes peace with Prussia and declares war on the Dane in order to retake Holstein. The Russian army had already reached Lübeck when his consort had him deposed and murdered and made peace with Denmark, so ensuring the freedom of millions.

IV.ii THE CHURCHES

IV.ii.1.1 CATHOLICISM

Spiritual goods are the foundation of the chair of Rome; the church's rights are separated from the state, spiritual courts from their secular counterpart.

The king swears always to protect the rights of the church; the ceremonies, the pomp, the processions, the songs, music, candles and holy water, blind a superstitious peasantry.

Divine service is sung morning and evening, mass at noon, then there is also matins, evensong and high mass.

Churches, churchyards, bells and chasubles are consecrated and dedicated; everything is blessed and crossed, the making of signs being a new kind of magic.

Saints are introduced by those superstitious enough to believe the pope; the churches preserve relics of the holy cross, the Virgin Mary's hair, John's head, Eric's bones; saints are credited with preposterous miracles.

The forgiveness of sins is purchased through a letter of indulgence, through selling the body of our Lord to the sick, through assiduously attending church to hear an absurd mass, through peregrinating to a wonder-working cross or an image of Mary.

The world is filled with fictions, dreams, monks' tales.

The cathedral chapter is founded in order to defend the rights of the church; the tithe intended for the poor is appropriated for the support of the clergy.

There are the houses of the Benedictines, the Bernardines, the Dominicans or blackfriars, the Franciscans or barefoot, mendicant, grey friars, the Brigittines of St. Bridget, the Hospitallers who serve pilgrims, these being the original institutors of the Knights of Malta, also known as soldiers of Christ, sworn to defend the religion with life and goods.

Punishment conciliated the king, the community, the officialdom; monks are introduced in order to conciliate God – hence the spiritual punishment of confession, the adulterer's stool, the stocks, the denial of holy communion, excommunication, the refusal of burial in consecrated ground, the paying of fines to the church.

One wears a hairshirt, flagellates oneself, lives on bread and water, denies oneself a mattress, goes on a pilgrimage to Rome or to Christ's tomb, recites Our Father and Hail Mary so many times, lies on the ground, refuses banquets, eats off the floor.

Testaments to the church make Christ the heir. Taxes such as Peter's pence went to Rome, the tithe of all kinds of corn and of young livestock had to be paid by Easter or one was not worthy of the Lord's table.

The dean was entitled to twopence from the farmer.

Payment was required for visitations, procurations, celebrations, palliations, authorizations, exemptions, indulgences, fines, dispensations, escheat, the priest's tithe, offerings, predial, the tithe on dairy produce, as well as for marrying, christening, churching and burial.

The origin of the churching of women was to get the candles paid for, of the requiem to get the cost of burial paid for.

Only the priest had any learning and he was practically illiterate.

Many were unable to read from the Bible. The places of learning were the monasteries, where one learnt to recite *Pater noster*, *Ave Maria*, *Credo*.

Should clerics pay tax to the temporal power? Should the servants of God be summoned before a temporal court? Should the sheep shear its own master?

To give to Christ was to adopt Christ as one's heir, to exchange what is earthly for what is heavenly. This implies contempt for the world, sheer fear of God.

There is an outward service of God, one repudiates one's spouse, abrogates one's marital duty, watches, fasts, gives alms and offerings, observes the times of prayer.

The pope is the vicar of Christ, his word is God's word, the church is God's house.

The monks said mass for the dead, consecrated bells and dedicated churches, enslaved and despised the labourers in the fields.

God's blessing is on the celebrating of holydays, the hearing of requiems, on watching, praying and fasting. Sins can be expiated with money.

Diseases are the punishment of God, and to cure them is to sin, to oppose divine punishment. But the monks did employ exorcism, "perish or live in accordance with thy fate".

Whenever the bishop travelled, there was free conveyance, a subsistence allowance, a supply of mead, for both him and all the clerks accompanying him.

IV.ii.1.2

In Lisbon every All Hallows day a festive bonfire is lighted, to which the papal inquisitors, in the name of religion, lead the unfortunate sinners who are to die by burning. There is no more horrendous or cruel crime on earth than this parading of the devil's work in the name of God. It was on All Hallows day in 1755 that the earthquake, flooding and conflagration struck Lisbon, that the full force of God's punishment was visited upon these obdurate sinners. Half the earth trembled: God revealed that He could sense, hear and feel pity for the unfortunates, despite their being heretics.

IV.ii.1.3

Damiens

Damiens, servant to a gentleman, when his master is ill, gives him poison instead of medicine and steals from him; the gentleman survives.

A few days later, when the gentleman is convalescing, he has to have an injection; Damiens administers poison again, killing his master.

It was by such means that he acquired his resources and his wealth.

He turns pious, devoted to the Jesuits; thinking he can obtain blessedness by killing his sovereign Louis XV, he knifes him while he is surrounded by his life guards, wounding him slightly in the ribs.

He comes to a frightful end, torn apart by horses.

IV.ii.1.4

Lord lieutenant Lillienberg to Henrik Benzelius: "Uppsala cathedral was built for Catholics, not Lutherans; their mass takes one hour, ours takes three".

IV.ii.2.1

The authorized copper mines in Norberg were ruined because Lady day fell on the Tuesday after Easter.

The mine was waterlogged and had to be pumped at least every third day; no work was done on holydays.

Since Lady Day fell on the third day after Easter, no pumping took place for four days; the water got so out of hand that it proved impossible to reduce it to the required level, and the mine had to be abandoned.

Tidström.

IV.ii.2.2

Controller

A controller in Gothenburg sleeps with the wife of a ship's cargoman, who is away in India. When the husband returns, his wife has a child. He divorces her.

The controller wants to marry her. They both try to arrange this, but the clergy and the lawyers oppose it. The case eventually comes before parliament. Three of the four houses grant permission to marry, but not the clergy.

On the day on which this permission is granted, the controller and his whore fall out. They are prayed for in church, but prove irreconcilable.

An employee of the Gothenburg East India Company travels to the east. While he is away, someone else sleeps with his wife. The employee returns and divorces her.

The adulterer wants to marry the woman. Faced with opposition all round, he appeals to parliament and wins the case in three of the four houses, the only opposition coming from the clergy.

On the day on which the case is decided he falls out with her. Although they marry as intended, they remain at odds, fighting and squabbling.

IV.ii.2.3 Melander

Melander, professor of theology at Uppsala, always played his politics hard in the university consistory. On a certain occasion we had been sitting all day until gone six o'clock in the evening, dealing with some tedious business. There he was, on his high horse, stirring things up in the interest of an eminently unjustified cause, when his head suddenly twisted backwards and he slid under the table. He was carried home and never really recovered. All were chastened as they dispersed to their homes; they had seen that the Lord is aware of our intrigues.

IV.ii.2.4 Mathesius

Mathesius, a master of philosophy and subsequently a professor at Uppsala.

Norrelius is married to Margareta Benzelia, who is unfaithful to him, fornicating with the students. When he applies for a divorce she engages Springer as her solicitor.

Norrelius is supported by professor Nesselius and professor Ullén. Springer therefore has occasion to inveigh against Nesselius in the consistory. Nesselius undertakes to cede the professorship to Mathesius if he will arrange to have Springer beaten up.

Late one evening Springer is going home to Roddens in Uppsala when he is surprised by two attackers, who throw him to the ground and stab and slash him in the stomach and face. Mathesius gets the professorship.

Mathesius pays court to the daughter of alderman Axberg and the banns are bid twice.

Ryman, a postmaster in Stockholm, is engaged to Miss Axberg and forbids the banns. After proceedings in the council chamber, the court of appeal and the senate, Miss Axberg wins the case.

In the meantime, however, Mathesius has found her mother more attractive, and the partiality is mutual.

Miss Axberg gets married to Piper, councillor of the board of domains, who is deeply in debt. When he brings financial ruin on his mother-in-law, who even when Axberg was still alive had made money by fornicating, Mathesius breaks off his relationship with her.

In 1758 Mathesius proposes to the daughter of an innkeeper by the name of

Uddbom, a girl born on the day the banns were bid for him and Miss Axberg, and is accepted.

Miss Uddbom is smitten with secretary Stenhammer, who gets her pregnant. She wants to break off her engagement with Mathesius, but nearly all her friends advise her against this. It is against her will that she eventually marries Mathesius.

Six months later she gives birth to a child. This is an eye-opener for Mathesius, since he knows he cannot be the father, but he has to keep quiet and pay back a long-standing debt.

The provostship of Uppsala cathedral becomes vacant. In the normal course of events the position should have gone to Mathesius, but since professor Asp is annoyed with him for having selected Annerstedt and not Floderus for a curacy, he so arranges things that it is given to Hydrén.

IV.ii.2.5 Ullén

Ullén, professor of theology at Uppsala, is pre-eminent in the consistory, where he overrides everyone by his mastery of intrigue.

Ihre, Skyttean professor at Uppsala, had published a pamphlet to which the theologian had taken exception. The matter is remitted to the academic consistory. Ihre provides his explanation. Ullén arrives in the consistory during the afternoon, after having taken part in a doctoral ceremony. He is tipsy, and attacks Ihre in the most disparaging terms. Ihre says nothing.

During the next consistory meeting, Ihre makes his position clear in respect of Ullén, adding that he had said nothing during the previous meeting on account of Ullén's having had too much to drink, and asking for this to be entered in the minutes.

Ullén is vexed. He, a theologian, has been accused of having been drunk. This so upsets him that he never again feels sound and well; he falls into decline and dies.

IV.ii.2.6 Annerstedt

Daniel Annerstedt, master of arts, a quick-witted, lively, talkative teacher.

A curacy falls vacant at Uppsala. Annerstedt enters the running, and gets the support of Mathesius, professor of theology, who throws himself energetically and wholeheartedly into backing him.

Professor Carl Asp, Mathesius' best friend, puts in a word for professor Floderus, a decent fellow and a scholar. Asp cannot persuade Mathesius to change his mind, however, since Mathesius knows that Asp wants to marry Floderus to his niece and make him his heir. The outcome is mortal hate between Asp and Mathesius.

Annerstedt gets the curacy and marries the daughter of bishop Humble, despite having been secretly engaged for many years to the daughter of fencing-master Porath and having boarded with her widowed mother. When she hears of the marriage the girl has a fit and remains permanently depressed. How do you think it all turns out?

The provostship of Uppsala cathedral becomes vacant. Mathesius is the obvious choice and the whole consistory votes for him. When it comes to confirming the decision, however, the entire body does an about turn and votes for Hydrén, who is appointed. Asp arranged this in order to spite Mathesius, and he did so with the help of his bosom pal university treasurer Julinschöld, to whom most of the professors were in debt.

1769: parliament is called and Annerstedt is a member of it. Mathesius requests to be relieved of his teaching duties but not of his membership of the cathedral chapter. In spite of the help Annerstedt has received from Mathesius, he so arranges things that the request is granted by the council but altered when it is confirmed: Mathesius is relieved of all his duties with no right of appeal; the point being that Annerstedt wants his professorship.

Mathesius, who had helped Annerstedt and failed to get the important cathedral provostship because he had done so, was dismissed against his will without right of appeal.

How do you think it all turns out once God takes the matter in hand?

Those who survive will see.

Before very long Annerstedt has more than enough on his plate – extensive debts and another child every year. Mathesius had accounts enough to settle with God, but since he had given me a helping hand I would never have chosen to be his scourge.

Annerstedt died of consumption on the first of May 1771. Right up to the year of grace he was deeply involved in debts countersigned by his wife. He had already ruined his mother-in-law, who had died some years before. When prices had peaked he bought a farm for 24,000 dollars which no one else in the city would have given 10,000 for. The creditors therefore took the lot: the widow and her numerous children do not know where the next meal is coming from.

It is at this price that one parts for ever from wife and children.

IV.ii.2.7 Florinus

As soon as provost Florinus of Kimito has handed the parish over to his son, he turns out to be ungrateful in the extreme. He has undertaken to keep his father for the rest of his life but is excessively hard on him, allowing him nothing more than a shack to live in. It is the consistory which has pity, granting him a minor chaplaincy and then the tiniest parish on Åland. The old provost sighs to God on account of the ungratefulness of his son.

A student by the name of Peldan is engaged as tutor and chaplain in the house of the young Florinus. As Peldan is leaving Åbo his horse goes down on its knees gasping for breath. An elderly person sees this and says: "Don't go on the journey. If you do, you will never return". Peldan replies that if every tree were a devil he would still go.

Florinus takes a fancy to the district commissioner's wife and she to him. With the help of his maid he administers poison to his wife while she is in childbirth and kills her. He invites the district commissioner and his wife to a Christmas party, and after getting his guest dead drunk has him dragged away to a separate room.

Peldan is incited into striking the old fellow on the forehead with an axe while he is asleep, a girl holding the candle.

The manservant mounts the dead man on his horse and leads it to a bridge on the way to his farm; there he allows the corpse to tumble off and the horse to trot on home.

Passers-by discover the dead man. The story is that he had ridden home drunk and must have killed himself when he slumped off at the bridge.

A farmhand who was hoping to catch a glimpse of the Christmas guests had been moving around in the dark outside the house; he saw the light in the room and Peldan strike the sheriff on the forehead.

Two years later, after this farmhand has suffered a thrashing at the hand of Florinus, he reports what he knows.

In 1722, Florinus, Peldan, the wife, the maid, the girl and the manservant are executed.

Peldan was well received by the pastor; he was engaged to the maid, who was related to Florinus.

Proverbs 19: 26: He that wasteth his father, and chaseth away his mother, is a son that causeth shame, and bringeth reproach.

IV.ii.2.8 Malung and Kvikkjokk

The pastor's wife in Malung fornicates and is unfaithful to her husband. The daughter becomes a boozer and adopts the habits of her mother. The sons fight a duel; one kills the other and flees to Norway.

Christina Groth, wife of the pastor in Kvikkjokk and Lule Lappmark, fornicates with regimental quartermaster Kock. The pastor, in despair, takes to the bottle. The daughter turns strumpet, and is tumbled by a Lapp.

IV.ii.2.9 Muræus

Muræus, pastor in Kristinehamn, on account of something he had done wrong, the significance of which had been exaggerated by his enemies, was removed from office and replaced by court chaplain Carlberg.

When Carlberg was due to deliver his inaugural sermon, Muræus preached from the aisle, willing damnation upon him for the part he had played in the intrigues.

Crispin Flygge, son of the wealthy general inspector of Kristinehamn, dies soon after marrying Sigrid Ekehielm, leaving his wife pregnant; soon after her husband's death, she leaves to have her child in Stockholm.

Carlberg and the mayor of Kristinehamn accuse her of having practised deception, of having put on a false stomach and acquired someone else's child in order to retain the inheritance. There is a court case and she proves that the child is hers.

Carlberg's honour is declared forfeit and he is condemned to Marstrand. His brother the bishop of Gothenburg finally gets him absolved, released and re-instated, but he dies before he is able to return to Kristinehamn.

The widow holds open house for Muræus for ten years.

Carlberg's hymn number 293.

IV.ii.2.10 Boëthius

Provost Boëthius of Mora preaches against sovereignty: woe to the country whose king is a child. He also inveighs against the prime minister for misuse of power and bad government.

Piper, Charles XII's favourite, has him prosecuted, and condemns him to imprisonment for life in the fortress at Nöteborg.

The Russians take Nöteborg and are victorious at Poltava. Boëthius is released and comes back to Sweden: Piper is captured and incarcerated in the same prison, where he passes away miserably.

IV.ii.2.11 Kihlmark

Kihlmark is a young army chaplain, intrepid and obstinate. By putting his colonel, Dellwig, on a charge, he brings about his dismissal.

When the revolt planned by Brahe is discovered, and announcements are made in the churches, provost Kihlmark comments on the matter from the pulpit, maintaining that Brahe had done nothing wrong – it was earl Brahe who had the advowson of Kihlmark's living.

Dellwig's son happened to be in the church, and together with the curate he sets the matter in its true light.

Kihlmark is arrested, removed from office, and dies in custody before his case is decided.

IV.ii.2.12 Fahlander

Gezelius was an old grey-haired clergyman at Grangärde. Fahlander, an army chaplain directly under the command of king Charles XII, gets the old man brought to the king's notice as difficult to get on with.

Fahlander is put in charge of the parish of Grangärde, and travels back home from the war to take up his new duties.

Gezelius, with tears in his eyes as he takes his leave tells Fahlander: "God will also order your departure".

In 1743 a grey-haired and grey-bearded Fahlander joins the rebellious Dalecarlians. When they are defeated, he is sentenced to imprisonment for life in Marstrand gaol, where he dies in misery.

Fahlander prepared a tomb for himself at Grangärde, but it was never to be his; he lies buried elsewhere.

IV.ii.2.13 Serenius

When Serenius bishop of Södermanland was a student he lodged in Siwert's cellar. He comes into his room on a certain occasion and finds an East Goth sitting there. "What sort of vermin has found its way into my room", he exclaims. "This sort", retorts the East Goth, slashing off his nose. It had to be stitched back on again.

Despite professor Hermansson's having taken such an intense dislike to Serenius, he eventually had him as his son-in-law.

As pastor in London Serenius so threw himself into his work that he won everyone's heart.

He returned home to become provost at Nyköping. He compiled his magnificent English dictionary, always supported the English lobby, and faced the consequences of doing so.

He was bishop already in 1765, and one of the most effective members of parliament.

Hildebrand died unmarried, by his will leaving everything to his mistress. His brother, cavalry captain Hildebrand, starts legal proceedings against the mistress in respect of the inheritance. When he loses the case in the crown court, he engages Serenius in order to prosecute it in parliament.

Faced with the problem of winning over the house of peasants, Serenius gets his son-in-law to arrange a get-together for the members on Munkholm. Twelve whores are made available, and are only to entertain those members who have given their word that they will vote for Hildebrand. Between twelve and fourteen members get dead drunk and catch gonorrhoea, with the result that Serenius' religion is much discussed in the city.

Serenius was second on the list when Beronius became archbishop. It will be interesting to see if he succeeds him.

Hildebrand lost his case in parliament.

IV.ii.2.14 Benzelia

One of archbishop Spegel's daughters was married to archbishop Jöns Steuch, the other was betrothed to privy councillor Düben.

When Düben was being held prisoner in Russia, it was common knowledge that Erik Benzelius, who was later to become archbishop, was fornicating with Düben's betrothed, and doing so at the archbishop's palace. Late one evening, when Steuch was leaving Spegel's place, he was badly beaten up by students from East Gothland who thought that he was Benzelius.

Düben eventually arrives back, and since he could so easily have heard about the relationship, she marries him the same evening.

Benzelius subsequently marries the daughter of bishop Swedberg, a foolish woman who fornicated with Mörner, later to become lord lieutenant.

The piglets suffer for the piggery of the porker.

Benzelius' eldest daughter Greta, who turns out to be utterly wanton, gets married to librarian Norrelius, a learned dryasdust.

Greta Benzelia, daughter of bishop Benzelius: as her father violated Düben's right, so her mother violated her father's right.

Although Norrelius, librarian at Uppsala, makes her his wife, she goes out at night with the students, cannot stand her husband, encourages undergraduates to beat him up, plays havoc with his property. The subsequent divorce case drags on for years.

Olaf Rudbeck the younger, angry with Norrelius for having published under his own name a manuscript he had entrusted to him in order to get it published in Holland, takes the woman into his house. She faints in a fit of hysteria; doctor Rosén is called, and next morning is found in her bed. She gives birth, and maintains that Gerdessköld, later to become president of the court of appeal, is the father of the child. He denies this and witnesses are called for. She says that there can be no witnesses to such a relationship. She is condemned as a whore and Norrelius is rid of her.

When the Russian officers arrive in 1743 she lives among them as a common whore and they have her whipped.

She moves on to Norway, where she lives in misery.

Her son is quick-witted, earns twenty-four and thirty-six dollars a time reproducing our banknotes in Copenhagen. Extradited from Denmark and sent to Stockholm, he is hanged during his mother's lifetime.

IV.iii LIVE IRREPROACHABLY

IV.iii.1.1

Live irreproachably, God is near.

IV.iii.1.2

Though you honour the age for its new insights,
There is constant flux in the views of men,
What one age esteems another decries.
Your memorials follow you into the dust,
Your name is forgotten, there is scarcely a trace
   Of what once was admired.

Open time's book and see what names you find.
Most owe their fame to their vices.
Those who gained life's goal by quiet virtue
   Are not recorded.

Your life never achieves its purpose,
For in eternity it dwindles to a point.
How many detours have bedevilled the short course
Since you first set out to plan your way.
One false step, and your whole life is remembered
   With shame and suffering.

I leave the pleasures of the town to whoever seeks them –
The bone-shaking coach on the cobblestones,
The frenzied pursuit of place and money,
The robber masquerading in a lordly wig,
The sheer folly to be found in the best of all worlds.

Though there is a cheap respect for virtue,
Unless virtue can be paraded around
It is out of place in a world like ours.
Conscience, virtue, courage, three old Gothic words,
No longer understood, are marketed around.
Small souls are bought and sold. Where then are virtue
And love of country? Not here among these slaves.

It is to this that discord has brought the unfortunate citizenry.

IV.iii.1.3

NATURALISM

Rather than concerning themselves with petty heretics, the theologians ought to concentrate upon refuting the freethinkers, whom I have come across in all periods of the world, and who are in agreement on the following points:

1. All believe that there is a God, creator and preserver of the universe, and that this is made apparent to them by the whole of nature. They do not believe in the other two persons of the Godhead, however, for they say:

2. that nature provides no evidence at all of these two further persons, whereas it would do so were they part of the creation;

3. that God's indulgence of Christ is in imitation of the creation of gods by Jupiter;

4. that a divine creator of this magnitude, not to be contained by the whole world, could never have been born of Mary;

5. that Christ was a holy man, who taught morality in an exemplary manner.

6. They either attempt to give a natural explanation of all his miracles, or they call in question the account given of them.

7. They believe neither in the sin of our first parents nor in the atonement, and say that such a lifting of the punishment would be the end of death.

8. They say that the Holy Ghost was accepted at a certain council, but that a single vote would have decided the matter the other way.

9. They say that it was idolatry as monstrous as that of the heathen to limit the power of the Omnipotent by foisting assessors upon Him.

10. They deny the immortality of the soul, and believe that it passes away like the flame of a candle.

11. They believe neither in the kingdom of heaven nor in hell, neither in angels nor in devils.

12. They therefore deny that there is any punishment or reward in an afterlife.

13. They do not believe in the sacraments as such.

14. They should pray to God to grant the faith to believe what is impossible.

There are many such propositions, which in the interest of blessedness, of protecting the simpleminded from being led astray, ought to be refuted succinctly by the theologians.

IV.iii.2.1

PRIDE

Pride is fortune's bastard.

Pride is the first step toward becoming a fool.

Jupiter turns mad those he will destroy.

Fortune makes fools of those whom she spoils.

Fortune exalts the wicked, that their fall may be the greater.

Fortune breeds audacity, none is saved from pride by pride's being obnoxious to others.

As you bear your fortune, Celsus, so shall we bear with you.

All hate the proud, and if they are courageous enough, make enemies of them. The proud are praised and hated by all.

It is rash and foolhardy to put out to sea simply because of a favourable wind.

Everyone is hostile to the unfortunate.

As you bear your fortune, Celsus, so shall we bear with you.

Drunk with the sweetness of rapid good fortune, you have wandered far from life's straight and narrow.

Proverbs 16: 18: An haughty spirit goeth before a fall.
Proverbs 18: 12: Before destruction the heart of man is haughty.
Tobit 4: 13: Let not pride rule, since it is an initiation of punishment.
Ecclesiasticus 10: 7, 12, 13: Pride is hateful before God and man. The beginning of pride is when one departeth from God, and his heart is turned away from his Maker. Therefore the Lord brought upon them strange calamities, and overthrew them utterly.
Proverbs 29: 23: A man's pride shall bring him low.

Dwell on what is your own.
You have arisen from a frothing drop of detestable lust.
You have emerged from a nasty hole, between excrement and urine.
You swell with the content of your bowels and void it daily: a libidinous shitbag.
Life hangs by a thread, cobweb fine, nothing more fragile.
Life is a beautiful bubble.
You are full of yourself when prospering, but a beggar in adversity. Kierman.
When dead, an abominable corpse.
Death shows what an empty bubble man is.
How great I think I am, but it is all a fiction.
Dwell on what is your own: you will then see how meagrely your house is furnished.
What is magnanimity when the wheel of fortune turns?
What is wisdom? To be aware of your ignorance.
What is power? Leadership among fools.
What is wealth? To be the treasurer for other fools.
What are clothes? A pantomime costume to make the kiddies gape. For we often find a good-for-nothing in velvet, a whore in silk, a rogue in a gown, a poltroon in armour. Children take those who are masked for what they appear to be.

Daniel 5: 20, 21: Nebuchadnezzar's mind was hardened in pride, he was deposed from his kingly throne, and his heart was made like the beasts.
Daniel 5: 22, 23, 25: Belshazzar did not humble his heart and drank in God's vessels. Written over against the candlestick: Mene, Mene, Tekel, Peres.
Luke 12: 20: Thou fool, this night thy soul shall be required of thee.

God sees and hears all. He is near.

Daniel 4: 17, 35: The most High ruleth in the kingdom of men, and giveth it to whomsoever he will. He doeth according to his will, and none can stay his hand.

It is only by degrading means that ambition will take you to the top.

He who has not been exalted by chance will not be toppled by fortune. Seneca.

IV.iii.2.2

Beware of sins writ large.

There is no atonement for the sin, except the deed be undone.

1. Beware of blood, especially murder.
2. Beware of blood-guilt.
3. Beware of disdain for God.
4. Beware of ingratitude to parents.
5. Beware of ruining the well-being of another.
6. Beware of harming the helpless.

IV.iii.2.3

When fortune is granted by the grace of God, a thousand hands are there to prevent misfortune. A murderer encounters another murderer (Spegel). The one expected is delayed; others are there instead.

A hundred circumstances frustrate the adulterer and his desired one.

Divine wisdom is at play in human affairs.

They keep acquiring property fortuitously; they inherit.

God grant my son good fortune, learning is not necessary.

Great and precocious gifts seldom mature.

Go gently, it is the gentle who inherit the earth. The mother of the retiring seldom has occasion for tears. Fulfil your office as well as you can, let the world go its own way, remain on good terms with him who is set above you.

IV.iii.2.4

Job 15: 8: Hast thou heard the secret of God?
Job 8: 3: Doth God pervert judgment?
1 Samuel 16: 7: For man looketh on the outward appearance, but the Lord

looketh on the heart.
Ecclesiasticus 17: 15: Their ways are ever before him, and shall not be hid from his eyes.

Wisdom of Solomon 2: 21: Such things they did imagine, and were deceived: for their own wickedness hath blinded them.
Job 4: 7: Who ever perished, being innocent? or where were the righteous cut off?

A just cause triumphs.
Psalms 37: 25: I have been young, and now am old; yet have I not seen the righteous forsaken, nor his seed begging bread.
Proverbs 12: 14: The recompence of a man's hands shall be rendered unto him.
Job 34: 11: For the work of a man shall he render unto him, and cause every man to find according to his ways.

Job 4: 8: They that plow iniquity, and sow wickedness, reap the same.
Galatians 6: 7: Whatsoever a man soweth, that shall he also reap.
Wisdom of Solomon 11: 16: That wherewithal a man sinneth, by the same also shall he be punished.

IV.iii.2.5 Klingspor

When the Dalecarlians were defeated at Norrmalmstorg in 1743, colonel Klingspor was a captain in the guards. The insurgants had laid down their arms and were begging for mercy when he rode at them like a madman, slashing down and massacring the poor souls, including one who was there on his knees, reaching out his hands, begging for mercy. Even his comrades found it objectionable.

In Pomerania in 1761 he was so frightfully cut about the head and face by the Prussians, that he remained an unsightly spectacle for the rest of his life, suffering infinite hardship and pain.

IV.iii.2.6 Kierman

Gustaf Kierman, mayor of Stockholm, began his life in the city as a poor boy from Askersund in the service of his predecessor in marriage. The wife was unfaithful and is said to have had a secret crush on Kierman. The predecessor, a godfearing person, was found lying in his room one morning, slashed and murdered with a penknife. Kierman gets the widow, but does not remain faithful to her.

He becomes wealthy, a councillor, mayor, speaker of the house on numerous occasions, has immense influence.

He has no mercy when Brahe is sentenced to death; he is godless but frank, everyone is delighted with him and he makes many friends.

In 1762 he arranges for his son-in-law captain Duwall to be proposed for the privy council. Rudbeck gets the same number of votes for the position, but loses to Kierman's son-in-law in the final vote.

In 1765 Rudbeck becomes speaker of the house of lords. Kierman is put under arrest and everything he owns is confiscated.

Kierman is sentenced to a month on bread and water, followed by life-imprisonment in Marstrand. All his possessions are auctioned off, the proceedings going to the crown; he dies in Marstrand the same year.

The poor fellow could hardly have envisaged this two parliaments before, when Brahe was sentenced and he declared that there was to be: "No mercy".

None is blessed before his death.

Ecclesiastes 8: 9: There is a time wherein one man ruleth over another to his own hurt.
Job 12: 23: He increaseth the nations, and destroyeth them: He enlargeth the nations, and straiteneth them again.

In 1756, while showing some bigwigs over his shipyard, he treated them to some truly magnificent entertainment. As merry as the rest of them while the banquet was under way, he regaled them with an account of his life:

as a lad in Askersund he had peddled for Borås;

he had traded in nails when he first came to Stockholm;

his matrimonial predecessor had taken him on on account of his quickwittedness;

he had slept with his patron's wife;

his patron had been murdered and the skulduggery that had ensued;

the widow had accepted him as the alternative to becoming a common cowhouse whore;

he had slept with Forsberg's wife, and infected her as he had many others, with the French pox;

he had slept with his present wife while her husband was still alive;

and a hundred and one other things.

"The outcome is that you see me here now with 200,000 dollars in gold under my belt. Spend the whole lot of it on a gibbet, and it would not make up for all the jiggery-pokery I have been involved in".

In 1763, when the tax-assessments for the inhabitants of Stockholm were being drawn up, he is reputed to have said: "Tax the devils to the hilt, the system has to be maintained, the rabble can eat barkbread". It was this remark which gave rise to the unappeasable hostility which toppled him during the next parliament, which led to his being stripped of most of his possessions and dying in Marstrand prison.

IV.iii.2.7 Buskagrius

Buskagrius, professor of Greek at Uppsala, though erudite, quotes an author incorrectly during a disputation. The opponent declares that the text has been cited inaccurately. There is a vituperative exchange of words.

The professor prays to God in the heat of the moment, that if he is indeed in the wrong he may never lecture again.

When he gets home he discovers his error, falls into a melancholy, and dies two years later. He never did lecture again.

IV.iii.2.8 Sinclair

Captain Sinclair, while in captivity, knifes to death another non-commissioned officer by the name of Lood. Though he is tried for murder when he gets home, he is not convicted.

He hated Russians so intensely, that he said he had no desire to go to heaven if he was likely to meet any of them there. Artedi hated the Dutch in a similar manner, and was drowned in Amsterdam.

Sinclair is dispatched to Turkey to incite the Turks against the Russians.

Münnich, prime minister in St. Petersburg, acquires his portrait and arranges for four officers to ambush him at Ingerstedt in Germany. Küttler was the one who murdered him.

IV.iii.3.1

To be king is not to fear,
And all may have the power.
Let him who will, in pride of place,
Try not to slip and fall;
I choose peace and quiet;
Let me unnoticed go my way,
Enjoy the rest, the calm alone;
Let my life slip quietly by
To the world at large unknown.
Then, when my days have run their course,
Free from the taint of idle strife,
As man, no more, I would pass away.
The end is hard if when we leave,
Though known to all in public life,
We are strangers to ourselves.

IV.iii.3.2

GLADNESS

Do what is good and be glad.

Eat, drink and be merry, after death there are no such pleasures.

Ecclesiasticus 30: 22: The gladness of the heart is the life of man, and the joyfulness of a man prolongeth his days.
Ecclesiastes 11: 9: Rejoice, O young man, and let thy heart cheer thee in the days of thy youth.
Ecclesiastes 3: 12, 22: I saw that there is nothing better than for a man to rejoice, and to do good in his life. There is nothing better, than that a man should rejoice in his own works; for that is his portion.
Ecclesiastes 2: 24: There is nothing better for a man, than that he should eat and drink, and that he should make his soul enjoy good in his labour. This also I saw, that it was from the hand of God.
Ecclesiastes 3: 13: That every man should eat and drink, and enjoy the good of all his labour, is the gift of God.
Ecclesiastes 8: 15: Then I commended mirth, because a man hath no better thing under the sun, than to eat, and to drink, and to be merry: for that shall abide with him of his labour, which God giveth him.
Ecclesiastes 5: 18, 19: Behold that which I have seen: it is good and comely for one to eat and to drink, and to enjoy the good of all his labour that he taketh under the sun all the days of his life, for it is his portion. Every man also to whom God hath given riches and wealth, and hath given him power to eat and drink thereof, and to take his portion, and to rejoice in his labour; this is the gift of God.
Ecclesiastes 9: 7, 9: Eat thy bread with joy, and drink thy wine with a merry heart. Live joyfully with the wife whom thou lovest all the days of the life of thy vanity: for that is thy portion in this life, and in thy labour.

If you are successful, you must know that it will not be for long; if you are unsuccessful you must know that you are not so unless you consider yourself to be.
Ecclesiasticus 21: 20: A fool lifteth up his voice with laughter; but a wise man doth scarce smile a little.

IV.iii.3.3

When I contemplated revenge everything went wrong for me. When I changed and put everything in the hands of God (1734), everything went well.

IV.iii.3.4

Genesis 50: 19, 20: Fear not, for I am under God. You thought evil against me, but God has turned all to good.

IV.iii.3.5

O for the grace
   to share in Thy counsel
and secret decree.

Raise me aloft
from the dust to Thee
   and let me see
in a moment of time
   how the wide world turns,
reveal the prime cause
   of all that happens
so strangely here.

A rose withers
While the nettle thrives
   and flourishes on.
The rich are laden
   with still more wealth
while the poor see
   food become scarce
and their goods dispersed.

Mid falsehood and wrong
mid art and trickery
mid civil murder
mid fair-speaking friends
mid fair-sounding words
   I wilt and wither;
unless you sustain me
   I fall and perish.

You marked my fate
when still I lay there
   in darkness alone.

You set my clockwork
   and give me my bread;
almighty Lord
   will you now turn from me
and cast me off?

I have built my house
   by the Creator's grace,
   and sleep secure.

IV.iv THE JUDGEMENT OF GOD

IV.iv.1.1

O mighty Ruler, how mighty is Thy rule,
And how unfathomable are Thy decrees.
Though the fateful power which forms things may be seen,
Of that which moves invisible and alone, which brings
All that seems the work of human hands under the sway
Of God's sceptre, we know and perceive nothing.

IV.iv.1.2

Ecclesiastes 9: 2: There is one event to the righteous, and to the wicked.

IV.iv.1.3

FATE

Fate is the judgement of God and there is no escaping it. Philosophers maintain that given man's freedom of choice there can be no fate; they completely deny its existence, and assert that each is the author of his own fortune, something which even the Devil would not dare to trumpet around.

A parable can help to show how freedom of choice may be reconciled with the inevitability of fate.

A man can hang himself, drown himself, cut his own throat, and can also choose not to do so. But if for any reason he is condemned to death by the supreme judge, he no longer has any choice in the matter, his execution is an unavoidable necessity.

A man is therefore at liberty to commit or not to commit a crime, but not to avoid the issue once he has committed it and been condemned.

Man does have freedom of choice, is able to do what he wants, and God sees and hears all; consequently, if man acts wrongly and is not brought to book, if he stands accused before God of the injury, God orders the natural course of the inevitability of fate.

The judgement of God is, therefore, fate.

IV.iv.1.4

It is often a long time before a sinner is brought to book.
Job 33: 29: Lo, all these things worketh God oftentimes with man.

Ecclesiasticus 16: 17: Say not thou, I will hide myself from the Lord: shall any remember me from above? I shall not be remembered among so many people: for what is my soul among such an infinite number of creatures?
Ecclesiasticus 23: 18: I am compassed about with darkness, the walls cover me, and no body seeth me. The most High will not remember my sins.

Psalms 94: 9: He that planted the ear, shall he not hear? he that formed the eye, shall he not see?
Ecclesiasticus 23: 19: Such a man knoweth not that the eyes of the Lord are ten thousand times brighter than the sun, beholding all the ways of men, and considering the most secret parts.
Proverbs 15: 3: The eyes of the Lord are in every place, beholding the evil and the good.
Job 34: 21: For his eyes are upon the ways of man, and he seeth all his goings.
Jeremiah 16: 17: For mine eyes are upon all their ways: they are not hid from my face, neither is their iniquity hid from mine eyes.
Proverbs 5: 21: For the ways of man are before the eyes of the Lord, and he pondereth all his goings.

Psalms 73: 11; *Galatians* 6: 7: Be not deceived; God is not mocked.
Wisdom of Solomon 16: 15: But it is not possible to escape thine hand.
Job 12: 14: Behold, he breaketh down, and it cannot be built again.
Ecclesiastes 3: 9: What profit hath he that worketh in that wherein he laboureth?
1 Corinthians 3: 7: So then neither is he that planteth any thing, neither he that watereth; but God that giveth the increase.
Ecclesiastes 9: 11:

> The race is not to the swift;
> nor the battle to the strong;
> neither yet bread to the wise;
> nor yet riches to men of understanding:
> nor yet favour to men of skill;
> but time and chance happeneth to them all.

Amos 3: 6: Shall there be evil in the city, and the Lord hath not done it?
Let it go as it goes, it goes as God wills.
Ecclesiasticus 18: 4: To whom hath he given power to declare his works?
Ecclesiasticus 22: 22: For there may be a reconciliation: except for upbraiding, or pride, or disclosing of secrets.
Ecclesiasticus 21: 22: A foolish man's foot is soon in his neighbour's house: but a man of experience is ashamed of him.
Ecclesiasticus 4: 30: Be not as a lion in thy house, nor frantick among thy servants.

IV.iv.1.5

> I have often wondered if the heavenly powers
> Are concerned with the Earth, if a steersman is in charge,
> If mortal fate is anything but uncertain chance.

Rufinus' punishment finally freed me from such doubt,
And acquitted the gods.

Claudianus

IV.iv.i.6

Boatman

When the Reverend Collin was studying with me in Växjö, his mother sent him a supply of food by a pack-horse, ridden by one of the servants.

A boatman up from the country also comes riding along, and accompanies the servant for a while. That evening, however, just as they are passing the gallows, the boatman draws his knife on the servant, knocks him off the horse and takes what he wants of what is being carried.

The servant revives after this assault, and reaches the town. The bailiff is sent out and captures the offender, who is condemned to death.

An aged official looks at the offender's hands, and declares that he will never die a violent death.

The offender dies while under arrest, two days before the death-warrant arrives.

If God has delineated our fate, before it comes to pass, in our hands, we ought to commiserate the unfortunates whose fate is as it is.

IV.iv.1.7

A Norwegian

There was a murder in Norway: three people were in the company of a fourth, who died of a stab wound. Since the evidence remains undecisive, they are obliged to cast lots on who should be beheaded. The lot falls to one of the innocent parties, the king despatches the death-sentence.

The one sentenced appeals, maintains by all that was, is and will be, that he is innocent. One of the most eminent barristers is obliged to take up the case, and demonstrates as clearly as two plus two are four, that the one sentenced cannot possibly be the murderer.

The case is therefore referred back to the king. Though occupied with something else, he recalls the name, and without reading the investigation declares that since the person has been sentenced to death he should be executed.

Utter desperation when the prisoner is informed. The barrister goes along to see him and says: "It is now clear to me that the judgement of God is upon you. Though you may be innocent in this case, you must be guilty of someone else's blood".

This caused the prisoner to pull himself together: "The righteousness of God's judgement is now apparent to me. When that fellow was murdered and no one was brought to book five years ago, it was I who was guilty. In the present case, however, I am innocent".

Professor Strömer recounted this.

IV.iv.2.1

FORTUNE

God grants to whom he will.

There are times when one person is unfortunate in everything he undertakes, while another has all his wishes fulfilled.

One of the farmers in Hammarby is generous and brisk, has enough of everything, a happy home, is sitting pretty. Another is miserly and niggardly, is generally miserable and never gets on top of his work.

In 1770 the scene changes. The second farmer is now sitting pretty, cheerful, prosperous; from one day to the next everything begins to go wrong for the first farmer.

One person works and never gets anywhere; someone else, like Broman, simply drifts into precisely the position he wants.

A lazy Jack wafts to the top of the tree, a busy Tom slips into a rut and gets nowhere.

The old woman's wish: God grant my son fortune, he has no need of knowledge.

Fortune's turns are fickle, but the pace is steady.

Order in the greatest confusion, as in the realm of nature.

The Devil would not dare to maintain that anyone hammers out his own fortune.

God's will has to be obeyed, the whole of nature contributes to the bringing about of misfortune. As with dogs, all are hostile to the unfortunate. All hurry along the wagon of misfortune. As can be seen from the case of Kyronius, neither heaven nor earth can help. Pride leads the way, making enemies of all. No calamity comes alone. Our role is to make the rod for our own back. Jupiter turns mad those he will destroy.
Proverbs 21: 30: There is no wisdom nor understanding, nor counsel against the Lord.

When fortune is granted by the grace of God, a thousand hands are there to prevent misfortune. A murderer encounters another murderer (Spegel). The one expected is delayed; others are there instead.

A hundred circumstances frustrate the adulterer and his desired one.

Divine wisdom is at work in human affairs.

They keep acquiring property fortuitously; they inherit.

God grant my son good fortune, learning is not necessary.

Great and precocious gifts seldom mature.

Go gently, it is the gentle who inherit the earth. The mother of the retiring seldom has occasion for tears. Fulfil your office as well as you can, let the world go its own way, remain on good terms with him who is set above you.

IV.iv.2.2 A rabbinical story

When Moses spoke with God on mount Sinai, he asked why he who is just should permit injustices in His world, why the just should so often be so extremely unfortunate and the unjust get by so very comfortably. God replied: "You judge in accordance with what you see and experience, I in accordance with my omniscience. Cast your eyes down to the base of the mountain, to that well-spring".

Moses did so and saw:

1. A wild and blustering soldier, galloping along on a horse, who dismounts at the spring, drinks, drops his purse without realizing it while remounting, and thunders off.

2. A ragged and sweating boy, who comes along, drinks, notices the purse, discovers and pockets the money, and goes happily on his way.

3. A grey-haired old man, exhausted, weary and puffing, who reaches the spot, drinks, and lies down to rest.

The soldier comes careering back, and demands that the old man should return the purse, vowing to run him through with his sword if he refuses to do so. The old man denies ever having seen the purse, swears that he is telling the truth, and shows everything he has in his possession. Regardless, the soldier slashes away at him, and a frightful murder is committed.

When he sees this, Moses cries out: "O God of justice, why should such a worthless soldier be allowed to murder such an innocent old man!"

God replies: "That is how you see it, but I have ordered it to be so.* Eight years ago, there in that wood, the old man strangled the boy's father because he thought he had money on him. Since then, the fatherless boy has begged his keep from door to door. I have given him the money, which the soldier acquired unjustly. I have used the soldier in punishing the wicked old man. The soldier has committed many crimes, and in due course he too will be punished".

* *1 Samuel* 16: 7: Man looketh on the outward appearance, but the Lord looketh on the heart.

IV.iv.2.3 Asp

The court steward, such an efficient manager of the household and farms, marketing over a thousand tons of grain a year, that he inspires confidence in everyone.

He comes a cropper in his phaeton, sustains injuries to the head which make him feebleminded.

Creditors move in on five hundred thousand dollars in gold, and it bankrupts him.

Both brothers-in-law, who had invested heavily in him, are at their wits' end.

His mother-in-law had left her property to the children in exchange for a regular income.

His only daughter, married to Post, justice of the court of appeal, dies of small-pox after a confinement.

His son, his only son, is wasting away with consumption, constantly gasping for breath.

No calamity comes alone.

If God intends to punish, no misfortune is to be avoided.

It is now in 1769 that God's judgement is being executed upon this family.

I have not discovered what they have done.

I know that God never judges without cause.

IV.iv.3.1 NEMESIS DIVINA

Nemesis divina is talion, exact retribution.

Friess, F.C., *The Divine Law of Talion*, Stockholm, 1763, 8:vo, 105 pages, translated from the Danish into Swedish, has a little to say on this.

IV.iv.3.2

Psalms 119: 137: Righteous art thou, O Lord, and upright are thy judgments.

There is no doing away with the sin, except the deed be undone.

Proverbs 12: 14: The recompence of a man's hands shall be rendered unto him.

Galatians 6: 7: Whatsoever a man soweth, that shall he also reap.

Wisdom of Solomon 16: 15: But it is not possible to escape thine hand.

Amos 3: 6: Shall there be evil in a city, and the Lord hath not done it?

Let it go as it goes, it goes as God wills.

Psalms 127: 2; *Ecclesiastes* 3: 9: It is vain for you to rise up early, to sit up late, to eat the bread of sorrows.

Romans 9: 16: It is not of him that willeth, nor of him that runneth, but of God that showeth mercy.

1 Corinthians 3: 7: So then neither is he that planteth any thing, neither he that watereth; but he that giveth the increase.

Psalms 127: 2: God giveth to his own while they sleep.

Psalms 37: 25: I have been young, and now am old; yet have I not seen the righteous forsaken, nor his seed begging bread.

But God will be avenged upon children and children's children. The sons of heroes bear the punishment. The piglets suffer for the piggery of the porker.

God help us to do well on our way through this wicked world.

Psalms 73: 11; *Galatians* 6: 7: The Father will not, God is not mocked.

Let us not charge through the world, let us slip through quietly, lest Nemesis should hear us.

Isaiah 10: 5–15: Shall the axe boast itself against him that heweth therewith? O Assyrian, the rod of mine anger. I will send him against an hypocritical nation, to tread them down like the mire of the streets. But since he saith that by his hand he hath done it, I shall punish the stout heart of the king of Assyria.

There is no one greater than a truly honest man.

Ecclesiastes 3: 12: Rejoice and do good in life.

Live irreproachably, God is near. God hears and sees all.

Psalms 94: 9: He that planted the ear, shall he not hear? he that formed the eye, shall he not see?

He who would govern must do so gently.

See to it that you do no great injury to those who are both powerful and miserable; though plundered, they still have their arms. Juvenal.

He who has not been exalted by chance will not be toppled by fortune.

To be very active is to submit to fortune, usually an experience best avoided.

Beware of building your welfare on another's ruin. Fate is a tyrant, and can topple you too.

There is nothing so sublime and above all perils, that it is not under God and subject to Him.

See the section on Christian II's attitude to Slagheck.

We are punished with that whereby we sin.

Wisdom of Solomon 11: 16: That wherewithal a man sinneth, by the same also shall he be punished.

It is rash and foolhardy to put out to sea simply because of a favourable wind.

As you bear your fortune, Celsus, so shall we bear with you. Horace.

The willing are led by fate, the unwilling are dragged.

Everything will help nature, so that its work may be completed.

Call no one happy till his death. Solon.

Life will work out in the way that you live it.

IV.iv.3.3

NEMESIS

Retribution is visited upon him. Everything runs counter. No calamity comes alone.

For some, everything they undertake goes wrong, for others, regardless of their stupidity, everything goes swimmingly.

Whole families are unfortunate. The children are taught; act completely contrary. Hell bent on their own destruction. Heaven and earth can neither help nor save.

Many always do badly after a certain day. Everything then began to go wrong. Charles XII's first nine years went well, then nine years of misfortune.

One misfortune follows another. Wherever he turns, whatever he does, it goes wrong. Someone else is sitting in the lucky chair.

One dies after the other, once God begins to settle the accounts. No calamity comes alone. The house burns down, everything goes wrong. God's retribution has now been visited upon the house.

Call no one happy until his death, not even Croesus.

Gunnar Gröpe murdered St. Sigfrid's three nephews; no natural death among his descendants, the Ulfsax family, for twelve generations.

The unfortunate born of wicked parents. The piglets suffer for the piggery of the porker.

Psalms 37: 25: I have not seen the righteous forsaken, nor his seed begging bread.

Misfortune pursues him wherever he turns.

When I contemplated revenge everything went wrong for me. When I changed and put everything in the hands of God (1734), everything went well.

IV.iv.3.4 Gallus

Tschepius, the parson at Soldau in Prussia, often has occasion to suggest indirectly to George Gallus, an old irreligious linen-weaver, that he ought to mend his ways. On Sunday February 23rd 1755, just before the bells are rung for the second time, Gallus sends a message to the priest by the sexton, requesting that he should come and administer the sacrament to him.

The priest dons his vestments, but is prevented from going by a nosebleed. Gallus panics, falls on his knees before the altar, wounds himself in the throat. People rush forward; he is bound, admits that on account of the priest's repeated warnings he had resolved to kill him, and then do away with himself.

IV.iv.3.5 Joseph

A person I knew well was extremely libidinous. Although he finally got engaged and resolved to behave himself, he continued to fornicate; caught gonorrhoea which developed with cancer. After any amount of pain and worry he was eventually cured.

After a number of years he got married; continued to copulate with the girls. Admitted to having slept with more than a hundred females, most of them servant-girls.

His wife was getting on a bit, and was called away on an urgent journey for a period of three months. He fornicates, eventually shows signs of incipient gonorrhoea. Before the disease has the chance to develop, however, he undergoes a rapid cure through the administering of mercury. He has some excuse, since his wife was away longer than she had said she would be.

Finally, his wife sets out on an eight-day journey but stays away for three weeks. In order to get his own back, he sleeps with a married woman, catches gonorrhoea badly and is in danger of his life.

His conscience troubles him. He realizes that God is not only right to have punished him for all his flagrancies, but that He ought also to have brought his life to an end, that the righteousness of the Almighty demands that he should die. He finds it difficult to believe that God's grace can prevail over His righteousness. He did become well again, however, and although God in His infinite mercy spared him, he was troubled in spirit from then on.

IV.iv.4.1

Virgil

To what will the hearts of men not be driven by the cursed lust for gold?
The secret wound still rankles in the heart.
May sword flash against sword among their children's children.
Since fortune constrains, let us follow,
And steer our course whither she calls.
Unhappy soul, what wild madness has possessed you?
Unhappy race, what destruction has fortune reserved for you?
Let happen what will, any fortune can be mastered by bearing it.
Cease to hope that prayer may change the will of the gods.
Each, ere long, will face his own fate.
The human mind knows neither fate nor the future.
What is the fortune that disturbs your calm,
Invoking once more the wars of the past?
Many an ill has been repaired by time, and by the shifting toil
Of the changing years; the fickleness of fortune has mocked many a man
Before setting him on the firm ground again.

IV.iv.4.2.

Be bold and valiant in time of stress.
Be wise and take in sail when the wind sets fair.

Though the harshness of winter is Jupiter's gift, so too is the spring;
Though we are faring ill today, things are bound to change.

God plays about at pleasure with all that we do. Seneca.

IV.iv.4.3

BLESSING

He will be with them in trouble!
suffer them to be honoured!
satisfy them with long life!
shew them his salvation.

IV.iv.4.4

Raise me aloft
from the dust to Thee
and let me see
in a moment of time
how the wide world turns,
reveal the prime cause
of all that happens
so strangely here.

The rich are laden
with still more wealth
while the poor see
food become scarce
and their goods dispersed.

IV.iv.4.5

I have built my house
on the Creator's will
and sleep secure.

IV.iv.4.6

Genesis 50: 19, 20: Fear not, for I am under God. You thought evil against me, but God has turned all to good.

IV.iv.4.7

Thanks be to Thee, great and almighty God,
For all the goodness shown me in the world.

THE CORRELATION OF THE TEXT

Classification	*Title*	*Petry*	*Uppsala ms.*	*Malme-ström*	*Lepenies*
	Title page	83	1	33	43
I.i	My only son	86	2b	35	45
I.ii	*Ecclesiastes* 9: 2: There is one event	87	1b	36	46
I.iii	I have often wondered	87	1b	36	46
I.iv	There is nothing so sublime	87	32b	50	61
I.v	Thou mighty Ruler	87	1b	37	63
II.i	*What is life?*	89	12b	58	71
II.i.1.1	It is an axiom	90	22b	94	115
II.i.1.2	Dwell on what is your own	90	42b	101	124
II.i.1.3	*Rutting*	90	57	72	88
II.i.1.4	Where the free child of the air	91	5b	44	54
II.i.1.5	What is life?	91	12b	58	71
II.i.1.6	I conceive of man	91	13	59	72
II.i.1.7	Voigtländer	91	78	194	233
II.i.1.8	Friesendorff	92	142	134	162
II.i.2.1	*Vanity*	92	41	105	128
II.i.2.2	Rudbeck	92	114	176	212
II.i.2.3	Netherwood	93	101	166	200
II.i.3.1	*Death*	93	55	77	94
II.i.3.2	*Passing on*	95	54	79	96
II.i.3.3	*Seneca*. Be assured	97	50b	85	103
II.i.3.4	*Wisdom of Solomon* 2: 2: The breath	97	50b	85	103
II.i.3.5	Yxkull	98	88	199	238
II.ii	*The self within me*	99	12b	58	71
II.ii.1	What is the God	100	12b	58	71
II.ii.2	Madam N.N.	100	173	165	200
II.ii.3	Lewenhaupt	100	201	157	190
II.ii.4	Ostermann	101	102	170	205
II.ii.5	Hamilton	101	192	140	169
II.iii	*Going out into the dark*	102	53	99	120
II.iii.1	*Haunting*	103	53	99	120
II.iii.1.1	Haunting is mentioned	103	53	99	121
II.iii.1.2	According to Holy Scripture	103	53	99	121
II.iii.1.3	Maja Hierpe	103	53	99	121
II.iii.1.4	Caesar's soothsayer	103	53b	99	121
II.iii.1.5	On the day when my mother died	104	53b	99	122
II.iii.1.6	But why do hauntings take place	104	53b	100	122
II.iii.2	*Shades*	104	51	75	91
II.iii.2.1	My room	104	51	75	91
II.iii.2.2	The evening before	104	51	75	91

Classification	*Title*	*Petry*	*Uppsala ms.*	*Malme-ström*	*Lepenies*
II.iii.2.3	In 1728 I was lodging	105	51	76	92
II.iii.2.4	At twelve o'clock at night	105	52	76	93
II.iii.2.5	Professor Dahlman's daughter	105	52	76	93
II.iii.3	*Spellbinding*	105	15	65	79
II.iii.3.1	It is the custom	105	15	65	79
II.iii.3.2	An executioner, incognito	106	16b	46	57
II.iii.3.3	At Uppsala	106	16b	47	57
II.iii.3.4	An executioner in Trondheim	106	16b	47	58
II.iii.3.5	The chancellor of Denmark	106	16b	47	58
II.iii.3.6	A German	107	8	47	59
II.iii.3.7	A country lad	107	15	65	80
II.iv	*You too are here*	108	24	94	115
II.iv.1	*Fortune-telling*	109	44	97	119
II.iv.1.1	A woman	109	44	97	119
II.iv.1.2	My brother Samuel	109	44	98	119
II.iv.1.3	In Gävle	109	44	98	120
II.iv.1.4	The astrologer in Rome	109	44b	98	120
II.iv.1.5	An astrologer told Domitian	110	44b	98	120
II.iv.1.6	Boatman	110	156	122	149
II.iv.1.7	Tiliander	110	93	192	230
II.iv.2	*Portent*	110	46	91	111
II.iv.2.1	It is generally said	110	46	91	111
II.iv.2.2	Laodamia	111	48	91	111
II.iv.2.3	Löfling	111	48	92	112
II.iv.2.4	Colonel Freidenfelt	111	48	92	112
II.iv.2.5	A student, Peldan	111	48	92	112
II.iv.2.6	On the very day	111	21	92	112
II.iv.2.7	Princess Elizabeth's revolution	112	21	92	112
II.iv.2.8	Antonius Saturninus	112	21b, 23b	93	113
II.iv.2.9	Dean Risell	112	22	93	114
II.iv.3.1	Everyone is supposed	113	22b	94	114
II.iv.3.2	How is it	113	22b	94	115
II.iv.3.3	It is an axiom	113	22b	94	115
II.iv.3.4	Caesar	113	167, 24	123, 94	149, 115
II.iv.3.5	General Carl Cronstedt	114	24	95	115
II.iv.3.6	Some country people	114	16	45	56
III.i	*The sapping of affection*	117	10b	61	74
III.i.1.1	Joseph	118	175	147	178
III.i.1.2	Myhrman	118	103	165	199
III.i.1.3	*Onanists*	119	172	89	109
III.i.1.4	Brahe	119	127, 127b	117	142
III.i.1.5	Schmidt	120	120	180	216
III.i.1.6	Backman	120	95	110	134
III.i.2.1	*Incest*	121	62	68	82

Classification	*Title*	*Petry*	*Uppsala ms.*	*Malme-ström*	*Lepenies*
III.i.2.2	Lagerbladh	121	184	156	188
III.i.2.3	The vow of chastity	122	154	114	139
III.i.2.4	Christina Juliana Thun	122	77	191	229
III.i.2.5	Welin	123	82	196	236
III.i.2.6	The gardening man	123	90	192	230
III.i.3.1	*Adultery*	123	10, 62b	60, 71	73, 87
III.i.3.2	Sjöblad	124	40b	57	69
III.i.3.3	Dörnberg	125	136	130	158
III.i.3.4	Meldercreutz	125	194	162	196
III.i.3.5	Hökerstedt	125	170	144	174
III.i.3.6	Bergqvist	126	180	112	137
III.i.3.7	Controller	126	180, 106	127	154
III.i.3.8	Canutius	127	166	151	183
III.i.3.9	Wertmüller	127	83	197	236
III.i.3.10	Nietzel	127	99	166	201
III.i.3.11	Urlander	128	81	194	233
III.i.3.12	Psilanderhielm	128	108	173	208
III.i.3.13	Jan Jansson	128	176	147	178
III.i.3.14	A Brålanda farmer	129	124	121	147
III.ii	*Thy children should seek to thee*	130	12b	58	70
III.ii.1.1	O thrice and four times blest	131	12, 12b	57	70
III.ii.1.2	Ribbing	131	113	174	210
III.ii.1.3	Solander	132	71, 152	184, 214	221
III.ii.1.4	Lindberg	132	183	156	189
III.ii.1.5	Såganäs	133	92	190	227
III.ii.1.6	Karl Jansson	133	165	124	151
III.ii.2.1	*Wife*	133	11	103	126
III.ii.2.2	Ihre	135	187	145	175
III.ii.2.3	Von der Lieth	135	199	159	192
III.ii.2.4	Wallerius	136	85, 85b	196	235
III.ii.2.5	An Uppland farmhand	136	91	194	232
III.ii.3.1	*Parents*	137	60b	90	109
III.ii.3.2	A Parisienne	137	171, 171b	171	205
III.ii.3.3	Sånnaböke	138	118	179	215
III.ii.3.4	Leijel	138	200	158	191
III.ii.4.1	*Retribution*	138	28	88	107
III.ii.4.2	Krabbe	139	186	154	186
III.ii.4.3	Appelbom	140	96	109	133
III.ii.4.4	Wrangel	140	87	198	238
III.ii.4.5	Morga	140	195	163	197
III.iii	*The right common to all*	141	56b	75	91
III.iii.1.1	*Seneca*. Customary submission	142	39	55	68
III.iii.1.2	To the unfortunate	142	5b	44	54
III.iii.1.3	Though all can be swept	142	37	53	65

Classification	*Title*	*Petry*	*Uppsala ms.*	*Malme-ström*	*Lepenies*
III.iii.1.4	The mineworker	142	140	138	167
III.iii.1.5	A Scanian farmhand	142	178	182	219
III.iii.1.6	Strutz	143	73, 117b	188	226
III.iii.1.7	Per Adlerfelt	143	158	106	129
III.iii.1.8	Planting	143	107	173	208
III.iii.1.9	Wallrave	143	86	195	234
III.iii.1.10	Tavaste-bo	144	177	190	228
III.iii.1.11	Cronhielm	144	162	127	155
III.iii.1.12	Tordenskjold	145	76	192	230
III.iii.2.1	*Greed*	145	19, 19b	62	76
III.iii.2.2	Broberg	145	155, 155b	119	145
III.iii.2.3	Kyronius	146	185, 185b	154	187
III.iii.2.4	Hauswolff	147	191	141	170
III.iii.2.5	Odhelius	148	111	170	205
III.iii.2.6	Schleicher	148	66	183	220
III.iii.3.1	*Envy*	148	61b	70	85
III.iii.3.2	Bréant	148	125	119	145
III.iii.3.3	Westrin	149	84	198	237
III.iii.3.4	Krabbe	149	186	154	186
III.iii.4.1	*Ingratitude*	149	60, 69	69, 126	84, 153
III.iii.4.2	Cicero	150	69	126	152
III.iii.4.3	Hallman	150	196	140	169
III.iii.4.4	Engberg	151	138	132	160
III.iii.5.1	*Malice*	151	56	73	89
III.iii.5.2	Mrs. Dahlman	152	134	129	156
III.iii.5.3	The farmhand at Jönköping	152	174	150	181
III.iii.5.4	Erik Grubbe	153	144	137	166
III.iii.5.5	Dahlberg	154	133	128	155
III.iii.5.6	King Christian II	154	68	125	152
III.iii.6.1	*Friendship*	154	20	62	75
III.iii.6.2	There is the tale	155	69	126	153
III.iii.6.3	Sedlin	155	121, 121b	181	217
III.iii.6.4	Uggla	156	79	193	231
III.iii.6.5	A Swedish captian	156	192	189	227
III.iii.7.1	*Diligence*	156	61	69	83
III.iii.7.2	Maja Hierpe	157	182	142	171
III.iii.7.3	Grizell	157	193	137	165
III.iii.8.1	*Law*	158	58	72	87
III.iii.8.2	The more ardent the moralist	158	40	56	69
III.iv	*Distributing wealth*	159	2	39	49
III.iv.1.1	Fortune stands accused	160	2	38	48
III.iv.1.2	Plutus, accused before Jupiter	160	2	39	49
III.iv.1.3	The coronations of kings	160	2	39	49
III.iv.2.1	*Poverty*	160	45	90	110

Classification	*Title*	*Petry*	*Uppsala ms.*	*Malme-ström*	*Lepenies*
III.iv.2.2	The voice of Abel's blood	161	56	74	90
III.iv.2.3	Sohlberg	161	67	183	220
III.iv.2.4	Funck	161	143	134	163
III.iv.2.5	Do not regard	161	41	105	128
III.iv.3.1	*Riches*	161	17, 17b	64	78
III.iv.3.2	Anders Jöranson	163	67	107	130
III.iv.3.3	Håkansson	163	169	144	174
III.iv.3.4	Broman	163	123	120	146
III.iv.3.5	Charles XI	164	192	124	151
III.iv.3.6	Meldercreutz	164	194	162	196
III.iv.3.7	Julinschöld	164	164, 164b	148	179
III.iv.3.8	Hultman	166	188	143	173
III.iv.3.9	A Danish sheriff	166	179	130	157
III.iv.3.10	Norrelius	166	97	168	203
III.iv.3.11	Brauner	167	128	118	144
III.iv.3.12	Gyllenkrok	167	150	139	168
III.iv.3.13	Nordenflycht	168	100, 100b	167	201
III.iv.3.14	Råfelt	169	116	178	214
III.iv.3.15	Cederhielm	169	160	125	151
III.iv.3.16	Wallerius	170	85, 85b	196	235
III.iv.3.17	Stade	170	74, 74b	185	222
III.iv.3.18	Blackwell	172	131, 131b	115	140
III.iv.3.19	Only the good fortune	173	12	57	70
IV.i	*The Nations*	175	31	51	63
IV.i.1.1	*Job* 12: 23: He enlargeth	176	31	51	63
IV.i.1.2	It is to this that discord	176	40	56	69
IV.i.1.3	A traveller	176	157	174	209
IV.i.1.4	Stobée	176	72, 72b	186	224
IV.i.2.1	*Parliament*	178	25	175	211
IV.i.2.2	Herkepaeus	178	190	141	171
IV.i.2.3	Skeckta	178	119	179	216
IV.i.2.4	Springer	179	65	184	221
IV.i.2.5	Brahe	179	127, 127b	117	142
IV.i.2.6	Rogberg	180	115	176	211
IV.i.2.7	Renhorn	180	112	173	208
IV.i.2.8	Pechlin	181	110	172	207
IV.i.2.9	Gylling	181	149	140	169
IV.i.2.10	Antonsson	181	159	108	132
IV.i.2.11	General Rudbeck	181	117	176	212
IV.i.3.1	Slagheck	182	68	183	220
IV.i.3.2	Jöran Persson	182	109	151	182
IV.i.3.3	Griffenfeldt	182	145	136	165
IV.i.3.4	Görtz	183	148	135	164
IV.i.3.5	Münnich	183	102	164	199

Classification	*Title*	*Petry*	*Uppsala ms.*	*Malme-ström*	*Lepenies*
IV.i.3.6	Ekeblad	183	135	131	158
IV.i.3.7	Horn	184	189	142	172
IV.i.4.1	Marlborough	184	105	160	193
IV.i.4.2	Byng	184	129	113	137
IV.i.4.3	Buddenbrock	185	122	122	148
IV.i.4.4	Klingspor	185	202	153	185
IV.i.5.1	The coronations of kings	185	2	39	49
IV.i.5.2	*Judges* 1: 6, 7: Judah defeated	185	38	54	66
IV.i.5.3	*Judges* 9: 52–57: Abimelech	186	8	48	59
IV.i.5.4	*Emperors*	186	23, 23b	145	176
IV.i.5.5	Caesar	187	167	123	149
IV.i.5.6	The emperor Justinian	188	38	53	66
IV.i.5.7	Birger	188	154, 153	114	138
IV.i.5.8	King Gustavus	189	147	138	167
IV.i.5.9	*Daniel* 5: 1: King Belshazzar	189	147	139	167
IV.i.5.10	Adolf Frederick, king of Sweden	189	49	106	129
IV.i.5.11	Ziegler	190	89	199	239
IV.i.5.12	The Stuarts	190	75	188	266
IV.i.5.13	Louis XIV, the great	190	171b	159	192
IV.i.5.14	Peter I, emperor of the Russians	191	47	172	207
IV.i.5.15	Elizabeth	191	137	131	159
IV.i.5.16	Russians	192	43, 43b	177	213
IV.i.5.17	The king of Prussia	193	7, 7b	44	55
IV.ii	*The Churches*	194	6	41	51
IV.ii.1.1	*Catholicism*	195	6, 4	41	51
IV.ii.1.2	In Lisbon	196	56b	74	90
IV.ii.1.3	Damiens	197	163	129	157
IV.ii.1.4	Lord lieutenant Lillienberg	197	40b	57	69
IV.ii.2.1	The authorized copper mines	197	9	48	59
IV.ii.2.2	Controller	197	180, 106	127	154
IV.ii.2.3	Melander	198	197	162	195
IV.ii.2.4	Mathesius	198	198	161	194
IV.ii.2.5	Ullén	199	80	193	231
IV.ii.2.6	Annerstedt	199	152, 152b	107	130
IV.ii.2.7	Florinus	200	141, 141b	133	161
IV.ii.2.8	Malung and Kvikkjokk	201	181	160	193
IV.ii.2.9	Muræus	201	104	164	198
IV.ii.2.10	Boëthius	202	126	117	142
IV.ii.2.11	Kihlmark	202	130	113	137
IV.ii.2.12	Fahlander	203	139	132	160
IV.ii.2.13	Serenius	203	168	182	218
IV.ii.2.14	Benzelia	204	132, 94	111	135
IV.iii	*Live irreproachably*	205	31	51	63
IV.iii.1.1	Live irreproachably	206	26b, 31	51, 86	63

Classification	*Title*	*Petry*	*Uppsala ms.*	*Malme-ström*	*Lepenies*
IV.iii.1.2	Though you honour the age	206	3b	40	50
IV.iii.1.3	*Naturalism*	207	50, 50b	84	102
IV.iii.2.1	*Pride*	207	42, 42b	100	122
IV.iii.2.2	Beware of sins writ large	209	5	43	54
IV.iii.2.3	When fortune is granted	209	27b	97	118
IV.iii.2.4	*Job* 15: 8: Hast thou heard	209	31b	52	64
IV.iii.2.5	Klingspor	210	202	153	185
IV.iii.2.6	Kierman	210	64, 64b	152	183
IV.iii.2.7	Buskagrius	212	70	122	148
IV.iii.2.8	Sinclair	212	161	126	153
IV.iii.3.1	To be king is not to fear	212	37	52	64
IV.iii.3.2	*Gladness*	213	59, 59b	70	86
IV.iii.3.3	When I contemplated revenge	213	28	89	108
IV.iii.3.4	*Genesis* 50: 19: Fear not	214	13b	59	72
IV.iii.3.5	O for the grace	214	3	37	47
IV.iv	*The Judgement of God*	216	14	66	80
IV.iv.1.1	O mighty Ruler	217	31	51	47
IV.iv.1.2	*Ecclesiastes* 9: 2: There is one	217	31	51	63
IV.iv.1.3	*Fate*	217	14	66	80
IV.iv.1.4	It is often a long time	217	29b, 32, 32b, 33b	49	60
IV.iv.1.5	I have often wondered	218	31	51	63
IV.iv.1.6	Boatman	219	156	122	149
IV.iv.1.7	A Norwegian	219	98	169	204
IV.iv.2.1	*Fortune*	220	63, 27b	67, 96	82, 118
IV.iv.2.2	A rabbinical story	221	27	95	116
IV.iv.2.3	Asp	222	151	110	133
IV.iv.3.1	*Nemesis Divina*	222	26, 1b	86, 36	104, 46
IV.iv.3.2	*Psalms* 119: 137: Righteous art thou	222	26, 26b	86	104
IV.iv.3.3	*Nemesis*	223	28	88	107
IV.iv.3.4	Gallus	224	146	135	163
IV.iv.3.5	Joseph	224	175	147	178
IV.iv.4.1	Virgil	225	39	54	66
IV.iv.4.2	Be bold and valiant	225	40	56	69
IV.iv.4.3	*Blessing*	226	18	63	77
IV.iv.4.4	Raise me aloft	226	31	51	63
IV.iv.4.5	I've built my house	226	5b	44	54
IV.iv.4.6	*Genesis* 50: 19: Fear not	226	5b	44	54
IV.iv.4.7	Thanks be to Thee	226	203	199	239

PART THREE

NOTES AND APPENDICES

NOTES

Title-page:

Uppsala ms. 1. Latin. The lay-out indicates that Linnaeus had thought of the collection as a publishable book – probably as a result of having read Friess, F.C., 1763.

Nemesis: see Bacon, F., 1609, no. 22; Zedler, J.H., 1732/1754, 23: 1689–1692; Hederich, B., 1770, 1701–1707.

Talion: the definition is taken from Friess, F.C., 1763, §1, 1758 ed. 7–8, who took it from Ravanel, P., 1660/1663, 2: 1236–1237; cf. *Luke* 6: 38; Kittel, G., 1933/1979, 4: 637.

Autopathy: also taken from Friess; cf. *Polybios-Lexikon* 1: 261; Migne, J.-P., *Patrologiæ Græco-Latine* II: 440, sect. 389, line 5.

Live irreproachably: Ovid *Ars amatoria* I: 640; also inscribed over the door of Linnaeus' bedroom at Hammarby; cf. IV.iii.1.1.

I.i:

Uppsala ms. 2b. Swedish and Latin. 'Author' (werden) in line two, might be more accurately translated as 'Ward', recalling the opening line of Caedmon's hymn. On the last line of stanza three, which is in Latin, see Walther, H., 1963/1986, no. 8952.

Carl von Linné the younger (1741–1783) was Linnaeus' eldest child and only son. Despite being matriculated in the medical faculty at the age of nine, becoming a senior member of the university Småland association at the age of twelve, titular professor at the age of eighteen, full professor at the age of thirty-six, he never really attempted to follow in his father's footsteps, either personally or professionally. He was well-known as a ladies' man, an acquaintance observing in 1770 that "the young gentleman amuses himself for the most part by enquiring after the nymphs rather than Flora. He holds his head high, dresses and powders himself in a fashionable manner, and is generally to be found in the company of pretty women". When he was congratulated on being appointed his father's successor at the university, he retorted that he would prefer to be anything else, even a soldier.

He was bullied by his mother, who used to box his ears, even when he was a grown man. Linnaeus almost certainly began to collect material for the *Nemesis Divina* long before he realized that it might be useful as instruction for his son. This dedication seems to have been drawn up when he copied out his original collection of material during the autumn of 1765.

SBL 23: 715–718; Blunt, W., 1984, 175; *Proverbs* 5–7, *Micah* 7: 5.

I.ii:

Uppsala ms. 1b. Swedish. Quoted without the Biblical reference. Cf. IV.iv.1.2.

I.iii:

Uppsala ms. 1b. Latin. Cf. IV.iv.1.5. Claudius Claudianus (c. 365–c. 408) , the last of the Roman poets, published his epic on the downfall of Rufinus, the unworthy minister of Arcadius at Constantinople, in 396. In the opening section of the work, from which this quotation is taken, Claudianus contrasts the divine regularities of the world of nature with "the impenetrable mist which surrounds human affairs, the wicked being happy and long-prospering, the good without comfort". His treatment of the theme evidently influenced Tasso and Milton, and Bayle's discussion of it in the *Dictionary* article on Rufinus, by catching Leibniz's attention (*Theodicy* I §16, II §146), helped to set the scene for the wider eighteenth-century debate.

Rufinus was eventually assassinated and dismembered by his own troops, in the presence of the emperor, on a parade-ground outside Constantinople in 395.

I.iv:

London ms. 12; Uppsala ms. 32b; cf. Uppsala ms. 26b, IV.iv.3.2. Latin. Ovid *Tristia* IV.viii.47.

I.v:

Uppsala ms. 1b. Swedish. Cf. London ms. 34, Uppsala ms. 31, IV.iv.1.1, where the wording, punctuation and spelling are slightly different. Unidentified, but not likely to be by Linnaeus.

II.i:

Cf. II.i.1.5.

II.i.1.1:

Uppsala ms. 22b. Latin and Swedish.

According to Aristotle, "all things that move are moved by something else", the ultimate origin of all movements being an eternal mover which is itself unmoved: *Physics* 258b. For his discussion of theories concerning the way in which living being initiates motion, see *On the Soul* 404–405.

II.i.1.2:

Uppsala ms. 42b. Latin; cf. IV.iii.2.1. The Malmeström–Fredbärj edition (p. 101) omits the first line; cf. appendix C, section 1; Linné, C. von, 1961, 13.

"Man is but a bubble" is a Greek proverb; cf. the opening paragraph of Jeremy Taylor (1613–1667) *The Rule and Exercises of Holy Dying* (London, 1651).

Gustaf Kierman (1702–1766), mayor of Stockholm, see IV.iii.2.6; the sale of his ill-gotten gains was much publicized after his death: Brolin, P.-E., 1953, 405.

II.i.1.3:

Uppsala ms. 57. Latin. There is a large space between paragraphs two and three, and signs of further attempts at paragraphing in the final section.

Linnaeus deals most extensively with subject matter of this kind in *Diæta Naturalis* (1733) and *Lachesis Naturalis* (1907): its relevance to the central theme of the *Nemesis Divina* is indicated in the *System of Nature* 1766/1768[12] I: 28 – the footnote on physiology.

Tycho Brahe (1546–1601) co-habited for years with a commoner from Scania by the name of Kirstine Barbara (d. 1604), who bore him three sons and five daughters; a fact which may have been associated in Linnaeus' mind with the folk-belief that each year has thirty-two unlucky 'Tycho Brahe' days: Mogens Knudsen Hosum (1681–1745) *Inauspicatus dies Jovis, auspice Jova illustratus* (Hafniæ, 1706) 3.

The reference to Newton is even harder to account for, unless one takes into account the reputation of his niece Catherine Barton (1679–1740): "I thought in my youth that Newton made his fortune by his merit. I supposed that the court and the city of London named him Master of the Mint by acclamation. No such thing. Isaac Newton had a very charming niece, who made a conquest of the minister Halifax. Fluxions and gravitation would have been of no use without a pretty niece": Voltaire *Dictionnaire philosophique* (1757). Cf. Needham, G.B., 'Mrs. Manley: an eighteenth-century Wife of Bath', *The Huntington Library Quarterly* XIV (1951) 259–284.

II.i.1.4:

Uppsala ms. 5b. Swedish. The handwriting appears to indicate that it was originally followed by IV.iv.4.5, and that III.iii.1.2 was inserted between at a later date.

Cf. Linnaeus' conception of the elements and of childhood: *Introduction* II.i; III.iii.1.

II.i.1.5:

Uppsala ms. 12b. Latin and Swedish. This follows III.ii.1.1 after a large gap.

It is evidently a reference to a passage in: 'Sur les vaines terreurs de la mort et les frayeurs d'une autre vie', an epistle in French verse by Frederick the Great (1712–1786), addressed to his field marshal James Keith (1696–1758):

> De l'avenir, cher Keith, jugeons par le passée, ...
> Non, rien n'est plus certain, soyons-en convaincus,
> Dès que nous finissons, notre ame est éclipsée.
> Elle est en tout semblable à la flamme élancée,
> Qui part du bois ardent dont elle se nourrit,
> Et dès qu'il tombe en cendre elle baisse & périt.

Oeuvres du Philosophe de Sans-Souci (Potzdam, 1760): 202–213, 'Épitre XVIII au Maréchal Keith'; (Potsdam & Amsterdam, 1760): 209–220; (Berlin, 1764) I: 224–236, 'Imitation du Troisieme Livre de Lucrece', with improved punctuation: 231.

II.i.1.6:

Uppsala ms. 13. Swedish.

The origin and elaboration of the simile is evidently Biblical: *Proverbs* 20: 27, *Luke* 11: 36; *Job* 29: 3; *Psalm* 18: 28, *Jeremiah* 25: 10; *Revelation* 22: 5; *Job* 21: 17, 18: 6, *Proverbs* 24: 20, *Revelation* 18: 23; cf. II.iii.1.5.

II.i.1.7:

London ms. 12, no. 11; Uppsala ms. 78.

Johan Eberhard Voigtländer (1698–1759) was member of a medical family long-established at Uppsala. The date of his posting to Pomerania must be erroneous, since in a letter written at Uppsala on September 10th 1759, Linnaeus notes that he has just died of an apoplexy: Annerstedt, C., II: 248, III: 309; *Letters* III: 73, no. 997.

Daniel Gustaf Cedercrona (1706–1752), public prosecutor for the House of Lords 1741, herald of arms 1743, assistant judge of the Göta court of appeal 1748; for his contact with Linnaeus, see Fries, Th.M., 1903, I: 271.

Nils Kyronius (c. 1700–1784), purveyor of a popular beer at Uppsala, see III.iii.2.3, Carl Michael Bellman (1740–1795), matriculated 1758, *Fredmans Sånger* (1791) no. 28.

Voigtländer's case-history has been placed here on account of the emphasis Linnaeus lays upon his purely, *animal* drives and sensations.

II.i.1.8:

Uppsala ms. 142. Swedish.

Greta von Friesendorff (1661–1727) died on March 24th, her sister Elisabeth (1664–1727) on April 2nd. They were buried together in the parish churchyard at Danmark on April 26th. It was over thirty years later that Linnaeus purchased the Hammarby estate, see his letter of December 22nd 1758, *Letters* V: 57, no. 982, and Fries, Th.M., 1903, II: 379.

The main point of the case-history seems to be the bringing of human differences under the universal sway of mortality.

II.i.2.1:

Uppsala ms. 41, 41b. Swedish, apart from the final sentence in Latin, which is the only entry on the second page.

The same theme is dealt with in the *Lachesis Naturalis* 1907, 242: "There is the story of Charles XI's having asked a certain huntsman's boy what his wages were. 'All I get is my clothing and my keep', was the reply. 'We have nothing more', said the king. And it was a very great truth, for our food and our clothing are all we actually possess. There was a gentleman who spent a great deal of money on his house and paintings etc. When all was finished, he said that although it had cost him a pretty penny, he got a lot of pleasure out of it. On hearing this his servant also looked around the house, and said that although it had cost him nothing, it also gave him great pleasure".

II.i.2.2:

Uppsala ms. 114. Swedish.

Olof Rudbeck the younger (1660–1740) began to teach medicine and botany at Uppsala in 1692, and ten years later succeeded his illustrious father as professor. He was helpful to Linnaeus when he first arrived as a student, giving him lodging in his home and engaging him as tutor to his son Olof. Linnaeus eventually took over part of his university teaching

Rudbeck had twenty-four children – three sons, including Olof, by his third wife Charlotta Rothenberg (1691–1761) , who was the daughter of a gamekeeper: Fries, Th.M., 1903, I: 54, 64.

Linnaeus therefore knew of this case from first-hand experience. It is placed here on account of the elemental nature of the wife's ill-humour and the excruciating physical pain with which it was requited.

II.i.2.3:

London ms. 11, no. 3; Uppsala ms. 101. Swedish. Linnaeus spells the name 'Netterword', which is very close to the present-day official Swedish pronunciation of it: H.C. Claës Uggla 'Om rätt uttal av adliga namn', *Arte et Marte* 50 (1996) 3–8; 52 (1998) 21.

Johan Vilhelm Netherwood (1699–1749) came of a Scottish family which had first settled in Småland during the late sixteenth century. Following in the footsteps of so many of his forebears, and more particularly of his father Vilhelm Netherwood (1664–1735), who like Linnaeus had been educated at Växjö School, he began his career in the army, becoming a sergeant in the Kronoberg regiment in 1717, taking part in the siege of Fredrikshald in 1718, rising to the rank of full lieutenant in 1721, but quitting the service in 1724: Hyltén-Cavallius, G., 1897, 127; Elgenstierna, G.M., 1925/1931 no. 446, 417–421.

His promotion to the rank of full lieutenant was due in part to his having married well earlier the same year – Maria Wickman of Växjö (d. 1743), daughter of Johan Wickman, receiver of the revenues for the district of Kronoberg, by his first wife Catharina Wient. It was not Netherwood's father, therefore, as Linnaeus thought, but his father-in-law, who got into trouble for failing to balance the books. Johan Wickman recommended him for a position in the Växjö tax-office soon after he left the army on June 17th 1724, relinquished his post to him in 1727, and more than a decade later was still involved in legal wrangling in Stockholm: Kammarkollegiet records Oct.–Dec. 1725; Dec. 5th 1738.

Between 1722 and 1742 Netherwood's wife bore him thirteen children, eight of whom were still alive when he committed suicide in Stockholm on June 4th 1749. In 1757 the eighth of these children, Vilhelm Netherwood (1732–1809), married Anna Catharina Rothman (d. 1777), the daughter of Linnaeus' revered teacher Johan Stensson Rothman (1684–1763), and subsequently carved out a distinguished banking-career for himself in Växjö and Stockholm: II.ii.2; Anrep III: 25–28; Hedlund, E., 1936, 69. It was almost certainly through this network of acquaintances and family connections that Linnaeus gathered his knowledge of the case.

It has been placed here primarily on account of Netherwood's having chosen death through his own free will.

II.i.3.1:

Uppsala ms. 55, 55b. Latin. All these passages, except nos. 1, 9, 21, 25 are also to be found in the *Lachesis Naturalis* 1907, 5–10 – although not always in precisely the same form, and without the numbering. Together with those in II.i.3.2, they are grouped there under the general headings of Life and Death.

This re-grouping and numbering of the passages reflects a restructuring and a clarification of Linnaeus' thinking. The passages in this section are concerned mainly with the way in which fate or nature brings about the death of the individual, and they therefore find their logical conclusion in no. 31.

Since the great majority of the passages in the two sections are taken from works by Seneca, it has been found convenient to use the following abbreviations in identifying them: Ad Lucil. *Ad Lucilium epistolæ morales*, Ad M. *De Consolatione ad Marciam*, N.Q. *Naturales Quæstiones*.

1. *De Providentia* V.7.1; 2. Ad M. XXII.3.10; 3. Ad Lucil. XLIX.6.8, Ad. M. X.4.7; 4. Ad M. XI.1.1; 5. Ad M. XIX.5.1; 6. Ad M. XX.1.1; 7. Ad M. XXI.1.5; 8. Ad M. XXI.1.9; 9. Ad M. XXI.6.4; 10. Ad. M. XXIII.3.6; 11. N.Q. II.35.2; 12. N.Q. II.59.6; 13. N.Q. III præf. 12; 14. Ad Lucil. XCI.21.7; 15. N.Q. III.27.2.2; 16. N.Q. VI.1.8.6, VI.1.9.8; 17. N.Q. VI.1.10; 18. N.Q. VI.2.6.9; 19. N.Q. VI.2.7.4, VI.32.4; 20. N.Q. VI.32.12; 21. *Ludus de morte Claudii* XI.6.4, Catullus III.12; 22. Virgil *Aeneid* IV.653, rest unidentified; 23. Ad Lucil. CXXI.18.9; 24. Ad Lucil. LXIX.6.6; 25. Ad Lucil. LXX.5.8; 26. Ad Lucil. I.2.3; 27. Ad M. XVIII.8.2; 28. Ad M. XVIII.8.6; 29. Ad M. XX.21.9; 30. *Hercules Oetæus* 122; 31. *De Beneficiis* II.31.3, Ad Lucil. XI.85.37.

II.i.3.2:

Uppsala ms. 54, 54b. Latin, with some Swedish (nos. 35, 46, 48). All these passages, except nos. 40, 46, 50, are also to be found in the *Lachesis Naturalis*.

They are concerned mainly with the way in which the individual regards the fate or the natural course of things to which all are subject. No. 35 indicates that it was the events in Uppsala in 1765/1766 which were the occasion of Linnaeus' drawing up this classical commentary on the inter-relationship between physiology, diet and pathology: *Letters* V: 143–145, nos. 1064, 1065; *System of Nature* 1766/1768[12] I: 28.

32. Ad M. XXI.5.1; 33. *Ecclesiasticus* 9: 28; 34. Ad. M. XXI.6.4, Virgil *Aeneid* X.472; 35. *2 Chronicles* 34: 28; 36. *Ad Polybium De Consolatione* 1.7; 37. Ad. M. XI.5.3, Virgil *Aeneid* VI.714–715; 38. Ad Lucil. LXXVII.12.6, Virgil *Aeneid* VI.376; 39. Ad Lucil. LXXVII.11.3; 40. Ad Lucil. LXXVII.11.7, Ad M. XXI.7.4; 41. Ad Lucil. XXX.10.3; 42. Ad Lucil. XXX.10.6; 43. Ad M. XV.4.6; 44. N.Q. II.59.7.6; 45. Ad M. XVII.1.5; 46. N.Q. I præf. 4.8; 47. N.Q. III.27.2.5; 48. N.Q. VI.32.1.5, Ad M. XX.4.2, N.Q. VI.32.12.4; 49. N.Q. VI.1.8.6, VI.1.9.1, VI.1.9.7, in the *Lachesis* also, Linnaeus changes Seneca's 'tumultus' into 'tumulus'; 50. N.Q. VI.1.10.1; 51. N.Q. VI.2.1.4, Virgil *Aeneid* II.354; 52. N.Q. VI.2.3.5; 53. Ad Lucil. XXX.10.3; 54. N.Q. VI.2.7.5; 55. N.Q. VI.2.7.7.

II.i.3.3:

Uppsala ms. 50b. Latin. This is a garbled version of Seneca *De Consolatione ad Marciam* XIX.4.

II.i.3.4:

Uppsala ms. 50b. Swedish. Cf. II.i.2.1.

II.i.3.5:

Stockholm ms. 2 no. 2; Uppsala ms. 88. Swedish.

Johan Casimir Leijonhufvud (1583–1634) was member of a family which had risen to power during the sixteenth century and become notorious on account of the rapacity with which it administered its Finnish estates. His marriage with his cousin Sidonia Grip (1587–1652) of Vinäs took place in the castle at Nyköping on October 6th 1616. Towards the close of his career he was appointed governor of the estates of the widowed queen Eleonora (1599–1650), mother of queen Christina (1626–1689). Like so many of his family he made elaborate preparations for his interment, the final pomp and ceremony taking place in Uppsala cathedral nine months after his death, on May 16th 1635. The couple's epitaph is also to be found in Ed church in Småland. Linnaeus refers erroneously to another of the family's scions, Abraham Leijonhufvud (1583–1618), whose monument is in Lillkyrka church near Enköping.

Axel Didrik Meyendorff von Yxkull (1720–1798), military man, knight of the Order of the Sword, was related to the Leijonhufvuds through his paternal grandmother, his mother and his wife: Anrep II: 665, 886–887; SBL XXII: 464–474, 601–611.

The case-history has been placed here on account of Yxkull's sovereign indifference to death and the past.

II.ii:

Cf. II.ii.1.

II.ii.1:

Uppsala ms. 12b. Latin.

Descartes had reasoned from the self within him, by means of Anselm's ontological argument to the existence of God: *Meditations* (1641) no. 3.

The camera obscura was an instrument consisting of a darkened chamber or box, into which light was admitted through a double convex lens, so forming an image of external objects on a surface of paper or glass: Dellaporta, G.B., 1558, IV.ii.

II.ii.2:

Uppsala ms. 173. Swedish.

Linnaeus evidently heard this either from Johan Stensson Rothman (1684–1763), his teacher at Växjö, or from his son Jakob Gabriel Rothman (1721–1772), who studied under him at Uppsala, and when he first arrived in the city in 1741 lodged with him for a while: *Letters* I: 64, no. 30 (11.5.1753).

It has been placed here on account of Madam's showing scarcely any awareness of her maid's individuality.

II.ii.3:

Uppsala ms. 201, 201b, the last line being the only one on the second page. Linnaeus wrote "1748" in the penultimate line. Swedish.

Olof Johannson Dagström (c. 1678–c. 1760) fought with Charles XII at Poltava, and when he eventually returned to Sweden in 1719, he settled on his wife's estate in Scania. He was converted to pietism, and on May 23rd 1728 was arrested for having compiled a treatise declaring that Frederick I had usurped the throne. During his trial, his sanity was investigated by Kilian Stobaeus (1690–1742), professor of natural philosophy at Lund, from whom Linnaeus probably acquired his information. Dagström was eventually imprisoned for life in Malmöhus: SBL IX: 532–538.

Charles Emil Lewenhaupt (1691–1743), speaker of the Estates 1734, 1740/1741; given command of the Swedish land forces in Finland September 3rd 1741; executed August 4th 1743. Deemed to be lacking in commonsense by Johan Ihre (1707–1780), professor of rhetoric and politics at Uppsala: cf. III.iii.4.4.

The case is evidently meant to illustrate the shortcomings of the ways in which we assess others.

II.ii.4:

London ms. 15 no. 30, under the general heading of Nemesis; Uppsala ms. 102. Swedish.

Heinrich Joachim Friedrich Ostermann (1687–1747), the Russian statesman, was born at Bochum in Westphalia. He was matriculated at the University of Jena, not Halle, on September 9th 1702. During the following year, when drunk, he killed a fellow student who had laughed at his strange manner of dancing, by plunging his rapier into him. He fled to Holland, where he was recruited by Peter the Great's agent Cornelis Cruys (1657–1727). As a Russian civil servant, he distinguished himself by his self-control and soundness of judgement under the most trying and difficult of circumstances. He rose rapidly in the Czar's service, was highly successful in negotiating the Peace of Nystad with Sweden in 1721, and continued in high favour throughout the reigns of Anne and Ivan VI. Falling from power when Elizabeth, with the help of Sweden, became empress in the November of 1741, he was banished to Berezov in Siberia: *Westfälische Lebensbilder* VI: 37–59; IV.i.3.5.

Linnaeus may have heard of the dramatic conclusion to Ostermann's university career from Jacob Flachsenius (1683–1733), who took his master's degree at the University of Jena in 1711, and from 1723 until 1725, that is, soon after the signing of the Peace of Nystad, taught Linnaeus systematic theology at Växjö School; *Introduction* III.iii.3.

It was by distancing himself from his environment that Ostermann managed to control it, until his environment took its revenge upon him.

II.ii.5:

Uppsala ms. 192. Swedish.

Hugo Johan Hamilton (1668–9.1.1748) served in the French and Dutch armies, and throughout the whole of Charles XII's campaigning. He was in command of a number of cavalry regiments at Poltava, was captured, and did not return to Sweden until 1722. He married into the possession of extensive estates, mainly in the vicinity of Stockholm and Örebro. He was appointed field-marshal in 1734. In June 1743 he helped to deal with the Dalecarlian march on the capital, and it was probably then that he burnt down the guest-house at Rotebro, which is on the main road to the north-west of Stockholm: SBL XVIII: 94–96.

Linnaeus evidently regarded the burning as a completely irresponsible act of gratuitous destructiveness.

II.iii:

Cf. II.iii.1.1.

II.iii.1:

Uppsala ms. 53, 53b. Swedish. These six sections have been kept in the order in which Linnaeus noted them down. It looks from the handwriting as though the last was added somewhat later.

II.iii.1.1:

Scriptural evidence of haunting: *Job* 4: 15, 26: 4; *2 Maccabees* 12: 44; *Matthew* 10: 1; *Mark* 5: 2; *Luke* 24: 37.

Bacon, F., 1609, set himself the task of expounding the hereditary wisdom of the second age by interpreting its mythological fables in modern terms: *Introduction* III.iii.1.

In the *Lachesis Naturalis* 1907, 150, when discussing the external senses and dealing with opinion, persuasion and fantasy, Linnaeus notes that "children are intimidated by means of the bogey-man standing there at the window in the dark, into not going out at night, into behaving themselves. It is thus that they develop their fear of the dark, of ever going out into it: their heart thumps, they tremble, terrified that someone will grab hold of them. There is no point in rationalizing. I myself, until I was twenty, never dared to go out into the dark alone, and even now, against my better judgement, certain rooms give me the creeps".

II.iii.1.2:

Scriptural evidence of guardian angels: *Matthew* 18: 10; *Acts* 12: 15. On the central significance of Linnaeus' conception of them, see *Introduction* III.iii.4.

Although he is discussing the matter here in a very basic context, the reference at the end of this paragraph could be to Carl Carleson (1703–1761), the brother of his disciple Edvard Carleson (1704–1767), who between May 1730 and October 1731 published a journal devoted to the discussion of moral issues: *Sedolärende Mercurius*.

II.iii.1.3:

Maja Hierpe (d. 1773); cf. III.iii.7.2.

Linnaeus mentions this outbreak of fever in a letter to his friend Abraham Bäck (1713–1795), written at Uppsala on March 12th 1773: *Letters* V: 204, no. 1124.

Myresjö is near Vetlanda in Småland, some two hundred and sixty miles from Uppsala. Maja Hierpe returned to Uppsala infected. She was left in the hospital there, and Linnaeus and his family moved out to Hammarby.

II.iii.1.4:

Plutarch, *Life of Caesar*: "The same night, as he was in bed with his wife, the doors and windows of the room suddenly flew open. Disturbed both by the noise and the light, he observed, by moonshine, Calpurnia in a deep sleep, uttering broken words and inarticulate groans. She dreamed that she was weeping over him, as she held him, murdered, in her arms".

II.iii.1.5:

Linnaeus' autobiography, 1888b, 10: "On June 6th 1733, at six o'clock in the evening, my dearly loved and most devout mother passed away, in my absence, to my unutterable sorrow, anguish and loss". His mother, Christina Brodersonia (1688–1733), who came of a predominantly clerical family long-established in the Stenbrohult area, married his father there on March 6th 1706: Fries, Th.M., 1903, I, append. ii: 9; Skrede, P.O., 1987, 316–322.

His father, Nils Ingemarsson Linnaeus (1674–1748), born in Jonsboda in the parish of Vittaryd, some twenty-five miles to the north-west, became rector of Stenbrohult in 1708. Jacques Soberant (1676–1724), the French dancing-master, arrived at the vicarage at the beginning of December 1724 and died there a few days before Christmas. Linnaeus' father notes in the burial register that he was a native of 'Autram' or Autrey-lès-Gray in France, converted to Lutheranism in 1714, and "left behind him a fine reputation for honest living – godfearing, diligent and temperate": Virdestam, G., 1928, 29.

On corpse candles, see II.i.1.6; Sacheverell, W., 1702, 20; Martin, M., 1716, 313. Linnaeus' second daughter Sara Magdalena, not Helen, was born on September 8th 1744 and died fifteen days later: Fries, Th.M., 1903, I, append. ii: 10.

II.iii.1.6:

As one would expect from the whole tenor of Linnaeus' theodicy, it is not the supposedly objective influence of the stars themselves, but the immediate fact of their being viewed by man, which he takes to be the basic datum of any rational enquiry into the significance of astrology.

Ludvig Holberg (1684–1754), who devotes one of his epistles to a consideration of hauntings, reaches the following conclusion: "The most reasonable of the many opinions concerning these phenomena, is that which derives them from what are called *astral*-spirits, beings between the body and the soul, which Paracelsus regards as consisting of air and fire. Spirits of this kind are conceived of as being *sensuous souls*; although they are mortal like the body, they hover about it once they have left it, retaining the same *passions* as they had when they were united with it, and so giving rise to the *phænomena* we associate with burial grounds. I do not subscribe to this opinion, I simply say that it is the most reasonable". *Epistler* 1944/1954 IV: 218.

II.iii.2:

Uppsala ms. 51, 52. Swedish. These five sections have been kept in the order in which Linnaeus noted them down. From the lay-out and the handwriting, it looks as though the last was added as an afterthought.

II.iii.2.1:

The preceding section is concerned primarily with haunting in general, and Linnaeus states quite definitely that he has never seen anything. This section is concerned more specifically with what has been heard, and includes phenomena which did fall within his personal experience.

II.iii.2.2:

Linnaeus' father-in-law Johan Moræus (1672–1742) died on his estate in Falun, Dalecarlia, at three in the morning on November 29th 1742. He had a great admiration for him, and compiled a succinct but detailed biography: Fredbärj, T., 1962.

II.iii.2.3:

Linnaeus began his studies at Lund on August 17th 1727 and concluded them on June 28th 1728. While he was at the university there he lodged at 111 Storgatan, the home of Kilian Stobæus (1690–1742), professor of natural philosophy and medicine, from whom he also took private lessons.

Cf. Fries, Th.M., 1903, 1: 29: "Stobæus had taken upon himself the function of physician to the gentry of Scania, and was never free from being consulted by them. On one occasion he called up to Linnaeus and asked him to come down and write out a letter prescribing something for a certain illness, but on account of Linnaeus' untidy hand, what was written turned out to be useless".

II.iii.2.4:

This incident is also recorded in the London ms. 18, where it must have been noted down just prior to Linnaeus' having copied out the whole of the *Nemesis Divina* onto octavo sheets: *Introduction* I.i.

Carl Alexander Clerck (1709–1765) was a tax-inspector in Stockholm. With Linnaeus' help, and after battling for years with ill-health and lack of funds, he eventually saw through the press two outstanding entomological works: *Aranei Suecici* (Stockholm, 1757), and *Icones insectorum rariorum* (2 pts. Stockholm, 1759): Vries, Th.M., 1903, II: 247; *Letters* V: 265–307.

II.iii.2.5:

Lars Dahlman (1701–1764) was appointed professor of moral philosophy at Uppsala in 1748. On the general history of his family, see Wikström, L., 'Gammal adelsätt återinsatt i sina rättigheter', *Arte et Marte* 47 (1993) 19; cf. III.iii.5.2.

When attempting to date this occurrence, it should be remembered that Linnaeus' four surviving daughters were born in 1743, 1749, 1751 and 1757.

II.iii.3:

Uppsala ms. 15. Latin and Swedish. All seven of the following instances involve the exercizing of a positive influence – the general progression in their arrangement being from the physical handling of an animal to awareness of an inter-personal involvement.

II.iii.3.1:

Uppsala ms. 15. Latin.

II.iii.3.2:

Uppsala ms. 16b. Latin.

The inn at Diö, a village in the parish of Stenbrohult on the main road between central Sweden and Denmark, dated from the late sixteenth century, a royal decree of 1561 having given rise to the establishment of many such hostelries along the country's major highways. During the Great Northern War (1700–1720) it became the focal-point for army officers in an extremely busy transit area: during the January of 1710, for example, more than seven thousand troops passed through Diö on the way to active service: Stille, A.G.H., 1903; Virdestam, G., 1928, 16; *Stenbrohult sockens historia* (Kalmar, 1956) 253–255, published by the Stenbrohult local history society.

The proprietor of the inn was therefore a person of considerable local importance, and certainly the social equal of the rector of the parish. In 1721 the running of the establishment was taken over by Peter Bexell (1675–1724), a graduate of the University of Åbo who had previously been employed as the superintendant of a copper mine, the manager of a nobleman's estate and the supervisor of a tax office. In 1719 his daughter Christina Elisabet had married Jonas Törnqvist (1699–1746), an army orderly who was eventually (1730) promoted to the rank of sergeant. In 1727 this couple took the inn over from Christina's stepmother and her new husband Johan Svanstedt, and continued to run it together for almost twenty years. After her husband's death and prior to her re-marrying in 1750 Christina Törnqvist was the sole proprietor, and it is therefore quite likely that the incident recorded by Linnaeus occurred during this period. If the sentence was carried out locally, the execution took place at Alvesta just west of Växjö: Selling, O.H., 1960, 77, 78, 125.

There were many contacts between Linnaeus and his family and the proprietors of the inn. When Peter Bexell was buried early in April 1724, Linnaeus' father noted in the church book that "he loved both God and virtue, always sought to further orderliness, faced fortune and misfortune ever rejoicing in God's will". On February 7th 1726 Linnaeus stood godfather to the Törnqvists' daughter Dorotea Christina. When he visited his family on his way to the Netherlands in the spring of 1735, he saw a lot of the Törnqvists. On the Easter Monday, April 7th, the family was entertained at the rectory. On the Tuesday and the Wednesday the compliment was returned, the Linnaeus family being entertained by the Törnqvists at the inn in Diö. When the party finally left Stenbrohult for Hälsingborg on April 15th, Jonas Törnqvist accompanied them as far as Ryfors: Linné, C. von, 1929, 196–197.

On August 13th 1741 Linnaeus paid another visit to the inn at Diö, noting in his journal that it "had a blast-furnace in which ore dredged from Möckeln and other lakes is smelted into iron". But his recording of this curious case of intuitive foresight would seem to indicate that the hostelry was more firmly associated in his mind with his earlier visit, and perhaps also with the passage from *Romans* chapter nine which his father quoted in Latin together with a reference to *Tobit* 5: 21, as they were taking their leave of one another: "It all depends not on human will or effort, but on God's mercy". Linnaeus, C., 1745, 320; *Minnesbok* 1919, April 14th 1735.

II.iii.3.3:

Uppsala ms. 16b. Latin.

II.iii.3.4:

Uppsala ms. 16b. Latin.

During the seventeenth century, there was a general humanization of penal law, helped on by the increasingly large share of the responsibility for the administration of justice transferred from local to central courts. Between 1635 and 1699, for example, the Göta Court of Appeal revoked nearly eighty per cent of the death sentences passed by lower courts. In 1699 it became illegal to recruit for the profession of hangman (bödel) by reprieving those who had been sentenced to death. The torture, branding, dismembering, hanging, burning, etc., carried out by such officials tended to be replaced by decapitation, carried out by an executioner (skarprätteren), and it is therefore this second term which has been used to translate Linnaeus' 'carnifex': Thunander, R., 1993.

II.iii.3.5:

Uppsala ms. 16b. Latin.

Peder Schumacher, count Griffenfeldt (1635–1699), arrested in the king's name March 11th 1676, tried in Copenhagen May 3rd 1676: see IV.i.3.3.

II.iii.3.6:

London ms. 26; Uppsala ms. 8, Swedish.

Linnaeus gives as his source: "Altona Mercurius 1764, n. 159", that is, the *Altonaischer Mercurius*, Anno 1764, No. 159, Thursday October 4th, p. 5, which carried the following extended account of an event first recorded in the *Nachlese einheimischer und auswärtiger Nachrichten* VI: 94 (Zittau, June 1764): "*Leipzig, September 27th.* On June 29th this year, in Türchau, a village a mile from Zittau, the following remarkable event took place: a seventy-two year old man, resident in the area, some forty-two years before, had so beaten up a young fellow that shortly after the event he had given up the ghost. The man was taken into custody by the authorities, but by taking a corporal oath he seemed to establish his innocence, and after being under arrest for six months he was set free. God visited numerous plagues on him in a variety of ways, and in his old-age he was eventually reduced to beggary. He remained so impenitent, however, that on numerous occasions he made a jest of the matter. "O I know what will happen", he used to say, "the dogs will lap my blood on the spot where I did him in". On the above-mentioned day, this man was walking along the street where he had committed the murder, when two dragoons burst out of the smithy, their horses bolting out of control. One of the horses collided with the old beggar so violently, that soon afterwards he bled to death – on precisely the spot where he had beaten up his fellow man. "Whoso sheddeth man's blood, shall also his blood be shed". It is thus that God maintains this and all his holy laws. This also provides proof of the incomprehensible patience and forbearance of God, of His not removing the sinner immediately, but bringing him to book ever a long period of time".

It is apparent from the village chronicle of Türchau that the person in question was Michael Schmied (1692–1764): Gottfried Hinke, *Chronologische Nachrichten oder Chronik des Zittauischen Raths-Dorfes Türchau* (Zittau, 1804) : 16; cf. *Nachlese einheimischer und auswärtiger Nachrichten* XI: 168–169 (Zittau, November, 1764) ; Adolf Schorich, *Aus unserer schönen Heimat* (Zittau, 1932) I: 35–38. I am greatly indebted to Uwe Kahl of the Christian-Weise-Bibliothek Zittau for having documented Schmied's background.

II.iii.3.7:

Uppsala ms. 15. Swedish.

This account is entered at the foot of a largely blank page, as if Linnaeus intended to use it as it is used here – as a highly complex case rounding off the whole section.

II.iv:

Cf. II.iv.3.4.

II.iv.1:

Uppsala ms. 44. Swedish.

These cases involve not merely awareness of the future, but also a conscious assessment of specific persons in future situations.

II.iv.1.1:

Uppsala ms. 44. Swedish.

This incident can be roughly dated from the fact that Linnaeus' mother died on June 6th 1733; his recording of it from the fact that the rectory at Stenbrohult burnt down during the night of April 20/21st 1746: Linnaeus, C., 1751, May 15th 1749: "Travelled from Virestad to Stenbrohult Church. I found the birds all gone, the house burnt down".

It is worth noting that Linnaeus would seem to be implying that his father, who experienced the fire and died on May 12th 1748, was not upset by the fortune-telling.

It is by no means certain that this fortune-teller is to be identified with the soothsayer mentioned in the following case. Ingeborg of Mjärhult had a family and lived on into an extremely active old age – there is no evidence that she ever attempted to seek advantage by passing herself off as being "poor and sickly". She evidently gave advice only when consulted personally, as far as we know was never "conveyed around and presented" by anyone, and certainly had no occasion to tout for custom.

II.iv.1.2:

Uppsala ms. 44. Swedish.

This incident can be roughly dated from the fact that Linnaeus left home for Lund on August 14th 1727; his recording it from the fact that his brother Samuel (1718–1797) became rector of Stenbrohult in 1749.

The fortune-teller was almost certainly Ingeborg Danielsdotter (1665–

1749), commonly known as Ingeborg of Mjärhult, well-known in the local area throughout this period. She was born in Uthövdan, a hamlet on the southern shore of lake Femlingen, a couple of miles west of Härlunda and some ten miles south-east of Stenbrohult. She married young, a certain Måns Gudmundsson (d. 1716), and settled at Mjärhult on the Grettasjön, a shallow lake or bog a mile or so to the west of her birthplace. The site of the home is now deserted. She had four children, two sons and two daughters, and ended her days some three miles north of Mjärhult, on her son Håkan Månsson's smallholding at Peaboda, being buried in the parish church at Virestad on July 23rd 1749; Nilsson, J., 1952/1960, III: 144–150; IV: 110–116; Ekstedt, O., 1988, 162–166.

Ingeborg seems to have begun to make her mark as a wise-woman at about the time that Linnaeus' mother consulted her concerning the future of her two sons. She went on to build up her extraordinary reputation not only or even mainly on account of her effectiveness as a fortune-teller, but as a result of her phenomenal success in recommending the most unlikely remedies for illnesses. To some extent her prescriptions simply consisted of applying plant-lore, of taking potions and putting on poultices. They also involved asking forgiveness of rocks through the offering of milk, however, frequenting streams flowing north, consulting reed-beds and copses, making water and falling asleep in particular places, paying particular attention, over a period of three weeks, to what one did on a Thursday evening. Probably the most remarkable feature of her successes was that they very rarely depended on her having any direct knowledge at all of the situations, characteristics or symptoms of her clientele. In most cases, all she asked for when consulted was a piece of the patient's clothing: Linnaeus, C., 1745, 312–314; Virdestam, G., 1930, 72–76.

Linnaeus evidently discussed these strange prescriptions with his mentor and friend county physician Johan Rothman (1684–1763), when he visited him in Växjö in August 1741, reaching the conclusion that their effectiveness had to be regarded as transcending all the tenets of normal medical science, and that they could only be assessed rationally as examples of curing by expectation. The local clergy were quite evidently disturbed, not only by the reputation she was building up but also by the conclusions people were drawing from it. It was brought to public notice that in 1715 she had been prosecuted in the district court for declaring publicly that her sister Sissa was a whore and a witch. The indictment had been disallowed due to lack of evidence, but it enabled her enemies to raise doubts concerning her general character. The moral and theological implications of her cures were brought to the notice of the bishop of Växjö Gustaf Adolf Humble (1674–1745), who interviewed her at the rectory in Virestad on July 2nd 1739. He put it to her that she was widely suspected "of standing in some relationship with unclean spirits, who had taught her the use of strange and unusual means for carrying out what she did". Ingeborg replied that she used no other means than those she had learnt from her mother, who had taught her when she was still a maid what had to be done on Midsummer Eve, what herbs, plants and grasses had to be gathered for which potions, medicines and plasters. The bishop responded by telling her that since she was unable to read she was particularly susceptible to being deceived by the Evil One, and that if she did not cease her present activities voluntarily she would have to be constrained legally. She said that as far as she knew she had never done anyone any harm, and thanked the bishop for his fatherly care and advice: Petersson, O., 1971, 110–113.

The relevance of Linnaeus' general assessment of Ingeborg's activities to the central principles of his theodicy has been indicated in the *Introduction* (III.iii.4). In his main account of them he noted that she "lived a godfearing life, attended church regularly, was friendly and courteous to all". When visiting his home area in May 1749 he made a point of searching out a tiny islet in the stream which flows north into lake Fanhult, not far from her home in Mjärhult. He did so because it was common knowledge that it was there: "of a morning, in front of a certain bush, that she used to fast and plead and seek advice of I know not whom, as a result of which the spot is known to the inhabitants of Virestad as Ingeborg's pulpit". He identified the "bush" as a particularly fine specimen of Royal Fern or Bog Onion (*Osmunda regalis*), a plant which is very rare in Sweden, and which in both Scandinavian and English folk-belief has medicinal properties as well as associations with Thor: Linnaeus, C., 1751, May 13th 1749; Vickery, R., 1995, 322; Hyam, R. and Pankhurst, R., 1995, 360.

II.iv.1.3:

Uppsala ms. 44. Swedish.

Much of the island of Gotland consists of limestone, which was burnt into quick-lime and exported to the mainland through the port of Gävle. Being involved in the hazards of this traffic evidently brought to the officer's mind the paradox of such a death.

II.iv.1.4:

Uppsala ms. 44b. Swedish.

Astrologers purported to foretell events by consulting the stars; cf. II.iii.1.6.

In 216 the Roman emperor Caracalla (186–217) ravaged Mesopotamia because the Parthian king had refused to give him his daughter in marriage, and after spending the winter at Edessa, went on a pilgrimage to the temple of the moon at Carrhæ. A desperate legionary by the name of Martialis, at the instigation of the minister Opilius Macrinus, stabbed him to death while he was making this journey: Dio Cassius 77; Herodian iii.10, iv.14; Gibbon, E., 1776/1788, ch. vi.

II.iv.1.5:

Uppsala ms. 44b. Swedish.

The Roman emperor Domitian (51–96) was the first to arrogate divine honours in his lifetime. After the revolt of Antonius Saturninus in 89, he was in constant fear of assassination: "There was nothing by which he was so much disturbed as a prediction of the astrologer Ascletarion and what befell him. When this man was accused before the emperor and did not deny that he had spoken of certain things which he had foreseen through his art, he was asked what his own end would be. When he replied that he would shortly be rent by dogs, Domitian ordered him killed at once; but to prove the fallibility of his art, he ordered besides that his funeral be attended to with the greatest care. While this was being done, it chanced that the pyre was drenched by a sudden storm and that the dogs mangled the corpse, which was only partly consumed; and that an actor of farces called Latinus, who happened to pass by and see the incident, told it to Domitian at the dinner table, with the rest of the day's gossip". Suetonius *De Vita Caesarum* VIII. Domitianus XV.3.

At the instigation of his wife, Domitian was stabbed to death in his bedroom shortly after this event.

II.iv.1.6:

Uppsala ms. 156. Swedish. Cf. IV.iv.1.6.

Linnaeus was at school in Växjö from 1716 until 1727. Johan Svensson Collin (1707–1766) was his contemporary there, married his sister Sophia Juliana (1714–1771) on March 8th 1737, and subsequently became vicar of Ryssby in the diocese of Växjö, a village some nine miles to the east of Ljungby. Linnaeus visited them there during the early August of 1749, while returning back to Uppsala from his tour of Scania.

The case has been placed here on account of the palmistry involved – the reading of fate from a purely physical phenomenon.

II.iv.1.7:

Uppsala ms. 93. Swedish.

Abiel Tiliander (1680–1724), vicar of Pjätteryd in the dioces of Växjö, a village some six miles to the west of Stenbrohult, on the other side of Lake Möckeln, was the son of Linnaeus' father's cousin Sven Tiliander (1637–1706), who was also vicar of Pjätteryd: Fries, Th.M., 1903, I. appendix 8.

He was certainly a troublemaker at school. In 1699 he was given three of the best on his bare bottom and put in the stocks for the day, for unruly behaviour in the classroom. The school authorities did not dare to put him in the lock-up, since he was in favour with those running the local off-licence, and his resources of beer and brandy would simply have turned the confinement into a drunken orgy.

Linnaeus would have heard of the case while he was still at the school. It was while Tiliander was attending the funeral of his father-in-law Samuel Unnerus, vicar of Rydaholm, a village some twenty-five miles north-east of Pjätteryd, that he fell down the well and was drowned. Just prior to this, his brother-in-law Johan Lyberg, the curate at Rydaholm, had been killed in a road accident. Three clergy were therefore buried in the village at one and the same time: Virdestam, G., 1928, 38.

The case has been placed here on account of the curse and its fulfilment.

II.iv.2:

Uppsala ms. 46. Latin.

Linnaeus lists the four cases following the account of his dream, in descending order of moral significance. This accords with the structure of the 1748 theodicy, reversed in the 1758 *Systema Naturae*: *Introduction* I.i.2; I.iii.3.

II.iv.2.1:

London ms. 18; Uppsala ms. 46. Latin.

In accordance with the central principle of his theodicy, Linnaeus' basic consideration when defining a field of enquiry such as this is his own immediate experience of it.

Since this note occurs in the London manuscript and it was during the autumn of 1765 that he transferred the whole of the *Nemesis Divina* onto octavo sheets, this reflection must have been noted down soon after the event. He evidently wrote in Latin so that his wife would not be able to read it and create more disturbances; *Introduction* I.i.2; II.iii.2; III.iii.1.

II.iv.2.2:

Uppsala ms. 48. Latin.

"I confess now, I would have called you back, and my spirit strove; but my tongue stood still for fear of evil auspice. When you would fare forth from your paternal doors to Troy, your foot, stumbling upon the threshold, gave ill sign. At the sight I groaned, and my secret heart said: 'May this, I pray, be omen that my lord return'." *Ovid Epistulæ Heroidum* XIII: 85–90.

The event was interpreted in a variety of ways: Protesilaos had offended the gods by not sacrificing to them before entering into marriage, Homer *Iliad*

II: 695; he was to be the first to land at Troy, and the oracle had warned that whoever that was would be the first to die, Lucian *Dialogues of the Dead* XIX; when Laodamia heard that he was dead, she implored Zeus for one more glimpse of him, and when it had been granted stabbed herself to death in order to be re-united, Ovid op. cit. XIII: 152; when Troy can be seen from the tops of the elms surrounding his grave, the trees wither and die, new ones springing up in their place, Pliny *Natural History* XVI: 88.

II.iv.2.3:

Uppsala ms. 48. Latin.

Pehr Löfling (1729–1756) went up to Uppsala in 1743 to study theology. Unbeknown to his parents, and as a result of Linnaeus' influence, he also signed on in the medical faculty, and in December 1750 left on a botanizing expedition to the Iberian peninsular. On October 15th 1754 he left Madrid for an extension of the expedition to Spanish America. He died of fever at the Merercuri mission station, Guiana, on February 22nd 1756, Linnaeus receiving the news on June 29th: Fries, Th.M., 1903, II: 40–47; *Letters* II: 210, no. 317; V: 28, no. 951.

Löfling was Linnaeus' favourite pupil. Aware, perhaps, of the significance of his having drawn him away from theology, he had a foreboding as he was leaving for Spain: "It was thus that I parted from him, familiar guest and dearest disciple as he was, leaving him in the hands of the Almighty, of Him who determines our fate, who rules with equal might in all countries". He published the letters and desciptions he had sent back to him in *Petri Loefling Iter Hispanicum* (1758), observing in the preface that: "Mortals as we are, we must rest content with the decree of Providence, but I can never forget my Löfling".

Pehr Forsskål (1730–1763) also went up to Uppsala to study theology, but took to following Linnaeus' lectures in the medical faculty. From 1753 until 1756, he studied oriental languages at Göttingen under Johann David Michaelis (1717–1791), the expert on Mosaic law. In 1759 he published a political treatise which Linnaeus, as vice-chancellor of the university, was obliged to censor. In 1760 Linnaeus arranged for him to join the ill-fated Danish expedition to the Middle East. In October 1761 he arrived in Alexandria, and subsequently sent back reports on the flora and fauna of Egypt, the Red Sea and Arabia. He died of the plague at Jerim in Arabia on July 11th 1763. Carsten Niebuhr (1733–1815), the only member of the expedition to survive it, published Forsskål's findings 1775/1776: Fries, Th.M., 1903, II: 60–63; Stafleu, F.A., 1971, 151; *Letters* VI: 110–169; III.ii.2.4.

II.iv.2.4:

London ms. 1; Uppsala ms. 48. Swedish.

Christopher Freidenfelt (1685–1743) joined Charles XII's army in 1701 and was with him in February 1713 during the uproar at Bender. In 1714 he became adjutant and in 1732 colonel of the Småland cavalry. On November 24th 1742, at Kemi on the eastern shore of the Gulf of Bothnia, he led one of the few successful attacks on the Russian forces in Finland. Promoted major-general in February 1743, he planned a full-scale offensive in April, but the mustering of the troops in Hälsingland did not go according to plan. On May 20th, while he and his company were crossing the Gulf of Bothnia from Piteå to Oulu, not while they were crossing Kvarken, the vessel was crushed to pieces by pack-ice and he was drowned: Malmström, C.G., III ch. 16.

Linnaeus heard the story of the horse from Magdalena Bostadia (1713–1789), wife of Carl Daniel Solander the vicar of Piteå, mother of his disciple Daniel Solander (1733–1782). He had been on friendly terms with the family since he had stayed with them at Piteå in June 1732, during his Lapland journey. Daniel worked in the British Museum and accompanied Captain Cook 1768/1771, during his circumnavigation of the globe. His post to his mother in Piteå was sent from London via Linnaeus in Uppsala: *Letters* I: 320–326; II: 334, no. 427, 28.2.1772; III: 293.

II.iv.2.5:

Uppsala ms. 48. Swedish.

The student Jacob Peldan and his uncle Henrik Florinus, vicar of Kimito in Finland, were broken on the wheel and beheaded outside the customs house in Åbo in May 1706: see IV.ii.2.7.

II.iv.2.6:

London ms. 17, no. 46b, Latin; Uppsala ms. 21, Swedish.

Lucius Licinius Lucullus (110–56), the Roman general, despatched to defend the province of Bithynia against Mithradates of Pontus and his son-in-law Tigranes of Armenia, inflicted a crushing defeat upon them near the town of Tigranocerta on October 6th 69. He followed this up in the late summer of 68 with an equally decisive victory over them in the valley of the Arsanias, the eastern Euphrates. In Rome, these successes aroused the jealousy of Pompey, and Lucullus was accused of unnecessarily prolonging the war.

Linnaeus' interest in this aspect of Roman history may have been due to Jacob Flachsenius (1683–1733), who taught him systematic theology at Växjö School from 1723 until 1725, and had graduated at Åbo in 1703 with a thesis on the sibylline oracles: *Oracula Sibyllina* (Aboæ, June 6th 1703) 98 pp.

He may well have seen parallels between the career of Lucullus and that of Charles XII: Plutarch *Lucullus*, 30–36; Ooteghem, J. van, 1959.

II.iv.2.7:

London ms. 17, Nemesis no. 46, Latin; Uppsala ms. 21, Swedish.

Linnaeus almost certainly had this account first-hand from privy councillor Anders Johan von Höpken (1712–1789), who was a founding member of the Academy of Science, chancellor of the university from 1760 until 1764, and a close friend. His interview with count Fredrik Magnus Stenbock (1696–1745) must have taken place in Stockholm during the evening of Tuesday November 24th 1741 (o.s.). Queen Ulrika Eleonora (b. 1658), sister of Charles XII and consort of Frederick I (1676–1751), had died earlier the same day: IV.i.5.16; *Letters* VII: 147–179; Fries, Th.M., I: 260–269, II: 314–316; Malmström, C.G., III, ch. 15.

During the War of Austrian Succession (1740/1748), France attempted to counteract Russia's support for Maria Theresa by encouraging Sweden to think of regaining the Baltic provinces, including Livonia, and by helping Peter the Great's daughter Elizabeth (1709–1762) to remove the pro-Austrian regency council and have herself proclaimed empress. A Swedish manifesto was drawn up in consultation with Elizabeth, and in February 1741 a French subsidy for a Swedish attack on Russia was guaranteed.

The regency council expected an invasion across the Baltic, and by midsummer had moved 100,000 troops into the former Swedish provinces. Sweden declared war on July 28th. About a month later, after the Russians had taken Willmanstrand, fresh troops were shipped into Finland and placed under the command of Charles Emil Lewenhaupt (1691–1743); cf. III.iii.4.4. Besides von Höpken, only two or three members of the Swedish privy council were fully informed of all the diplomatic and military moves involved in these developments.

Between November 19th and November 23rd, Lewenhaupt advanced his troops from Fredrikshamn to within twenty-five miles of the Russian fortress of Viborg. This necessitated the mobilization of forces in St. Petersburg and so precipitated Elizabeth's coup, which began at midnight and had been completed by the mid-morning of November 25th.

Dynastic considerations played a large part in determining policy. Charles Peter Ulrich of Holstein-Gottorp (1728–1762), as the great-grandson of Charles XI and the nephew of the childless Elizabeth, had a strong claim to both the Swedish and the Russian thrones, and played a key role in cementing the alliance between von Höpken's party, the French and the Russian revolutionaries. Since Ulrika Eleonora, his great-aunt, had always been extremely sensitive to this dynastic rivalry, her death must have added point and urgency to the von Höpken–Stenbock interview.

II.iv.2.8:

Uppsala ms. 21b, 23b. Swedish, apart from the final phrase. The second version has been separated from the material Linnaeus added at the end of IV.i.5.4. On Domitian (51–96), see II.iv.1.5.

On January 1st 89 Antonius Saturninus, governor of Upper Germany, seized the savings of the two legions wintering at Mainz and induced them to proclaim him emperor. He was supported by the Chatti, who advanced from the upper reaches of the Lahn to the right bank of the Rhine. Domitian received news of the rebellion on January 12th, collected forces and hastened north. By the time he had reached the middle Rhine, however, the crisis was over. Appius Maximus Norbanus, governor of Lower Germany, had remained loyal, marched south, and defeated the rebels on the plain near Andernach, between Bonn and Coblenz.

On January 25th, tidings of victory, heralded by rumour and prodigies, were being celebrated by the Arval Brethren: "Domitian learned of the victory through omens before he actually had news of it, for on the very day when the decisive battle was fought, a magnificent eagle enfolded his statue at Rome with its wings, uttering exultant shrieks; and soon afterwards the report of Antonius' death became so current, that several went so far as to assert positively that they had seen his head brought to Rome". Suetonius *De Vita Caesarum* VIII. Domitianus VI.2.

II.iv.2.9:

Uppsala ms. 22. Swedish.

Christopher Risell (1687–1762), vicar of Filipstad in Värmland, and his wife Anna Kalsenia (1692–1771), were the parents of Nils Risell (1720–1789), professor of constitutional law at Uppsala from 1761 until 1772. The younger Risell was retired on full pay as a result of the constitutional changes brought about by Gustav III's revolution: Annerstedt, C., III: 1, III: 2, appendix IV; *Letters* V: 201, no. 1121, 27.10.1772.

In the manuscript, Linnaeus indicates that it was from his colleague that he heard this account of the girl's death: "Profess. Riesel".

II.iv.3.1:

Uppsala ms. 22b. Swedish.

Although the context in which Linnaeus himself places these observations makes it essential that they should be classified as they have been here, it should not be overlooked that he is referring to the central principle of his theodicy: *Introduction* III.iii.3; III.iii.4; text IV.iv.2.1.

Psalms 34. 7 of the King James Bible of 1611 and the Revised English Bible of 1989, appears in the Lutheran Bible and the Swedish versions available to Linnaeus as *Psalm* 34: 8. In the Hebrew original, the angel of the Lord or the messenger of Jehovah is conceived of as the leader of the heavenly host – to some extent, therefore, as a collective unit or a corporate personality. This could be regarded as a justification for Linnaeus' translation of the verse, which fits in well with his general theodicy: *Introduction* III.iii.4, notes 81, 82.

In Luther's translation of the Psalms, as in the Swedish translations of 1536, 1541, 1618 and 1703, it is "the angel of the Lord" which is on guard round those who fear Him: Luther, M., 1959/1963 II: 58, 62.

II.iv.3.2:

Uppsala ms. 22b. Swedish.

This section is a summary of what has already been illustrated by examples: II.iii.1–II.iii.2.

II.iv.3.3:

Uppsala ms. 22b. Latin.

Cf. II.i.1.1. As has also been noted in respect of II.iv.2, the sequence of these four sections reflects the theodicy of 1748 rather than that of 1758. Linnaeus presents the Risell case on one side of the sheet, and on the other side speculates on its presuppositions.

II.iv.3.4:

London ms. 17, a much shorter version, Latin; Uppsala ms. 167, Latin and Swedish, cf. IV.i.5.5; Uppsala ms. 24, Swedish.

Julius Caesar (102–44), the Roman soldier and statesman. All the details of Linnaeus' main account are derived from Suetonius and Plutarch, see IV.i.5.6. It is typical of him that he should round it off by taking note of Suetonius' final observation: "Hardly any of his assassins survived him for more than three years, or died a natural death. They were all condemned, and they perished in various ways – some by shipwreck, some in battle; some took their own lives with the self-same dagger with which they had impiously slain Caesar": cf. II.ii.3; II.ii.4.1; IV.i.3.3, etc.

In general terms, what the soothsayer foresaw was the untempered outcome of the war of all against all, as rooted in the collective subconscious of mankind (cf. II.i.3.1, no. 27), hence this placing of the case-study.

II.iv.3.5:

Uppsala ms. 24. Swedish.

Charles XII began the siege of Fredrikshald, the Dano-Norwegian fortress by Halden, just across the border from Sweden in the south-eastern corner of Norway, on September 7th 1718. Three weeks later his troops stormed and took the main outwork of Gyldenlöve. By November 30th the sapping was so far advanced that a final assault was clearly imminent. Between 9 and 9.30 that evening, while on the crest of a parapet inspecting the approaching trenches, the king was shot in the head and died instantly.

There is every likelihood that this was simply a matter of effective enemy fire. Nevertheless, advantage clearly accrued to the later Frederick I (1676–1751), husband of the king's younger sister Ulrika Eleonora (1688–1741), his adjutant and private secretary the Frenchman André Sicre (d. 1733) was in the vicinity of the king just before he died, and was sent by him to Stockholm immediately after the event to brief the Council deciding on the succession. In 1723, evidently when in a state of mental derangement, Sicre asserted that he had been paid to kill the king, but when the matter was investigated by a government commission, the evidence was found to be inconclusive.

The Stockholm pietist preacher Erik Tolstadius (1693–1759) purveyed the idea that Carl Cronstedt (1672–1750), an artillery officer who had accompanied the king on most of his campaigns and was present at the siege of Fredrikshald, had incited Magnus Stiernroos (1685–1762), a corporal in the life guards who had been with the king at Poltava and Bender, to carry out the assassination, but there is no evidence for this: Paludan-Müller, C.P., 1847; Weibull, L., 1929; Jägerskiöld, S., 1941.

II.iv.3.6:

Uppsala ms. 16. Latin. Neatly written and paragraphed, giving the impression that it was copied from a prepared text.

It draws together themes already touched upon in this opening section on man (II.iii.3.1; II.iii.3.3; II.iv.3.3, etc.), and has been placed here as a counterbalance to the two case-histories which precede it; there are respects in which the animal world is not so vicious as that of man, and as Aesop showed, moral lessons can certainly be drawn from it.

III.i:

Cf. III.i.3.1.

III.i.1.1:

Uppsala ms. 175. Swedish.

Cf. IV.iv.3.5. On his return from the Netherlands, Linnaeus first made his mark in Stockholm by prescribing mercury ointment as a cure for venereal diseases: Blunt, W., 1984, 131. This account of the effectiveness of the treatment has been placed here at the beginning of this section, since he presents Joseph's choice of female partners as being completely indiscriminate and uncritical – as not even involving such a basic consideration as the danger of contracting a potentially fatal disease.

III.i.1.2:

Uppsala ms. 103. Swedish.

Since Christoph Myhrman (1712–1775), who was born in Filipstad, studied at Uppsala between 1729 and 1733, he was probably a personal acquaintance of Linnaeus'. Soon after graduating, he was employed by the ministry of mines. He eventually became a wealthy foundry magnate in Värmland and Dalecarlia, and in 1746 represented Filipstad in Parliament. He married Eva Christina Bratt (1720–1752) on November 30th 1738.

Johan Faxell (1713–1770), son of the vicar of Köla in Värmland, was the town-lawyer of Filipstad and a Hat politician, representing the town in Parliament 1751/1752, 1755/1756, 1760/1762, 1765/1766: SBL XXVI: 105–107; XV: 433–436; Brolin, P.-E., 1953, 170.

The Mrs. Fernholm case probably blew up during the late 1740s. Faxell eventually married, although there is no record of his having had children. It looks as though Linnaeus is implying that his lapse into melancholy resulted in his committing suicide.

The case-history has been placed here largely on account of Mrs. Fernholm, who was quite evidently an incorrigible disrupter of family affairs – the female equivalent of the Joseph described in the preceding case.

III.i.1.3:

London ms. 24; Uppsala ms. 172.

All the Uppsala material is derived from Friess, F.C., 1758, 96–98, who uses the Biblical account of Onan in order to illustrate God's punishing a sinner: "immediately, without using another's rod, avenging Himself by drawing His own sword as it were". Bredstedt is in Schleswig not Holstein, as Friess should have known. The triplets were evidently born in the hamlet of Kærbølling, in the parish of Bredsten some fifteen miles north of Kolding – the home parish of Friess's first wife Sophia Dorothea Bang (1726–1757), where her father was the clergyman: IV.iv.3.1.

The treatment of the subject in the London manuscript is much fuller, reference being made to the following works: Martial *Epigrams* IX, 41, 9–10; the German edition (Leipzig, 1736) of the much-published *Onania, or the heinous sin of self-pollution* (London, 1710, 1737[16]); *Genesis* 38 and *1 Corinthians* 6: 9; Hadrianus Beverland (1651–1712) *De Fornicatione Cavenda Admonitio. Sive Ad hortatio ad Pudicitiam et Castitatem* (London, 1698). Linnaeus also adds a marginal note, in Latin, on the interruption of coition.

III.i.1.4:

London ms. 16, Fate no. 44, Latin, cf. 30, no. 65; Stockholm ms. 2, no. 3, Swedish; Uppsala ms. 127, 127b, Swedish, by far the fullest account.

Eric Brahe (1722–1756), after a military and diplomatic career, married into money and in 1751 joined the court party, the main political objective of which was to secure more constitutional power for the crown: Malmström, C.G., IV, chs. 21, 22.

Linnaeus' colleague in the medical faculty at Uppsala, Nils Rosén von Rosenstein (1706–1773), was married to Anna Christina von Hermansson (1718–1782). Brahe was a student at Uppsala from 1732 until 1739.

Brahe married the wealthy Eva Katarina Sack (1727–1753) on December 12th 1745. On April 28th 1754 he married Kristina Charlotta Piper (1743–1800), the daughter of Karl Fredrik Piper (1700–1770), speaker of the house of lords.

Johan Puke (1726–1756), born on Gotland of an unknown mother, was almost certainly Piper's son. After a military career he joined the court party, on a certain occasion being entrusted with the queen's jewellery, which he pawned for her in Berlin. Though unmarried, he had a distinguished son, admiral Johan Puke (1751–1816).

Magnus Stålsvärd (1724–1756), made his name as a military engineer, publishing on the subject. When arrested for having plotted the revolt he made no bones about confessing: "Life is not worth all this bother, sooner or later death comes to every one of us". When being led to the place of execution, he brushed aside the attendant clergyman, threw his hat into the crowd, and leapt up onto the scaffold.

Gustaf Jacob Horn of Rantzien (1706–1756), marshal of the court, was married to Eva Margareta Gyllenstierna (1730–1753). On May 2nd 1756 Linnaeus wrote to him recommending one of the students for a scholarship: "partly on account of his quiet way of life, greatly appreciated by all, partly on account of his skills, especially in the preparation of optical glasses, which he has learnt from Klingenstierna". *Letters* VII: 128, no. 1530.

The case has been included here on account of Linnaeus' emphasis on the connection between sexual profligacy and political misdemeanour. It has also been included in its political context: IV.i.2.5.

III.i.1.5:

London ms. 17, Nemesis no. 46; Uppsala ms. 120. Swedish. Linnaeus writes "Morey" instead of Munster, makes him the mayor and not the son of the mayor, and gives "Torne" as the place where Schmidt died.

Johan Bernhard Munster (c. 1662–1714) came of a family which had recently taken its name from the hamlet of Munstis in the parish of Piikkiö just to the east of Åbo, where his father was born. The father, Berend Riggertsson (c. 1626–1696), after starting his career as a clerk, had inherited a couple of estates through his first wife (d. 1654), and had subsequently been elected mayor of Åbo, (1675) and member of parliament for the city (1676, 1682). Munster himself had an academic career, being matriculated at the academy in 1677, taking his master's degree and holy orders in 1688, being appointed professor of ethics and history in 1694, and rector of the university in 1700 and 1708. He married Christina Flachsenius (c. 1670–c. 1711), daughter of Jacob Flachsenius (1633–1694), professor of logic and metaphysics and dean of the cathedral, and elder sister of Jacob Flachsenius junior (1683–1733), who took his master's degree under his brother-in-law's supervision on June 6th 1703, with a thesis on the sibylline oracles. It was this Jacob Flachsenius junior who was later to teach Linnaeus systematic theology at Växjö school (1723/1725), and who was therefore almost certainly his source of information for the goings-on recorded in this case-history. Christina's elder sister Margareta (1666–1707) married David Lund (1657–1729) who became professor of theology at Åbo in 1697 and bishop of Växjö in 1711; the couple evidently took care of her younger brother after the father's death.

It looks as though the drunken party must have taken place about 1705. Christina Munster seems to have died in Åbo during the plague of 1711. In 1714, during the Russian occupation, when attempting to flee to Sweden with his son and two daughters, Munster was apprehended by the enemy on the island of Kumlinge and forced to return to Åbo, where he continued to perform his pastoral duties until June. The family was then arrested and earmarked for exile to Siberia. Munster died during the transportation at Helsingfors. The children did in fact spend some years in Siberia, before being repatriated at the end of the war: Stjernmans, A.A., 1719, 121; Strandberg, C.H., 1832, 54; Elmgren, S.G., 1861, 98; Wasastjerna, O., 1879, II: 109; Lagus, V., 1891, I: 169, 302, 340; Carpelan, T., 1950, 752; Ranta, R., 1977, 756. I am greatly indebted to Juoko Taimi of the City Library Åbo, for having supplied me with the information and documentation which has made possible this reconstruction of the family history.

Peter Schmidt (c. 1685–1714), son of a civil servant at Viborg in south-eastern Finland, joined the Åbo cavalry in February 1704, was promoted to the rank of second lieutenant in 1705, and to that of full lieutenant in December 1709. He married Helena Berends (c. 1687–1743), who after his death remarried with major Konrad von Creutlein (c. 1672–1737): *Genos: Tidskrift utgiven av Genealogiske Samfundet i Finland* 30 årg. (1959) no. 1 (117): 51. His brother Henrik Schmidt (c. 1686–1714) was matriculated at the University of Åbo on February 18th 1701, and in both 1703 and 1704 got into trouble in the town on account of his debts. He later joined the Åbo infantry, and in 1710 was appointed captain in the Björneborg infantry regiment: Lewenhaupt, A., 1920/1921, I: 603.

In August 1713 Carl Gustav Armfelt (1666–1736) was put in command of the army in Finland which at that time consisted of some six thousand men. At the beginning of February 1714 a Russian force of eighteen thousand men, on the express command of the czar, marched north out of Åbo in order to drive Armfelt's troops out of the country. The forces met at Storkyro, some twenty-three miles east of Vasa, on February 19th. In the opening move only Peter Schmidt's regiment of Åbo cavalry acquitted themselves well. The infantry in the centre, which included Henrik Schmidt's Björneborg regiment and was under the direct command of Armfelt himself, although it had some initial success, was eventually borne down by weight of numbers and harassed by the Cossack cavalry on its flanks. Henrik Schmidt was one of the two thousand Swedes to fall. The Russian losses amounted to about five thousand. Armfelt, forced to get away as best he could, gathered the remnants of his forces together again at Gamlakarleby, some sixty miles to the north. The Russians spent a couple of weeks plundering in the Vasa area, and then marched back south. In September Armfelt received orders from Stockholm to ship some of his troops across to Umeå, and to move the main body north, back into Sweden around the gulf of Bothnia via Torneå and Kalix.

It was in Torneå, on November 12th 1714, that Peter Schmidt met his end in the manner described by Linnaeus: III.i.3.1; Lewenhaupt, A., 1920/1921 I: 603; Mankell, J., 1870, 240–249.

It is certainly a matter of some significance that in the London ms. this case is followed by that of Henrik Florinus (IV.ii.2.7), executed at Åbo in 1706, not 1722 as stated by Linnaeus, and that although Linnaeus has a good grasp of the details of the Florinus case, his account of it contains inaccuracies which could well have originated from his having got to know of it by word of mouth, quite evidently also from Flachsenius. It seems reasonable to conclude from this that it was the way in which systematic theology was taught at Växjö which first aroused Linnaeus' interest in the wider significance of such cases.

III.i.1.6:

Uppsala ms. 95. Swedish. Linnaeus writes "Backmanson".

Carl Asp (1710–1782) – appointed professor of philosophy in 1755, rector of the university during the crown prince's visit in September 1767, retired 1770: Annerstedt, C., III: 1, 300; III: 2, 430.

Anders Borell (1693–1771) – alderman of Uppsala, his daughter Anna Maria married Linnaeus' colleague Johan Gustav Acrel (1741–1801) July 18th 1771: *Letters* V: 187, no. 1109.

Johan Backman (b. 1743) was tried in the Västerås town-court for misdemeanour at the toll-gate and the murder of the cartwright on March 4th 1774: Stadsdomstolarna, Västmanlands län, Västerås rådhusrätt och magistrat, Dombok 1774/1775, A II: 41: Landsarkivet Uppsala.

The case has been placed here on account of the casualness of Backman's attitude to the sexual relationship.

III.i.2.1:

Uppsala ms. 62. Latin, apart from the account of Mennander.

According to a theory which Linnaeus first developed toward the close of the 1740s and expressed in its most complete form in his *Clavis medicinæ duplex* (Stockholm, 1766; Swedish tr. 1967), all plants and animals consist of marrow, the female reproductive principle, bearer of the vital, internal generative powers of the organism, and bark, the male reproductive principle, the exterior nutritive aspect of its bodily form. Fructification and generation are therefore to be understood as the fusing together of marrow and bark.

Carl Fredrik Mennander (1712–1786) was a close friend of Linnaeus. They first met at Uppsala in 1732. Although Mennander was training for the priesthood, Linnaeus inspired him with a lifelong devotion to the natural sciences. In his subsequent academic career he became professor of physics (1746) and then of theology (1752) at Åbo; in his church career he became bishop of Åbo (1757) and archbishop of Uppsala (1775).

Mennander married his half-cousin Ulrica Paléen (1721–1742) in Åbo on August 25th 1741; she died in childbirth at Uppsala on November 13th 1742, and was buried in the cathedral there five days later. In 1747 he married the sixteen-year old Johanna Magdalena Hassel, who died two years later after bearing him a son – the archaeologist and art-expert Carl Fredrik Fredenheim (1748–1803): SBL XXV: 414–417.

III.i.2.2:

London ms. 21, Adultery no. 75, Latin; Uppsala ms. 184, Swedish.

Since Suno Lagerbladh (1717–1757) was curate at Tutaryd, just outside Ljungby, and originated from Skatelöv, just south of Växjö, it looks as though Linnaeus must have gathered this information through his clerical and family connections in the area: cf. II.i.2.3; II.iv.1.6.

Lagerbladh married Fredrika Margareta, daughter of captain Paul Eggertz (1687–1767) and his wife Anna Dorotea (1708–1758), sister of the Växjö tax-collector Johan Vilhelm Netherwood (1700–1749). The Lagerbladhs were eventually divorced, the wife subsequently marrying her army officer – Carl Gustav Berg: Malmeström, E., 1968, 229, 232, 236.

The wife's refusal to consummate the marriage on account of her suspecting her husband's previous involvement with her mother, makes the case a particularly apposite link between III.i.2.1 and III.i.2.3.

III.i.2.3:

Uppsala ms. 154.

All three of these instances of chastity are drawn from the family history of St. Bridget (c. 1303–1373).

Her father Birger Persson (d. 1327), one of the wealthiest Swedish landowners of the time, owned a large estate at Finsta near Norrtälje; her mother Ingeborg Bengtsdotter was the sister of Ramborg the wife of Eringisl Plata (d.c. 1285) of Svanhals, which is near Osby, just south of Stenbrohult.

St. Bridget's husband, to whom she was married when she was thirteen and with whom she had eight children was Ulf Gudmarsson (1298–1344) – knight, magistrate, privy councillor. The pilgrimage to the shrine of St. James at Compostela took place 1341/1342, and as Linnaeus implies, almost certainly involved sexual abstinence. He died two years after completing it, and was buried in the monastery at Alvastra.

St. Bridget's daughter St. Catherine of Vadstena (c. 1330–1381) married Egard Lydersson von Kyren (d. 1350) in 1348. He evidently treated her as a sister not as a wife, and when shortly after the wedding Bridget left for Rome, Catherine decided to follow her south. When she heard of her husband's death, she felt that she had been called to stay with her mother and work for the founding of the Order; *Letters* II.ii, 33–35, no. 290, May 5th 1754; *Swänska helgonets St. Britä lefwerne, utur åtskillige historier samlat* (Linköping, 1788); SBL IV: 447–462; *Kulturhistorisk Leksikon for Nordisk Middelalder* (1980) 8: 345–347.

III.i.2.4:

Uppsala ms. 77. Swedish.

Christina Juliana Löfving (c. 1723–1769), first lady of the bedchamber to queen Lovisa Ulrika (1720–1782), very probably the daughter, by his first marriage, of the Finnish partisan Stefan Löfving (1689–1777). It looks as though the chronology of the main events of her life must have been as follows: 1743 marries Anders Thun (1717–1803); 1745 leaves him to take up a position at court; 1759 her husband appointed to a position with the Board of Commerce in Stockholm; 1761 she meets Georg Gustav Wrangel (1728–1795), who had just been recalled to Stockholm after serving in the army in Pomerania 1759/1760 and then in a diplomatic capacity in Hamburg 1760/1761: BL XXI: 98–104; SU 17: 1065.

Linnaeus seems to have gathered his information through being consulted concerning her health: see his letter to his close friend Abraham Bäck (1713–1795), written March 9th 1769: "Poor Mrs. Thun; God help her. She was created for a completely different catheter": *Letters* V: 154, 163, nos. 1074, 1081.

In the manuscript, the following passage is deleted after the third paragraph: "In 1767 she contracted hæmorrhage of the uterus and two years later it had turned cancerous. That whereby one sins is that whereby one is punished".

Her case-history has been placed here in accordance with Linnaeus' own assessment of it – as an example of the flawed fulfilment of the sexual relationship.

III.i.2.5:

Uppsala ms. 82. Swedish.

Linnaeus knew of the death of Johan Welin (d. 1744), professor of logic and metaphysics at the university of Åbo, from his close friend Abraham Bäck (1713–1795), who wrote to him about it from Paris during the spring of 1744: "Something terrible happened here last Tuesday. A countryman of ours, professor Welin, together with two or three other tourists, was playing a game of quadrille at a good friend's place, when fire broke out in the flat below, so ravaging the place that the floor collapsed beneath them before they realized what was happening. All of them perished miserably, burnt to a cinder". *Letters* IV: 12, no. 630.

At the time of his receiving this letter, Linnaeus' disciple Pehr Kalm (1716–1779) was travelling in Russia with baron Sten Carl Bjelke (1709–1753), like Linnaeus a founder member of the Academy of Science. On April 4th 1744, Linnaeus wrote to Bjelke suggesting that Kalm might be approached with regard to the possibility of his filling the vacancy left by Welin's death. Kalm, not so sure that the decrees of providence could be read in quite that way, wrote as follows to Linnaeus from Moscow on May 3rd: "I can hardly regard the professor as being entirely in earnest when he suggests that I might take over professor Welin's position at Åbo: the science he was required to cultivate at the academy was metaphysics, which has always seemed to me to be a subtle brain-teaser, only to be indulged in by those not endowed by the Almighty with an interest and involvement in anything better, of more use to the general good. In fact I have to confess that if I were able to raise enthusiasm and interest for it, it would be as a result of something comparable to one of the miracles of our Lord – that, for example, by which he changed Nebuchadnezzar from a rational being into a dumb animal. What is more, I find that I am simply incapable of squaring my conscience with looking for a job wherever a position becomes vacant. It has always seemed to me that he who knows a little of everything and nothing in depth, remains very much of an all-round fool". *Letters* III: 191, no. 560; VIII: 13, no. 1589; *Daniel* 4: 33; cf. IV.iii.2.1.

III.i.2.6:

London ms. 12, no. 9; Uppsala ms. 90. Swedish.

This worker must be the Mr. Lindstedt appointed by Linnaeus in the spring of 1744 as assistant to Dietrich Nietzel (1703–1756), keeper of the botanic garden from 1739 until his death. Lindstedt's main task was carrying out the heavy work involved in running the hothouse or orangery – cutting and carting the wood, stoking, ventilating. He died in 1748: Fries, Th.M., 1903, II: 103–109; *Letters* I: 121–122, no. 67, January 1744; 143–145, no. 86, 1750; 200–202, no. 119, November 30th 1765.

When they were first appointed, Nietzel was paid 450 copper dollars a year, Lindstedt 60. As Linnaeus increased the scope and efficiency of the garden, the work-load increased, and he was fairly successful in getting their wages raised accordingly. He realized, however, that by and large they were underpaid.

He can therefore have taken no exception to Lindstedt's attempting to improve his financial position. As he saw it, retribution was visited upon him and his wife on account of the means they had employed to this end.

Lindstedt and his wife are assessed not only as individuals but also as a family unit – hence the placing of this case-study.

III.i.3.1:

London ms. 21, 24, Latin; Uppsala ms. 10, 62b, Latin and Swedish.

The London material differs widely from its Uppsala counterpart, and constitutes the first subsection of a general category headed *Sin*.

Linnaeus gives the wrong verse numbers for the quotation from *Ecclesiasticus*; the latter part of the observation concerning the two lovers is taken from Ovid *Remedia amoris* 444.

III.i.3.2:

Uppsala ms. 40b. Swedish. The anecdote was almost certainly part of Uppsala university tittle-tattle.

Johan Henrik Sparfvenfelt (1698–1768), bibliophile, lord chamberlain, crown equerry, was the son of Johan Gabriel Sparfvenfelt (1655–1727), bibliophile and benefactor of Uppsala university library, owner of an estate at

Åbylund near Vadstena, who had married Antonetta Sophia Hildebrand in 1695 and fathered ten children on her: BL XV: 153–155; Annerstedt, C., II: 2, 336f.

Pehr Johansson Sjöblad (1683–1754) was a general staff officer in the artillery, son-in-law to Carl Cronstedt (1672–1750), who was suspected by some of having instigated the assassination of Charles XII: II.iv.3.5; BL XIV.305.

III.i.3.3:

London ms. 25; Uppsala ms. 136. Swedish.

Johann Caspar baron von Dörnberg (1689–1734), councillor to Friedrich landgrave of Hesse-Cassel (1676–1751), was much in Stockholm after the landgrave became king of Sweden in 1720. He was ambassador at Regensburg in 1720 and at Soissons in 1728, and after 1728 president at Cassel. On August 22nd 1714 he married Sophie Charlotte von Heyden (1692–1738), by whom he had nine children over the next twelve years: *Stammtafeln der Althessischen Ritterschaft*: Freiherrn von Dörnberg, Tafel 11 (1888).

Henrika Juliana von Lieven (1710–1779), a lady of the court at Stockholm, married Linnaeus' patron the architect and art-expert baron Carl Hårleman (1700–1753) on May 29th 1748. Linnaeus first became acquainted with Hårleman in 1742 when he was given the task of redesigning the botanic garden at Uppsala. Mutual trust and admiration soon developed, and both were in the habit of describing their relationship, semi-jocularly, as being that of "father and son". It was a relationship which was extended when Hårleman married, since although his wife was three years younger than Linnaeus, he usually addressed her as his "mother". Fries, Th.M., 1903, I: 362–363; II: 534–535; *Letters* II: 100, 149, 165; VII: 131–147.

Dörnberg's death on February 6th 1734 had therefore been a crucial factor in Linnaeus' having acquired these "parents". Nevertheless, it is clear that the focal point of the narrative is Dörnberg's treatment of his wife, and it is this that has determined the placing of the case.

III.i.3.4:

Uppsala ms. 194. Swedish and Latin.

The primary career of Lars Molin (1657–1723) was as a theologian and churchman: padré in the life guards 1694, professor of theology at Pernau 1699, chaplain to the queen mother 1702, senior professor of theology at Uppsala 1705, close friend of archbishop Erik Benzelius (1632–1709). After 1710, however, he was also very active politically, concerning himself among other things with the financial aspects of the administration, with acquiring for his own benefit properties alienated from the nobility, and with accumulating a huge deposit in the national bank, the interest on which was to provide stipends for his descendants: SBL XXV: 645–647.

The primary career of his son Jonas Meldercreutz (1714–1785) was that of a mathematician and engineer: 1732/1734 he travelled widely throughout Europe with Anders Celsius (1701–1744), visiting scientific societies, investigating agricultural, mining and industrial undertakings; in 1736 he accompanied Maupertuis to Lapland on the expedition for measuring the length of a degree of the meridian; from 1751 until 1772 he was professor of mathematics at Uppsala. He was also very active economically and politically: in 1740 he began to develop the mineral and forestry resources of Lapland; in 1753 he obtained exclusive grazing, farming and forestry rights over a vast tract of country (1,500 km^2) between the rivers Kalix and Lule; he sat in all ten parliaments between 1738 and 1772. Towards the end of the 1750s, however, things began to go less well for him and he was forced to mortgage and raise bank loans; this financial crisis deepened during the 1760s and 1770s, and in 1781 he only just avoided complete bankruptcy: SBL XXV: 359–363; Annerstedt, C., III: 2, 272–275.

Meldercreutz's first wife Fredrika Yxkull was born at Arbrå in 1719, married him at Åbo on December 12th 1738 and died in Uppsala on October 24th 1768. On March 9th 1775, he married Magdalena Sophia Margareta Falkenburg (1755–1801).

Since George Buffon (1707–1788) was one of the most ardent opponents of the system of classification promulgated in the *Systema Naturae*, Meldercreutz's enthusiasm for his writing cannot have recommended him to Linnaeus.

Anders Philip Tidström (1723–1779) was one of Linnaeus' favourite disciples. He was appointed reader in chemistry at Uppsala on February 22nd 1758, despite determined opposition from Meldercreutz. On account of his wit and good humour, he was always a welcome guest at Hammarby. Fries, Th.M., II: 347, Bil. 14; Annerstedt, C., III: 1, 507.

The case-history has been placed here on account of Meldercreutz's treatment of his wife. It has also been given a placing under the general heading of *Riches* (III.iv.3.6).

III.i.3.5:

Uppsala ms. 170. Swedish.

Jacob Olofsson von Hökerstedt (1685–1757) was a mercantilist economist and civil servant, notable as one of the first in Sweden to take up the ideas of Charles Davenant (1656–1714) and advocate the advantages of a national census. When he put the idea of holding one to the state council in 1733, it was rejected with reference to the disastrous results of David's having numbered the people of Israel and Judah (*2 Samuel* 24). He was appointed lord lieutenant of Gotland in 1738, and in this capacity was evidently always ready to help Linnaeus with his researches: *Letters* II: 91, no. 234, January 9th 1747; III: 120, no. 518, June 18th 1752.

He married in Stockholm on November 4th 1712 – Hedvig Eleonora von Bleichert (1696–1742), daughter of the court jeweller.

During his journey through Gotland, Linnaeus visited Hökerstedt at his residence in Roma, staying there during the night of July 15/16th 1741. He has left the following account: "During the afternoon we arrived at *Roma* monastery. It suddenly became thundery, with lightning and rain, which quenched the parched earth, there not having been any rain since we arrived on Gotland. We remained there until the following afternoon. – *Roma* monastery, now the residence of lord lieutenant Hökerstedt, is uniquely well-situated and well-built, with a fine stone building and beautiful gardens. The view to the south is over extensive meadowland, that to the north over arable fields, both being enclosed by the thick evergreen woodland which surrounds the estate on all sides. Looking north from the residence itself, one sees Roma church beyond the fields, with a dead straight drive leading away to it".

Although there are several tales in Boccaccio's *Decameron* resembling that told here by Linnaeus (III: 6; VIII: 4; IX: 6), there is no real reason for regarding what he records as nothing but folklore. Conversation in the servants' hall at Roma may well have supplied him with reliable infomation: Linnaeus, C., 1745; BL XIX: 710–712.

III.i.3.6:

Uppsala ms. 180. Swedish, no heading.

Lars Bergqvist (d. 1768) was a master-cobbler in Uppsala.

The most distinctive feature of the case is the couple's apparently complete inability to sort out their differences themselves, the constant publicizing of their quarrels through the law and the church. It is on account of it that the case has been classified with that which follows.

III.i.3.7:

Uppsala ms. 180, 106. Swedish.

The Swedish East India Company was founded in Gothenburg in 1731 by Colin Campbell (1686–1757), after the rival English and Dutch companies had forced the collapse of a corresponding organization founded at Ostend in 1722. Although the charter granted was renewed in 1746 and 1766 and the main factory in Canton functioned quite effectively, it tended to be run by Scots rather than Swedes and always remained a somewhat precarious operation. Its first ships to the east were harassed by those of its rivals. Metal goods were exported and sold in Cadiz for Spanish silver piasters. Tea, porcelain and silk were imported, auctioned in the company's warehouse in Gothenburg, and bought mainly by the Dutch, who re-sold in the Baltic area.

Swedish divorce and re-marriage procedures were the result of changes in canon law and church organization introduced at the reformation. A cathedral chapter could grant divorce on grounds of adultery or desertion, and a civil court then decided on the legal consequences. In the normal course of events the guilty party was not allowed to re-marry, but this ruling could be overturned on appeal – hence the procedure mentioned here by Linnaeus.

In general, the legal reform of 1734 confirmed traditional practice. It also stipulated, however, that "if as a result of professional involvement in war, commerce, or some other necessary errand", the husband had been out of the country for a longer period of time, the wife was entitled to re-marry, on the understanding that if the husband returned he was entitled to "repossess" her, the later marriage then being automatically dissolved.

The main point of this case is in full accord with the central tenet of Linnaeus' theodicy: the legal proceedings of civil authorities and churches will always be ancillary to individual decision-making and conduct. Cf. IV.ii.2.2.

III.i.3.8:

London ms. 16, Fate no. 43; Uppsala ms. 166. Swedish.

In both accounts, Linnaeus gives the judge's name as "Rydén". Apart from the added information that his wife was "good-looking", the London text is reproduced almost word for word in the Uppsala version. Since Linnaeus passed through Klinte on July 14th 1741 while making his survey of Gotland, staying overnight a few miles to the north-east at Roma (III.i.3.5), it is to be assumed that it was at that time that he gathered his basic information concerning Canutius and his wife: Gullander, B., 1971, 112–113.

Olof Canutius (1700–1756) was the son of Israel Canutius vicar of Vamlingbo, a parish in the extreme south of Gotland, and his wife Christina Schmidt. His mother got him to study theology at Uppsala, but he soon left his studies and volunteered for the guards, seeing active service under Charles XII. Although he returned to the university after the war was over, he always regretted not having stayed with the colours. Later in life, when faced with the trials and tribulations of his calling as a clergyman, he was often heard to ask God to forgive his mother for having prevented him from becoming a soldier. After taking his degree and holy orders, he applied unsuccessfully for the living of När on the south-east coast of Gotland (1726), the headmastership of Visby school (1727), the living of Eskelhem just south of Visby (1729), and eventually, on December 18th 1731, was appointed vicar of Klinte, a village on the outskirts of Klintehamn, some twenty-three miles south of Visby.

Olof Brodén (d. 1737), the district judge for southern Gotland, lived in

Klintehamn and was therefore one of Canutius' parishioners. This was a somewhat controversial situation, since traditionally the judges had resided in the assize-village of Mästerby, which is some seven miles to the north-east. The house in which he lived had been built about 1660 by a Dutch merchant, with materials imported from Holland, whose family had sold it in 1688 to Nils Schmidt, a local businessman. The local population never really took to the assizes' being held there, and soon after Brodén's death in the spring of 1737 the location was moved once again. The local reputation of the house in Klintehamn remained tainted until well into the nineteenth century: a murder was reputed to have taken place there, and it was said to bring ill-luck on those who owned it: Snöbohm, Alfred Theodor 'Tre starke Gotländningar', *Gotlands nyaste tidning* April 21st 1864.

Brodén's "good-looking but promiscuous wife", immortal sweeper and duster, stewer of whips, was Anna Winbohm (d. 1757): David Gadd 'Herdaminneskommittén i Visby stift: excerpter, Klinte' 1950/1960, Landsarkivet i Visby. Canutius was a frequent visitor to the judge's house, and it had not gone unnoticed that he and Anna Winbohm had exchanged glances. It had also been noticed that Brodén was not unaware of the situation. Late one evening, when Canutius was returning to the vicarage from the judge's house, he was set upon, evidently by a group intent on administering rough justice. He proved equal to the situation, however, leapt from his coach, and set about his attackers with his Spanish cane, driving them off into the darkness. Once the judge had passed away the couple married. One of Canutius' shortcomings as a clergyman was that he made no attempt to keep the church books. There is therefore no record of the son mentioned by Linnaeus. The couple did have a child, however, a daughter by the name of Anna, who lived to the age of eighty (1739–1819), married Nils Qviberg (1720–1798), a later vicar of Klinte, and had eleven children.

Canutius was elected member of parliament in 1740 and again in 1742. He was therefore away in Stockholm when Linnaeus passed through the village, since the first parliament continued to sit until August 22nd 1741. The information concerning the horsewhipping and the stewing must therefore have been gathered at a later date: Lemke, O.W., 1868, 279–281.

III.i.3.9:

Uppsala ms. 83. Swedish.

Johan Ulrik Wertmüller (1712–1780) was member of a family long-established as proprietors of the Lion apothecary in Stockholm. They were known to Linnaeus as having anticipated by a couple of generations his own researches in Lapland, and he evidently purchased materials from them: *Letters* I: 4, no. 3; IV: 130, no. 701, October 9th 1750. The Wertmüller in question was appointed physician-in-ordinary to the king at Christmas 1762.

He married Maria Ravens (1720–1786). Their son, Johan Fredrik Wertmüller (1743–1770), after studying the theoretical background to the family trade under Linnaeus at Uppsala, returned to Stockholm and worked there as a "provisor" or qualified apothecary's assistant. The financial difficulties of the wife's admirer were probably associated with those of Peter Julinschöld (III.iv.3.7).

The case has been classified with that which precedes it as a further example of the wife's taking the initiative in the disruption of family life.

III.i.3.10:

Uppsala ms. 99. Swedish.

Dietrich Nietzel (1703–1756), keeper of the botanic garden at Uppsala 1739–1756, was son of a Hamburg gardener, from whom he learnt the rudiments of his craft. At the age of thirteen he left his family in order to widen his experience, taking gardening jobs at Gottorf near Kiel (1716), Herrenhausen at Hanover (1720), Salzthal near Wolfenbüttel (1721), Husum in Holstein (1722), Bremen (1726), in Jacob Ortmann's garden (1727), and in the earl of Peterborough's garden at Southampton (1731). In 1735 he moved to Hartecamp near Haarlem, to take charge of the rich garden collections of George Clifford (1685–1760), and it was there that he met Linnaeus. In 1738 Linnaeus returned to Sweden, and a year later arranged for Nietzel to be appointed at Uppsala: Jackson, B.D., 1915; Kuijlen, J., 1983, 35; Krol, J.L.P.M., 1982, 70–81; Fries, Th.M., 1903, I: 282; II: 104; *Letters* I: 145.

Nietzel must have become involved with the tanner's wife soon after arriving in Uppsala, since his daughter was ten years old when he died in 1756. Linnaeus took the child into his house when she lost her father, and arranged for her future care and upbringing. On September 27th 1756 he wrote to his friend Abraham Bäck (1713–1795): "My gardener is dead; he has left a silly wife, in debt, and a daughter only ten years old; where can I find a gardener?" *Letters* V: 13, no. 935.

The case has been placed here on account of the symmetry of the situation – just as Nietzel lacked proper respect for the tanner, so his wife lacked proper respect for him.

III.i.3.11:

London ms. 12, no. 12; Uppsala ms. 81. Swedish.

Harald Urlander (1709–1753), owner of a woollen mill in Norrköping, was married to Anna Christina Ehrenspetz (1718–1763). Linnaeus probably knew him through the Academy of Science, of which they were both members. Scarcely a week after he had passed away, Linnaeus wrote to his friend Abraham Bäck (1713–1795), asking if he knew what he had died of: *Letters*

IV: 243, no. 813, December 11th 1753.

Danviken, founded in 1527, was the Stockholm poorhouse and old people's home: formerly much used during witch trials, during the eighteenth century it housed unbalanced political dissidents, deranged ideologists, eccentric clergymen such as Boëthius (IV.ii.2.10).

The case has been placed here on account of the social implications of the adultery: the wife enjoyed living well but denied the husband the loyalty he was entitled to; she was therefore denied the comforts that would normally have accrued to her after her husband's death.

III.i.3.12:

London ms. 21, Sin, Adultery no. 74, Latin; Uppsala ms. 108, Swedish.

Peter Psilanderhielm (1695–1770), bibliophile, privy councillor, official at the royal treasury, member of a well-known Småland family, married Petronella Sofia Fries (1708–1748), daughter of a Karlskrona businessman in 1724. During the next eighteen years she bore him thirteen children, one a year between 1739 and 1742: Anrep III: 267–268.

Linnaeus probably knew of the case from his fellow Smålander Peter Jonas Bergius (1730–1790), who after studying medicine under him at Uppsala, became physician to the Psilanderhielm family (1751): *Letters* III: 128, no. 524, January 5th 1756.

The case is evidently meant to illustrate the dire psycho-somatic effect of an inverted conscience – the wife's lechery finds fitting retribution in the fatal tophus. In the first instance, however, the focus seems to have been on the "courtly gentleman", since in the London manuscript *Proverbs* 6: 29 is quoted: "So he that goeth in to his neighbour's wife; whosoever toucheth her shall not be innocent".

III.i.3.13:

Uppsala ms. 176. Swedish; the only entry on the page; evenly written, with what is presented here as the third paragraph added later.

Linnaeus purchased the two estates of Sävja and Hammarby in the parish of Danmark just south of Uppsala in 1758. As is evident from a number of his case-histories (III.ii.1.6; III.iii.6.3; III.iv.2.4), he took a close interest in what went on there: *Letters* V: 57, no. 282, December 22nd 1758; 176, no. 1095, October 26th 1770.

Jansson's wife evidently came from the village of Ängeby, between Storvreta and Vattholma just north of Uppsala. Her case-history has been classified with that which follows since it involves both adultery and incest (cf. III.i.2.3) – a combination of abuses which Linnaeus evidently regards as warranting the excruciating pain and suffering eventually meted out to her.

III.i.3.14:

London ms. 30, no. 73, in a listing beginning with no. 68, the preceding sheet having been lost. Uppsala ms. 124. Swedish. Cf. Linnaeus, C., 1747, 227, recording the events of July 24th 1746, when Linnaeus was crossing the heathland near Brålanda, not far from the western shore of lake Vänern, while on his journey through Västergötland. This original account was not noted down by Linnaeus' secretary Erik Gustaf Lidbeck (1724–1803), but was added to the manuscript later in Linnaeus' own hand.

Linnaeus quite evidently embellished his account of the case on the basis of hearsay. The farmhand Swen Persson (b. 1723) was twenty-eight years younger than the widow Gertrud Hemmingsdotter (1695–1755), whose mother had died many years before these events. The step-daughter Börta Persdotter (b. 1729) gave birth to an imperfectly developed male child on March 11th 1746, which was christened, and which was given a church burial three weeks later. There was, therefore, no question of infanticide.

The trial was for incest and complicity in incest, and was concluded in the Göta court of appeal on May 7th 1746. Persson was condemned to be broken on the wheel and beheaded, Persdotter to be beheaded and burnt, Hemmingsdotter to be fined eighty silver dollars or birched. The executions were carried out on June 25th 1746: Gillby, J., 1961, 74–75.

As Linnaeus well knew, the court which imposed these penalties was simply enforcing the law of the land as laid down in the statutes of 1608 and 1734 (*Introduction* III.i.3) – hence the placing of this case.

III.ii:

Cf. III.ii.1.1; *Ecclesiasticus* 33: 21.

III.ii.1.1:

Uppsala ms. 12, 12b. Latin and Swedish. The three Latin quotations are grouped together at the foot of a largely blank page, that from the *Apocrypha* at the top of the reverse side. There would appear to be no thematic connection between this material and that which precedes and follows it.

Virgil *Aeneid* I.94–96, Aeolus invokes the tempest; XI.158–159, Evander laments the death of Pallas; Ovid *Tristia* IV.x.81–82, the poet's autobiography.

III.ii.1.2:

London ms. 1; Uppsala ms. 113. Swedish.

Sten Ribbing (1730–1761) of Koberg near Trollhättan in Västergötland, a captain in the infantry, met Fredrika Pahl (b. 1729), daughter of a captain in the artillery, when posted to Finland; he subsequently served with distinction in the Pomeranian War (1757–1762), particularly during the opening carpaign, the defence of Fehrbellin. It was in 1764 that his brother Fredrik Ribbing (1721–1783), colonel of the Elfsborg regiment, married Eva Helena Löwen (1743–1813), daughter of Axel Löwen (1686–1772), governor-general of Pomerania and chancellor of the University of Greifswald. She was his daughter by his second wife Eva Horn (1716–1790). For a recent survey of the general history of the Ribbing family, see the four hundred page work by Magdalena Ribbing, obtainable from Börstorps Slott, Enåsa, Mariestad, Västergötland: *Arte et Marte* 49 (1995) 23.

Carl Fredric Piper (1700–1770) retired from the position of president of the entailed estate commission in 1756, subsequently devoting himself to the improvement of the family estate of Krageholm, near Ystad; his younger sister Sophia Carolina (1707–1732) was Löwen's first wife; his elder sister Ulrica Eleonora (1698–1754) was the mother of the two Ribbing brothers: SBL 24: 580–583; 29: 318–321; Anrep III: 386–387.

As Linnaeus notes, he gathered this information from Abraham Söderberg (1728–1803), with whom he had social contact in Uppsala (III.iii.5.2; III.iv.3.18). Söderberg evidently served in Pomerania during the war as surgeon attached to the Uppland regiment. He seems to have settled in Uppsala not so very long after the cessation of hostilities, and in 1774, after some arduous campaigning, he eventually managed to get himself appointed surgeon to the university: Annerstedt, C., III-1: 514–516; Bih. IV: 335.

III.ii.1.3:

Uppsala ms. 71, 152. Swedish. The final paragraph was appended to the account of Annerstedt (IV.ii.2.6) and then deleted.

Johan Hermansson (1679–1739), Skyttean professor at Uppsala, rector when Linnaeus was matriculated, head of the Småland union, married Anna Christina, sister of Jöns Steuchius (1676–1742), who in 1730 succeeded his father as archbishop of Uppsala: SBL 18: 691–692. On the way in which Hermansson reformed and controlled the university consistory, see Annerstedt, C., III-1: 29–32; III-2: 29–31; Bih. III: 154–156.

Daniel Solander (1707–1785) was the uncle of Linné's pupil of the same name (1733–1782); the chair he so set his heart on became vacant on the death of Johannes Reftelius (1659–1747); the girl he courted was Elsa Maria Hermansson (1719–1794), who subsequently married Jakob Serenius (1700–1776), bishop of Strängnäs (IV.ii.2.13); the woman he married was Anna Margareta Lambert, widow of Johan Lambert (1676–1738), university apothecary.

Solander's children evidently had little reason to seek to him – hence the placing of this case history, which there are good reasons for associating with one of the very few included in the London ms. (p. 13, no. 17), but dropped from the Uppsala version:

Major Cronhjort, after becoming lord lieutenant in Gävle, was removed from office for having furthered the interests of his relation. He had a mistress in Växjö, who bore him a whole lot of children. He completely abandoned them, married a rich wife, but had no children with her. During the 1761 parliament he was declared not guilty and subsequently granted a pension.

Carl Gustav Cronhjort (1694–1777) of Kläckeberga near Kalmar became a major in the Kronoberg regiment in 1731 and retired from the army in 1743. He married Catharina Sophia Lewenhaupt (1697–1772) in 1734 and in 1755 was appointed lord lieutenant of Wester-Norrland After his re-instatement, he became president of the war office; Anrep I: 491–492. It seems to have been the re-instatement and the granting of the pension which led to Linnaeus' dropping the case.

III.ii.1.4:

London ms. 7, Latin; Stockholm ms. 2, no. 1; Uppsala ms. 183. Swedish.

Before he was appointed to the chair at Uppsala, Linnaeus had made his mark in Stockholm by prescribing mercury ointment as a cure for venereal diseases, Blunt, W., 1984, 131 – hence his success in dealing with alderman Adam Lindberg (d. 1770). Linnaeus' pupil Fredrik Lindberg (1733–1779) was probably related to the alderman, *Letters* V: 133–134, no. 1055, 26.4.1765, and may well have supplied his tutor with some of the inside information purveyed here.

Like Lindberg, Justus Gottlob Goldhan (1717–1772) was member of the Uppsala city council. Isak Wikblad the tobacco manufacturer was almost certainly related to Olof Wikblad (d. 1734) the university printer: Elgenstierna VI (1918): 957–959; Annerstedt, C., III-2: 530.

The family disruption and the careers of Lindberg's sons have determined the placing of the case.

III.ii.1.5:

Uppsala ms. 92. Swedish.

The account is clearly and evenly written, without any deletions or alterations. Its absence from the London ms. would seem to indicate that when Linnaeus was copying his material out in the autumn of 1765, his mind went back afresh to the earliest experiences of his youth.

This case has been cited by Erland Ehnmark in support of the thesis that many of the basis conceptions of the *Nemesis Divina* derive from Swedish folk-belief. He instances a parallel case of an old woman on Gotland who had stolen something and when she was accused of having done so denied it, declaring that if she was guilty her fingers might rot from her hands – which is in fact what they did (1941, 36).

The plausibility of this thesis should not be taken to imply that what Linnaeus has recorded here did not actually take place. Kerstin Lindblom, Secretary of the *Linné's Råshult Trust*, has pointed out to me in private correspondence that at the expense of allowing for a modicum of inaccuracy in Linnaeus' narrative, the church books at Stenbrobult can be seen as yielding several possibilities for providing it with a firm basis in historical fact. The burial records make it evident that there was indeed a Jacob Carlsson (1673–1728) of Såganäs, who was married for twenty years and left nine children, and who had been ill for some time before taking to his bed on August 21st and dying six days later. Although there is no surviving record of his wife's having died by drowning in 1723, there is record of a Kjerstin Britta Truhliddotter (1678–1728) of Taxås, who was married for twenty-nine years and left eight children, and who on Sunday January 7th, "after having taken Holy Communion the previous day, fell and was drowned while hurrying to church over unsafe ice". Her husband's name was Germund, and there was indeed a Germund Bängtsson (1665–1735), who at one time lived very near Taxås, subsequently farmed in Såganäs, never prospered, and died on April 3rd 1735.

Linnaeus quite evidently sought to his parents (II.iii.1.5), as he realized he could never have done had he been brought up in the family in question in Såganäs – hence the placing of this case-history.

III.ii.1.6:

Uppsala ms. 165. Swedish.

Karl Jansson (1719–1773) was a farmer at Kyrkbyn in the parish of Danmark, just south of Uppsala. Linnaeus purchased the two estates of Sävja and Hammarby in the same parish in 1758, and quite evidently took a close interest in what went on there: III.i.3.13.

It was the watchfulness and attention of the community which eventually resolved the problems of the marriage, although it is not perfectly clear whether or not Linnaeus approved of the wiseman's solution: cf. II.iii.3.1; II.iv.1.1.

III.ii.2.1:

Uppsala ms. 11, 11b. Swedish. Linnaeus seems to have divided these quotations into groups by leaving gaps, and then inserted extra material: the divisions indicated here are partly his own and partly editorial. His numbering of the verses, which sometimes differs slightly from that now generally accepted, has been standardized.

III.ii.2.2:

Uppsala ms. 187. Swedish.

Johan Ihre (1707–1760) was one of the most distinguished of Linnaeus' colleagues at Uppsala, establishing a European reputation for himself in the field of Germanic philology, especially on account of his work on Gothic, Old Icelandic and Swedish etymology. After studying at Jena, Utrecht and Leiden, and visiting Oxford, London and Paris, he was appointed Skyttean professor at Uppsala in 1738, with responsibilities for teaching and research in politics and eloquence which he voluntarily extended to a whole range of broadly related subjects. His energy, outspokenness and popularity brought him into conflict with both the Uppsala theologians and the ruling Hat party during the 1740s, and as with Linnaeus in 1748, the university was obliged to discipline him: SBL XIX: 763–770; Lindroth, S.H., 1978, 601–610.

In 1738 he married Sara Charlotta Brauner (1714–1758), who came of a notable Småland family, her father being well-up in the civil service and a Latin poet of some distinction. In 1748 Ihre was able to purchase a civil service position for himself for considerably more than his annual salary as a professor. The running up of "a bit of a debt" when buying Sandbro was even more remarkable, since it was the second largest estate in the parish of Björklinge to the north-west of Uppsala, some 1,250 acres, just under half of which was arable.

Sara Charlotta died in Uppsala on September 17th 1758, and two days later Linnaeus wrote a letter of condolence to his colleague, addressing him in the third person in accordance with his position in the civil service. "It behoves me to implore you to bear this sorrow with the patience that becomes a Christian; but it is not for me to kindle the torch from which I have so often borrowed my little flame: it is easy enough to treat another's wound, not so easy to tend one's own. Should we not remember the vanity of our life-span; looking back on our time we find it nothing but a dream. That to come will certainly be no better; what advantage have we in being and dreaming longer; why complain at not being and dreaming the longest. Wife and children, all we have, are on loan from God and are not our own; He who has lent has the right to take back when He will, and who can complain at having been allowed to enjoy". *Letters* VII: 181, no. 1579.

Ihre married his second wife Charlotta Johanna Gerner (1728–1822) at Venngarn near Sigtuna on December 11th 1759.

III.ii.2.3:

Uppsala ms. 199. Swedish.

Nikolaus von der Lieth (1699–1772) was born on the family's hereditary estate at Elmelo near Bremen, in an area which had been ceded to Sweden by the Treaty of Westphalia in 1648. His father Christopher von der Lieth (1678–1741) had served in the Swedish army and been captured at Poltava before retiring to Elmelo.

Nikolaus also served in the Swedish army, rising to the rank of colonel, being naturalized as a member of the Swedish nobility in 1751, and retiring in 1760. In 1727 he married Anna Magdalena Tauscher (1703–1742), who bore him six children, the only one to survive infancy being a son Nils, born in 1741, who studied at Göttingen and died before his father. Nikolaus himself died at Uppsala on August 30th 1772: Anrep II: 680–681.

The girl he adopted was Catharina Lovisa Janssen (b. 1738), daughter of a captain in the merchant navy. He married her on March 30th 1760. Her reputed lover was Adolph Fredric Wedenberg (1743–1828), one of Linnaeus' pupils.

III.ii.2.4:

London ms. 25; Uppsala ms. 85, 85b. Swedish.

Nils Wallerius (1706–1764) was the elder brother of Johan Gottschalk Wallerius (1709–1785), professor of chemistry, metallurgy and pharmacy at Uppsala from 1750 until 1767. Although he came of a clerical family, his initial training also was in the natural sciences, more especially the statistical and mathematical approach to physics and meteorology. It was his tutor in physics Samuel Klingenstierna (1698–1765) who first introduced him to the general philosophical principles of Wolffianism, and it was as a result of his preoccupation with this aspect of his work, that the university appointed him professor of logic and metaphysics in 1746. His gifts as a polemicist, combined with his command of Wolffian ontology and logic and his clerical background, made him the ideal candidate for the chair of theological controversy, newly instituted and endowed by Andreas Kalsenius (1688–1750), bishop of Västerås.

Wallerius was appointed in 1754, and proceeded to promulgate the principles of the Lutheran orthodoxy of the established church with great gusto, the shrift given to the Hobbists, Spinozists and Bayleans on the one hand being as short as that shown to the Dippelites, Pietists and Moravians on the other. He adopted the highly successful tactics of moving straight in on any unorthodox doctrinal statement and rendering it untenable or ridiculous by a rigorous application of Wolffian logic. Ordinary churchmen were delighted, his voluminous expositions of the new scholasticized theology being widely read and cited: Wallerius, N. 1750/1752, 1754, 1756/1765; Frängsmyr, T., 1972; Lindroth, S.H., 1978, 357–358, 527–528.

Wallerius' second wife, pictured here so vividly by Linnaeus (cf. III.iii.5.2), was Anna Margaretha Boy (1712–1801): her financial dealings with the university are well-documented: Annerstedt, C., III-1: 398–399; III-2: 642–643.

It is possible that Linnaeus saw her difficulties as not unrelated to the theological stance adopted by her husband. Linnaeus' pupil Pehr Forsskål (cf. II.iv.2.3), in his *Dubia de principiis philosophiae recentioris* (Goettingae, 1756; Vinduae, 1760[2]), crossed swords with Wallerius, attacking his basic premisses with admirable verve and acumen, maintaining: i) that there is no compelling reason for accepting the principle of sufficient reason, that is, Wolff's postulate that contingent facts must be grounded in necessities, in the reason a perfect and omnipotent being has for actualizing one possibility rather than another; ii) that the Wolffian principle of non-contradiction, that is, that no conjunction of a proposition and its negation can be true, cannot be regarded as basically and universally valid, since it rests on the dubitable assumption that that which cannot be doubted is true; iii) that the individual freedom basic to law and morality is not to be derived from the fore-mentioned principles: Johan Dellner, *Forsskåls filosofi* (Stockholm, 1953).

In respect of these points, there are clear affinities between Forsskål's position and that of Linnaeus (*Introduction* II.i–iii). It is understandable, therefore, that while on his fatal journey to the Middle East, he should have written as follows to his mentor: "What's the situation in Sweden in respect of Parliament and the freedom of the press? And how has Dr. Wallerius reacted to the later edition of my disputation? Am I now marked there with the rest of his many heretics? And are the Uppsalians still promulgating my dangerous doctrines in the lecture-room? I'd be intrigued to receive a letter containing such a page torn from the disputations". *Letters* VI: 153, no. 1355, Alexandria, October 20th 1761.

III.ii.2.5:

London ms. 30, no. 74; Uppsala ms. 91. Swedish.

The province of Uppland contains Stockholm and Uppsala and has a total area of some five hundred square miles. It is not likely, therefore, that this case is simply another example of Linnaeus' interest in what went on in the parish of Danmark: cf. II.i.1.8; III.i.3.13; III.ii.1.6; III.iii.6.3; III.iv.2.4. In respect of its source, it probably has to be classified with II.iii.3.3 and II.iii.3.7.

It has been placed here on account of the contrast between the girls in respect of their attitude to their offspring.

III.ii.3.1:

Uppsala ms. 60b. Swedish.

The first two quotations are written at the top of the page, the last four at the bottom. It looks, therefore, as though Linnaeus left room for additions.

III.ii.3.2:

Uppsala ms. 171, 171b. Swedish.

As Linnaeus indicates, he derived this case-history from captain Carl von Gedda (1712–1794), probably about 1771, when as member of the university consistory he first became involved in the protracted legal proceedings Gedda initiated against the professor of theoretical philosophy Per Niclas Christiernin (1725–1799): Fries, Th.M., II: 225; Annerstedt, C., III-1: 453.

Niclas Peter von Gedda (1675–1758), the informant's father, was appointed secretary to the Swedish resident in Paris in 1702, and remained attached to the embassy there, in one way or another until 1736. His son served in the French army, rising to the rank of captain, married Marie Agnès Morell (d. 1763), the daughter of a French actor, and got an appointment at court when he returned to Sweden. Like his father, he also had strong British connections (Roberts, M., 1986, 144), and eventually emigrated permanently to England; Anrep I: 911.

The case-history has been placed here on account of the boy's relationship with his parents and guardian.

III.ii.3.3:

London ms. 6; Uppsala ms. 118. Swedish.

Sånnaböke is now the northernmost suburb of Älmhult on the eastern shore of lake Möckeln; Linnaeus knew it as "Sannaböke", a hamlet in the southernmost part of the parish of Stenbrohult.

The elder Måns Månsson, the hard man who had behaved so badly to his father, was also a wealthy man, and it seems to have been primarily on account of this that he was so often invited to stand as witness at christenings – that of Linnaeus' brother Samuel on May 1st 1718, that of the daughter of Peter Bexell (1675–1724), close friend of the Linnaeus family and proprietor of the inn at Diö on September 30th 1721 (II.iii.3.2), just as his wealthy brother Germund of Quarnatorp was invited to stand witness at the christening of his own daughter Ingebor in 1720. It was then common practice for such witnesses to make gifts or promises of considerable sums of money to the newly-christened child. Playing along with the practice did little to improve his general reputation, and when he died Linnaeus' father noted in the church book simply that he was a wealthy man and that "one would not want to blame his ashes for the manner in which he lived". Virdestam, G., 1928, 29; Kerstin Lindblom, Secretary of the *Linné's Råshult Trust*, private correspondence, September 1997.

Måns Månsson the younger and his brother Germund owned farms in the Diö area, on the lake to the north of Stenbrohult, which on August 3rd 1730 were purchased by members of the mining college for the development of iron-production: *Stenbrohult sockens historia*, 1956, 269, cf. II.iii.3.2.

As noted in the *Introduction* (III.ii.1), William Turner (1653–1701), vicar of Walberton near Arundel in Sussex, in his *A Compleat History of the Most Remarkable Providences* (London, 1697) I, ch. 99, no. 11, observes that he has "read of a Man, that was haled out of doors in a violent manner by his own Son, who cried out to him. Oh! pray, no further; for just so far I dragged my Father". Turner presents this as the eleventh of twenty instances of "divine judgments by way of retaliation". One can only assume that in the Stenbrohult area, as elsewhere, homiletic compilations such as Turner's were readily available, and that Måns Månsson's general reputation led to his being accredited with this particular action.

In his preface, Turner gives an account of his main sources, informing his readers that the foundation of his collection was laid some thirty years before by Matthew Poole (1624–1679), the presbyterian and biblical commentator. He emphasizes the broadening and humbling effect of studying these, "most remarkable providences both of judgment and mercy, which have happened in this present age": "Let those little Narrow-Soul'd Christians, that appropriate their Faith and Charity to a Canton, live in a little Corner of the World by themselves, they are hardly worthy to enjoy the Benefit and influence of an Universal Sun, and Gospel, and Government .... To be perfectly Wise, is the Property of God Almighty. For my part, I am very sensible of the Depths I have here taken upon me to fathom, and do declare openly to the World, That the Ways of God are unsearchable, and his Footsteps cannot perfectly be traced".

III.ii.3.4:

London ms. 3; Uppsala ms. 200. Swedish. In the London version Möhlman is said to have died within half an hour of having conferred his property on Gerdessk.öld.

The Leyel family originated from Arbroath in Scotland, a Henry Leyel serving in the Swedish army during the closing decades of the sixteenth century, and being granted property in Östergötland and Småland. Jacob Leijel (1612–1678) became burgher of Stockholm in 1638, and laid the foundation of the family's immense fortune by taking as his second wife Barbara Martha Dress (1633–1694), daughter of the owner of ironworks in the Örebro area. By making astute deals with his wife's co-heirs he assimilated most of the family's assets, and towards the end of his life acquired the lease of further foundries, hammering-plants and mines from the crown.

Adam Leijel (1658–1729), the member of the family mentioned by Linnaeus, was the son of Jacob and Barbara. Together with two of his sisters, he increased the family fortune still further, was ennobled in 1717, endowed a poorhouse at Hammarby near Nora, granted funds to the university of Uppsala, but died childless.

Adam's elder sister Maria (1657–1740) married Sven Möhlman (1636–1687), and had a son Jacob (1685–1761), who as Linnaeus notes, inherited the Leijel fortune when his uncle died.

Adam's half-sister Margaretha, daughter of Jacob Leijel's first wife Margaretha Eden (1627–1653), married the Stockholm banker Johan Gerdes (d. 1687). It was her grandson Johan Gerdessköld (1698–1768) who was president of the court of appeal, and who inherited the Leijel fortune in accordance with Möhlman's will.

Lars Salvius (1706–1773), the Stockholm printer, member of the Academy of Science, publisher of most of Linnaeus' Swedish works, married Charlotta Svedenstjerna, the granddaughter of Adam Leijel's second sister Eva Maria. David Schultz von Schulzenheim (1732–1823), the Stockholm physician and pupil of Linnaeus, in 1762, married Catharina Eleonora Svedenstjerna (1728–1797), Charlotta's sister: Anrep II: 619, 677, 942; III: 677; SBL 22: 448–451.

The case-history has been placed here as an example of the providential rectification of failed family relationships.

III.ii.4.1:

London ms. 15; Uppsala ms. 28. Swedish and Latin: placed here as an anticipation in family life of the right common to all; distinguished from IV.iv.3.3 by translating the title.

Charles XII came to the throne in 1697 and the Great Northern War began three years later; he defeated the Russians at Narva in 1700 and was defeated by them at Poltava in 1709; he did not return to Sweden from Turkey until 1714 and was shot at Fredrikshald in Norway in 1718; II.iv.3.5.

Croesus, last of the Memnad rulers of Lydia 560–546 B.C., already in his lifetime proverbial on account of his wealth, was reputed to have been visited by Solon, who reminded him that divine nemesis waits upon overmuch prosperity, and that one should "call no man happy until he is dead". Subsequently defeated and captured by Cyrus king of Persia, the story goes that Croesus was condemned to be burnt alive. When the pyre was lit he called out "Solon" three times, with lamentable energy. Cyrus asked him why, and was told of the conversation with the Athenian. This instance of the inconstancy of human affairs so struck the Persian ruler that he ordered Croesus to be taken from the pile, and became one of his most intimate friends; Herodotus I c. 26–29; Plutarch *Solon* VIII c. 24.

St. Sigfrid, a monk of Glastonbury (d.c. 1045), together with companions, was sent on a missionary journey to Scandinavia about the turn of the millenium. After founding a church at Växjö, he left it in charge of his nephews Unaman, Vinaman and Sunaman, and went on into Västergötland, where he evidently baptized Olof Skötkonung king of Sweden (d.c. 1025) at a spring in

Husaby, and founded the bishopric of Skara. While he was absent from Växjö, Gunnar Gröpe one of the twelve lords of Småland led a heathen reaction and murdered the three nephews, which led to his being exiled by the king. St. Sigfrid returned to Växjö and continued to work there until his death. His tomb is still to be seen in the cathedral there, and representations of him holding the heads of his three nephews in a basin are legion throughout Småland: Johan Magnus 1554 lib. 17, caps. 18–20, 560–564; Frondin, E., 1740; Klingspor, G.A., 1932; Rydbeck, M., 1957; Farmer, D.H., 1984, 357; Larsson, L.-O., 1991, 19–28; *Introduction* III.iii.4.

Gunnar Gröpe is said to have died at Tving near Karlskrona in Blekinge, then part of Denmark, and to have been buried in heathen ground. Linnaeus exaggerates the misfortunes of the Ulfsax family. By the end of the seventeenth century, the triple murder committed six hundred years before had simply left them with the reputation of not being able to rise to a higher military rank than that of captain. Of the three Ulfsax brothers captured by the Russians after the battle of Poltava, Carl Gustaf, a lieutenant in the Kronoberg regiment, died at Tobolsk in 1717, Zacharias, a captain in the Kalmar regiment, was murdered at Tomsk in 1721, and only Magnus, captain in the Kronoberg regiment, arrived back in Småland in 1723. Nevertheless, by the middle of the eighteenth century most of the main branches of the family were sixteenth-generation – hardly evidence of a less than normal longevity: BL 18: 70; Anrep IV: 482–500; *Arte et Marte* 43 (1989), 12–16.

On the piglets and the porker, see Walther, H., 1963/1986, no. 26060; cf. IV.i.4.2; IV.ii.2.14; IV.iv.3.2; IV.iv.3.3.

It was as a result of his differences with Nils Rosén (1706–1773), his rival for a permanent academic position at Uppsala, that Linnaeus, in 1734 "put everything in the hands of God": *Introduction* I.ii.3; Fries, Th.M., I: 171–191; Malmeström, E., 1926, 69–80; 1964, 51–62; IV.iii.3.3.

III.ii.4.2:

London ms. 16, no. 40; Uppsala ms. 186. Swedish. Cf. III.iii.3.4.

Scania was ceded to Sweden by Denmark at the Peace of Copenhagen in 1660.

Jörgen Krabbe (1633–1678) was owner of the estate of Krageholm, a couple of miles to the north of Ystad, acquired by his father Iver Krabbe (1602–1666) in 1642. Although the father had fought fiercely for his country in the two Swedish wars of 1657–1660, and was devastated by the ceding of his home province, he objected so strongly to the introduction of constitutional absolutism in Denmark (1661/1665), that he seriously considered throwing in his lot with the enemy. The son had a position in the Danish chancery prior to the wars, but he stayed in Scania when it became Swedish, took the oath of allegiance to Charles XI, and in 1664 became member of the Swedish nobility.

Göran Sperling (1630–1691) studied at Uppsala and then entered on a military career; he became vice-governor of Scania in 1677 and went on to hold governorships elsewhere in Sweden; he sickened and died very rapidly after being appointed field-marshal in 1690, but there would appear to be no evidence that he ended his days in poverty. He was married three times; four of his sons survived infancy and all of them took up military careers, three of them dying in the war against Russia, the fourth, Gustav (1667–1726), ending his career as colonel of an infantry regiment.

Sperling's daughter Helena (1668–1704) married Henning Rudolf Horn of Rantzien (1651–1730), who was appointed commandant of Narva in 1695. She died when the Russians took the town in 1704, and he was transported to Moscow with his six children, not returning to Sweden until 1715. While in captivity, four of his daughters married Swedish officers: Anrep IV: 86–87.

III.ii.4.3:

London ms. 28; Uppsala ms. 96. Swedish.

Axel Wrede-Sparre (1708–1772) of Sundby near Eskilstuna became colonel of the Västergötland cavalry in 1747.

Anders Appelbom (1724–1770) first joined the regiment in 1739; after a period of service abroad, he rejoined it in 1754 as a lieutenant. When the trouble with the governing Hat party blew up in 1756 and he was forced to go into exile, he joined the allied forces serving under duke Ferdinand of Brunswick (1721–1792) and George Sackville (1716–1785), and took part in the action at Bergen in the Hunsrück near Idar-Oberstein on April 13th 1759, when the allies were defeated and so failed to drive the French out of Hesse and Frankfurt.

It was in the course of this action that Carl Axel Sigge Wrede-Sparre (1737–1759) sustained the wounds which resulted in his death in Frankfurt on October 27th; Anrep I: 78–79; IV: 72.

III.ii.4.4:

London ms. 15, no. 25; Uppsala ms. 87. Swedish.

Erik Wrangel of Lindeberg (1686–1765) studied at Uppsala and then lived in England for a number of years. He probably travelled to France and Italy after finishing his job as secretary to the Swedish commissioner in London in 1714, and prior to his taking up the position of secretary to Georg Heinrich von Görtz (1668–1719) in 1718. In 1720 he married his relative Elisabeth von Rosen (1688–1751). He was member of the privy council from 1719 until 1731 and of the state council in 1739, and in 1743 was elected member of the Academy of Science. He had a fine private library, and concerned himself with agricultural improvement as well as with writing a couple of tragedies and an autobiography: BL 21: 112–122.

His daughter Ingeborg Wrangel (b. 1722) evidently had an affair with Josias Carl Cederhielm (1734–1795), married and was then divorced by Carl Hans Schrowe, a lawyer-in-training, and eventually married a bailiff or manager by the name of Tholin. She died in Christiania, Norway.

His son Erik Wrangel (1721–1753) had an affair with Catharina Frondin (1722–1789), the wife of Lars Braunersköld (1700–1753), secretary in Uppsala, later a district judge. He subsequently became one of the favourites of queen Lovisa Ulrica, but in February 1756 the ruling Hat party discovered that he had been attempting to stir up support for the crown in the army, and he was forced to flee via Norway to Hamburg, where he died on January 4th 1760: Anrep IV: 641.

III.ii.4.5:

Uppsala ms. 195. Swedish. In the heading and the last line of the main text, Linnaeus wrote "Wali", deleted it, and substituted "Morga". A case concerning a "farmer in Wallie", omitted from the Uppsala text, is to be found in the London ms. 29, no. 67. Valje near Bromölla in Scania is a possible identification for the place being referred to. The Morga case is also recorded in Linnaeus' unpublished papers on diet, where he dates it "the end of 1690". Linné, C. von, 1968, 219.

Alsike is a parish on the eastern shore of lake Mälaren, just south of Uppsala; a headland there is still known as Seven Neighbours' point, and traces of Morga farm were still in evidence in the 1880s. There are also local tales of farmer *Dager* of Morga, arrogant and avaricious, whose seven sons perished in the manner described when crossing the lake with their seven brides – daughters of a farmer who lived at Vreta in the parish of Dalby, on the western shore. There is, however, no record of the tragedy in the church books: Fredbärj, T., 1967.

Linnaeus has clearly got hold of a folk-tale reminiscent of the seven ewe lambs and the well which figure in the oath made between Abraham and Abimelech, the sevenfold sign on the troll-drum of the Lapps, etc. He may well have appreciated the irony of the way in which the tale turns upside down the normally lucky seven – hence the placing given to it here.

The final quotation is in Latin – a garbled version of Ovid, *Epistulæ ex Ponto*, I.3.61f.

III.iii:

Cf. III.iii.5.1.

III.iii.1.1:

Uppsala ms. 39. Latin.

Cf. note II.i.3.1: the sources of this pastiche of quotations are as follows: *De providentia* IV.15.2; *De ira* II.xi.3.7, I.xx.4.4; Ad M. XVIII.7.7; N.Q. III. præf. 7.6, 8.5, 12.3; Ad Lucil. CIII.1.8, 2.3, 2.9, 2.10.

In the preceding sections of the re-structured *Nemesis Divina*, consideration has been given:

1) to the nature of life, inwardness and the emergence of inter-subjectivity, and to a whole range of predominantly subconscious interactions (II.i–II.iv);

2) to the overcoming of the disruption of family life by unbridled sexuality, the ways in which children interact with their parents, the providential rectification of failed family relationships (III.i–III.ii).

The scene is therefore set for a consideration of the way in which man frees himself from "submission to the course of nature" by embodying ethical principles in a system of rights and law (III.iii), by curbing "man's delight in ruining man" and so opening up the possibility of humanizing economics (III.iv).

III.iii.1.2:

Uppsala ms. 5b. Latin. Cf. III.iii.6.1, IV.iv.2.1 section 3.

III.iii.1.3:

Uppsala ms. 37. Latin.

Seneca *Hippolytus* (*Phædra*) 1123–1127.

III.iii.1.4:

London ms. 30, no. 75, in a listing beginning with no. 68, the preceding sheet having been lost; Uppsala ms. 140, without the heading. Swedish.

Linnaeus' wife Sara Elisabeth Moræa (1716–1806), the eldest child of Johannes Moræus (1672–1742), town physician in Falun, Dalecarlia, kept in close contact with her family after her marriage. This contact was almost certainly the source of Linnaeus' information here.

He notes in his autobiography that when he first visited the gold and copper mines at Falun on January 26th 1734, they seemed to him to be "a pretty good replica of hell": 1888b I: 10.

III.iii.1.5:

London ms. 27, together with other cases from Friess, F.C., 1763, the date (1745) being included; Uppsala ms. 178. Swedish.

Friess (1758, 43) presents the case as an example of "the sword of civil authority" fulfilling the law of God.

The case was communicated to Friess about 1755 by "a respectable and pious" member of the Kolding Society, whose wife had been in service in that part of Denmark when the event took place. It has been placed here on account of the significance Friess attached to it, that is, as the civil counterpart of the purely providential course of events illustrated by the preceding case.

III.iii.1.6:

Uppsala ms. 73, shorter version 117b. Swedish.

Petter Strutz (1696–1746) studied at Lund, joined the army of Charles XII in 1715, saw service in Norway, and retired in 1723.

In 1719 he married Margareta Elisabeth von Friesendorff (d. 1736), and it was she who managed their estate at Bruskebo, not far from Enåker in Vestmanland. The market-town for this area is Sala, also known as Sala bergslag on account of its silver mine. On February 8th 1728 Strutz attended the market there, and while drinking in a tavern under the town-hall became involved in a heated exchange with a local tobacco-salesman by the name of Petter Bergqvist. He drew his rapier, lunged, and killed him. There are legal records of the court-case, the flight to Norway, and the wergild fine of one hundred silver dollars imposed in 1732.

Strutz had six children, and borrowed money to pay for the education of his sons at Uppsala. After his wife's death drink got the better of him, and he became completely incapable of managing the estate. He ran into trouble not only with his numerous creditors, but also with the local clergy, and in 1743 petitioned the crown for a resumption of his service in the life guards.

He died on November 14th 1746, and although no mention is made of the accident with the horse, the precise circumstances of his demise are particularly well-documented. It is not the case, however, that any of this documentation necessarily invalidates this particular element in Linnaeus' narrative: Tennemar, E., 1982/1983.

Linnaeus could have gathered his information through the university administration at Uppsala, the two sets of legal reports, or his von Friesendorff connections. As he records the case, the point of it seems to be that since Strutz's fatal violence was not properly countered by the wergild, further recompense had to be extracted from his fondness for the horse.

III.iii.1.7:

Uppsala ms. 158; the second, the chronological section, clearly added later. Swedish.

Per Adlerfelt (1680–1743) volunteered for the life guards in 1700, and served in all Charles XII's campaigns until condemned to death by court martial on April 23rd 1709.

The Swedish victory over the Danes at Gadebusch in Mecklenburg took place on December 12th 1712. The peace of Copenhagen, brokered by John Carteret (1690–1763), Britain's envoy to Sweden, and signed on June 3rd 1720, finally brought the eleven year war to an end.

In 1743 Adlerfelt was in command of the government troops opposing the Dalecarlian march on Stockholm. On June 22nd he was mortally wounded by shot from his own riflemen, who were purposely firing loose after having been forced by mutineers to target a cavalry attack he had ordered: SBL I: 159–140; Malmström, C.G., I c. 5, III c. 16.

Thirty-four years after not having faced the required consequences of the murder he had committed, Adlerfelt was mortally wounded as a result of the seemingly random consequences of his own decision-making.

III.iii.1.8:

Uppsala ms. 107. Swedish.

Georg Planting (1682–1747) was from Edeby in the parish of Rasbokil, just north of Uppsala. He became a staff sergeant in the local Uppland regiment in 1701, and ended his career in the same unit thirty-one years later as acting captain. Captured at Poltava, he was transported to Saranski, and returned home after the signing of the peace treaty in 1721. In 1723 he married his cousin's daughter Catharina Juliana von Post (1699–1757) of Stavby, a neighbouring village; cf. IV.iv.2.3. He died not at Christmas but on July 21st 1747, at his residence in Tibble, and was buried in Rasbokil church. His eldest son Gustaf Planting (1724–1799) was matriculated at Uppsala on February 15th 1737, his second son Arnold Fredrik Planting (1728–1767) on December 16th 1738.

There were, therefore, various possible sources for the details Linnaeus retails concerning his death, and the way in which it relates to what had occurred earlier in his career: Elgenstierna, G.M., 1911/1942 X: 755.

III.iii.1.9:

London ms. 20, no. 37; Stockholm ms. 2, no. 4; Uppsala ms. 86. Swedish.

Jakob Wallrave (1688–1739) began his career in the treasury at Stockholm, and after a study-tour in Germany in 1718, concerned himself extensively with the improvement of the national transport system. This phase of his life came to an end about 1724 as a result of the lantern-incident, which involved a lieutenant Pihlman.

Wallrave matriculated at the university of Franeker, not Harderwijk, for the study of medicine on September 20th 1725, and for the study of jurisprudence on July 10th 1726. He took his doctorate in medicine on September 29th 1725 with a dissertation on the circulation of the blood, and his doctorate in jurisprudence on July 30th 1726 with a dissertation on state authority and civil liberty: *Album Studiosorum* 1968, 321–322; *Auditorium Academiae* 1995, 473.

On February 24th 1728 he was appointed professor of Roman law at Uppsala, after an extraordinarily protracted selection procedure. He was well-known in the university as being hotheaded and contentious. Though popular with the students, he was markedly less so with his colleagues and the university administration: Bergius, P.J., 1758, 202.

Linnaeus seems to have got on well enough with him, and rightly or wrongly, credited him with having helped him financially during his undergraduate period: Linné, C. von, 1888b, I: 24; Fries, Th.M., I: 122, 197. After Wallrave's death, he sat on the committee which considered the possibility of converting his house into a museum of technology: Annerstedt, C., Bihang III no. 131. Cf. the maps of Uppsala by Lars Hoffstedt (d. 1723) 1702, and Jonas Brolin 1770.

Wallrave died in the manner described by Linnaeus on February 15th 1739, after having come to terms with the university treasurer Nils Rommel (d. 1754), see III.iv.3.7, and spent the evening in a student dive run by Gottfried Kähler (1699–1766) and his son. It is recorded that among those who knew him, his character and unusual career created the distinct impression of his being involved in some unfathomable but overriding pattern of events – hence the placing of his case-history at this preliminary level within the general sphere of social rights.

III.iii.1.10:

London ms. 25; Uppsala ms. 177. Swedish. The two versions are neatly and evenly written and very similar, the only peculiarity in the second being the later insertion of "in order to purchase seed", a phrase which is also present in the first.

Unfortunately, although the Häme County Library holds all the old legal documents relating to the area, an extensive search there has failed to trace the case: letter, April 11th 1997, from the County Librarian, Ms. Sinikka Sipila, Lukiokatu 2, 13100 Hämeenlinna, Finland.

Linnaeus' source of information may well have been the same as for III.i.1.5 and IV.ii.2.7, that is, either the family friend Peter Bexell (1675–1724), proprietor of the inn at Diö (II.iii.3.2), or Jacob Flachsenius (1683–1733), his teacher at Växjö School (*Introduction* III.iii.3). It is very likely, therefore, that the events took place soon after the turn of the century.

Tavastehus, some sixty miles north of Helsinki, has always been on one of the country's main north-south trading routes. It grew up around a fortress dating back to the late thirteenth century, and was regarded as on the "frontier" on account of its being situated on the easternmost limit of the Swedish-speaking area. The corresponding county was divided into seven jurisdictional districts, the supreme court being at Åbo.

III.iii.1.11:

London ms. 29, no. 69; Uppsala ms. 162. Swedish.

Axel Cronhielm (1699–1739) was the son of Salomon Cronhielm (1666–1724), appointed councillor of state and ennobled in 1719, and the nephew of Gustav Cronhielm (1664–1737), head of the Swedish civil service and chancellor of the university of Uppsala. The family originated from Saxony, Axel's great-grandfather being ennobled in 1675 after having entered the Swedish legal and administrative system during the middle decades of the century.

He had his seat at Hakunge in the parish of Össeby-Garn, some nineteen miles as the crow flies from the centre of Stockholm. When visiting the capital during the winter, he evidently travelled by sleigh over the lake to Åkersberga, and then across the frozen Värtan, which is in fact the open sea and not a lake. His end came when he was crossing the Värtan on February 1st 1739.

He had served for a while as corporal in the life guards and colonel of a Saxon regiment, and was master of the horse in the cavalry-training and horse-breeding establishments at Kungsör and Strömsholm.

This is remarkable among the cases recorded in the *Nemesis Divina*, since it is the only one in which the person experiencing the retribution of God indicates that he is aware that this is what he is witnessing.

III.iii.1.12:

Uppsala ms. 76. Swedish.

Peder Wessel Tordenskjold (1691–1720), the Danish naval hero, was born at Trondhjem in Norway. He first played a part in the Great Northern War in 1711, when he was given the command of a small sloop. He had a reputation for being flighty and unstable, and on two separate occasions was unsuccessfully court-martialed for criminal recklessness, for unnecessarily endangering his majesty's war-ships. He gained the confidence of the king, however, and on account of the audacity with which he attacked any Swedish vessel in sight, regardless of the odds, the unique seamanship by which he invariably managed to escape capture, he won the heart and admiration of the nation.

The climax of his career came on July 27th 1719, when he finally took the Swedish fortress of Marstrand and captured the Gothenburg squadron which had been interrupting communications between Denmark and Norway.

After the termination of hostilities in June 1720 he travelled down to Hamburg, where like Linnaeus fifteen years later he went to see the famous stuffed hydra with its seven heads. During the evening of November 9th 1720, while attending a large social gathering in Hanover, he got into conversation with a group of five or six persons which happened to include Jakob Staël von Holstein (c. 1670–1730), owner of the stuffed hydra, major in the Swedish army, half-brother of an officer who had served with great distinction in the defence of Gothenburg. Tordenskjold expressed the view that exhibiting a fake was a cheap way of making money. Von Holstein asked him to explain. Tordenskjold replied in a very loud voice that the sharks who were doing so ought to be dealt with. Von Holstein reminded him that there was no need to shout, and asked him to step outside and explain himself. When outside, Tordenskjold declared that he had no intention of parleying with a scoundrel such as von Holstein, and struck him with his cane. A duel with sabres took place at the village of Gleidingen near Hildesheim three days later. Tordenskjold was wounded under his right shoulder and died soon afterwards.

Poul Løvenørn (1686–1740), chief Danish negotiator at the recently signed peace, was an eye-witness of these events, and supplied the Swedish nobleman Nicodemus Tessin (1654–1728) with a convincingly impartial account. The inhabitants of Gleidingen regarded von Holstein as a murderer, however, and chased him from the village with sticks and stones. In general, the Danish view of the matter has of course been somewhat different from that of Linnaeus. Friess, F.C., 1758, 106–107, in a passage ignored by Linnaeus, mentions the Tordenskjold case as an example of God's employing means we can never understand in order to bring about retributive justice: "It is rumoured that it was with like for like that a Dane has since taken revenge of this Swede for Tordenskjold's death – a talion apparently foretold by the renowned Danish poet Reenberg in the following verses ...". Friess then quotes some significant lines from the *Epitaph on Tordenskjold* by Tøger Reenberg (1656–1742). Cf. Pontoppidan, E.L., (1698–1764) 1739/1741 I: 102–103; Rothe, C.P. (1724–1784) 1747/1750; Gjörwell, C.C. (1731–1811) *Den swänska Mercurius* (Stockholm, 1755/1765) October 1764, p. 745; Oehlenschlæger, A.G., (1779–1850) *Tordenskjold* (1832); BL XV: 204–205 (1848).

It is certainly understandable that Linnaeus' account of Tordenskjold should be somewhat perfunctory. Nevertheless, the shooting of the boy in the rigging, taken together with the duel, does bring out a marked and consistent feature of his general character.

III.iii.2.1:

Uppsala ms. 19, 19b. Latin and Swedish.

Cf. note II.i.3.1; Håkanson, L., 1982/1983. The quotations from Seneca in the first section are as follows: Ad Lucil. CVIII.9; N.Q. I præf. 6. The "is plagued by having acquired what he wants" in the last section is also taken from Seneca; *De Consolatione ad Polybium* IV.2.7.

The Lohe family originated from Jever near Wittmund in East Friesland. Johan Lohe (1643–1704) arrived in Stockholm in 1658 to work as a bookkeeper in the family firm. After establishing himself as an independent wholesale trader, he invested in mining, and in 1682 bought himself into the sugar industry, acquiring control of a very lucrative business based in Södermalm, south Stockholm. He then branched out into financing the shipping by means of which his mining and sugar products were exported throughout the Baltic area. He was ennobled in 1704.

He had eighteen children, eight of whom were still alive when his wife died in 1731. Linnaeus is referring here to his sole male heir Conrad Anton Lohe (1685–1763), who even in his lifetime had become legendary on account of his miserliness. At his death, his assets consisted of a substantial interest in his father's factory and blast-furnaces, and were valued at over a million copper dollars. By his will, they passed to Fale Henrik Burensköld (1739–1779), who was not related to him, and who returned much of the property to Lohe's relations: SBL 24: 91–94.

On the death of Lohe's son Johan Magnus (1739–1759) at Uppsala, see III.iv.3.13.

In the preceding section there has been a progression from cases in which the relationship between individuals is merely physical or thoughtlessly spontaneous, to those in which the injury inflicted is the outcome of having pursued a purpose or seen the victim in context. The further consequences of this increasing awareness of the significance of the other person are worked out in the structuring of the rest of the material included in this exploration of the foundations of "the right common to all". Civil society is made possible by mutual recognition within a framework of law.

III.iii.2.2:

Uppsala ms. 155, 155b. Swedish and Latin.

After the death of Dietrich Nietzel in 1756, Linnaeus had to manage the botanic garden at Uppsala without a qualified keeper: III.i.3.10. The routine work was carried out by the elderly Löfgren, who had succeeded Lindstedt in 1748: III.i.2.6, but the remuneration being what it was, replacing Nietzel proved to be no easy matter. In the first instance, Lars Broberg (c. 1740–1795) was taken on as a help for Löfgren, but he proved to be so quickwitted, capable and ready to learn, that it was eventually decided that he should go abroad for a while and qualify himself for the position of keeper by taking employment in a number of renowned Danish, German and Dutch gardens. The scheme proved successful, and on his return he was appointed keeper of the garden by the university consistory, taking an oath on March 17th 1764: "to do all in his power to ensure that the garden is run properly". On account of his very modest wages, he was empowered to take on helpers and train apprentices. When he was subsequently accused of making money illegally by selling seed, Linnaeus defended him in the consistory: 13th October 1770; Fries, Th.M., II: 109, 215; *Letters* I: nos. 109, 113, 116.

It could be the oath involved in Broberg's employment as keeper which provides the key to the significance of this case-history, which at first glance does not seem to have any real point. Broberg did not confirm it in writing until May 6th 1769: *Letters* I: 198.

Fredrik Hasselquist (1722–1752), Linnaeus' pupil, travelled to Palestine in 1749 to investigate the natural history of the area. Although he died in Bagda near Smyrna on February 9th 1752, his collections arrived back safely in Uppsala, and in 1757 Linnaeus published an account of them and of his travels.

Adonis capensis, a species of pheasant's eye from the Cape of Good Hope; *Bulbos capenses*, bulbs from the Cape of Good Hope; *Valerianam tetrandrum*, a four-stamened valerian; *Antholyza cepacea*, an onion with a divided or retrogradedly metamorphosed flower; *Geranium foliis peltatis*, its leaf stalk being attached to the centre of the leaf; *Bocconia*, named after the Italian monk and physician Paolo Boccone (1633–1703); *Kalmiæ*, named after Linnaeus' pupil Pehr Kalm (1715–1779); *Gardenia*, named after Alexander Garden (1730–1791), a Scot who lived in South Carolina; *Magnolia*, named after Pierre Magnol (1638–1715) of Montpellier.

III.iii.2.3:

London ms. 15, no. 31, Latin; Uppsala ms. 185, 185b, Swedish.

Nils Kyronius (c. 1700–1784) was the son of a clergyman of the same name who held a living at Öregrund, and then at Björklinge just north of Uppsala. He became a town clerk at Uppsala, then a pub-owner (II.i.1.7), and finally a city councillor. His influence in municipal affairs is apparent in his prosecution of Herkepaeus in 1741 (IV.i.2.2) and his election to parliament in 1746, his personal integrity in his financial dealings with the widow of the clergyman Achatius Gåse (d. 1739), mother of Carl Michael Bellman's friend Erik Gåse (1711–1780): Fries, Th.M., I: 128.

In 1727 he married Maria Elisabeth Holmberg (1710–1785), a skipper's daughter from Stockholm, who bore him eleven children, five of whom married. After his disgrace his wife reassumed her maiden name, as did his unmarried daughters and two of his sons. The love-letters were evidently written to the daughter of the Vaxholm skipper Lars Mattson Broms (1673–1735) and his wife Kerstin Steen (1679–1754): Elgenstierna, G.M., 1911/1920 VI: 164–166.

The trouble concerning the mayoralty blew up in 1747, the judge mentioned by Linnaeus being Henric Julius Voltemat (1722–1765), who as well as functioning in a legal capacity also taught history at the university; he was the son of a military man of the same name (1689–1764), who retired with the rank of colonel in 1744: Anrep IV: 653–634.

Linnaeus was close enough to Kyronius to receive embarrassingly pleading letters from him once he had fled from Uppsala; as late as March 11th 1773 he was writing to him in German from Bremen concerning natural history collections: *Letters* III nos. 555, 556, 557. He evidently gained his inside information concerning him from his close friend, Kyronius' son-in-law, Olof Celsius (1716–1794), professor of history at Uppsala, subsequently bishop of Lund, who married Catharina Charlotta Kyronia (1728–1765) in 1744.

III.iii.2.4:

London ms. 29; Uppsala ms. 191. Swedish.

The wider background to Linnaeus' interest in this case is quite evidently to be sought in the political situation during the summer of 1746, when the applications for the post in Åbo were being considered. As a result of the military disasters of 1741/1743, the ageing and childless king Frederick I (1676–1751) had been forced by the empress of Russia to accept her nephew as his heir, and a Russian army division of 12,000 men had been stationed in the heart of the Swedish mainland in order to enforce the acceptance of the settlement. The Cap administration, reverting to type, had simply capitulated to Russian interests.

By 1746, however, Linnaeus' patron the Hat politician Carl Gustav Tessin (1695–1770), had begun to strengthen his position within the party by emphasizing the importance of Swedish independence, and to create some sort of cross-party consensus by finding common ground with the moderate Caps: III.iv.3.18; IV.i.2.11; IV.i.5.10. It is by no means unlikely that it was Tessin who called Linnaeus' attention to the Hauswolff case: Uggla, A.Hj., 1961.

Henning Wilhelm Hauswolff (1712–1746) came of an eminent and well-connected family: Anrep II: 206–209. On his legal career, see Wilhelm Gabriel Lagus (1786–1859) *Åbo hofrätts historia* (Helsingfors, 1843).

The ship carrying Hauswolff from Stockholm to Åbo ran into trouble south of the Åland islands on November 5th 1746. He and five others, including the captain Daniel Cameèn, were in the sloop for five days before it came ashore in the parish of Fleringe in northern Gotland. His burial in Visby on November 21st is recorded in the church book (p. 217, no. 59) as follows:

On November 10th 1746 there died at sea here off Gotland an assistant justice of the court of appeal in Åbo, the honourable and well-born gentleman *Hindric Willhelm Hauswolff*, after he and five other persons, in very great distress for five days and nights, had drifted in an open boat from the Åland sea, where their ship had sprung a leak and threatened to sink. They had tried to save themselves, but Hauswolff died exhausted by the cold and hunger some six miles from land off the parish of Fleringe. He was buried here in the town church on the twenty-first of the month. Age: about thirty.

It seems that two of the five came ashore alive and survived to tell the tale – captain lieutenant Andreas Matthiæson (b. 1714), who died on November 15th and was buried in Visby a week later, and Daniel Cameèn, who died on December 6th and was buried in Visby three days later (Church book nos. 60, 64). There is evidence that the captain's funeral attracted great public interest in Visby, disrupting the routine business of the city, including the delivery of documents to the city court: information supplied March 11th 1998 by Tryggve Siltberg, Archival Director, Landsarkivet, Visborgsgatan 1, Visby.

III.iii.2.5:

Uppsala ms. 111. Swedish.

Jonas Larsson Odhelius (1712–1794), son of Laurentius Odhelius (1664–1721), rural dean at Flo near Trollhättan, Västergötland, in the diocese of Skara, was the uncle of Linnaeus' pupil, the well-known physician Johan Lorentz Odhelius (1737–1816): Fries, Th.M., II: 401; SBL 28: 54–57.

District judge in various parts of Uppland between 1747 and 1762, he was eventually appointed president of the court of appeal. In the parliament of 1771/1772 he was elected secretary to the house of the peasantry, and did all he could to influence decision-making in the interest of the Caps.

He was put under arrest when Gustav III took over the government in 1772, but Linnaeus is mistaken in thinking that he died in prison.

III.iii.2.6:

Uppsala ms. 66. Swedish. Linnaeus refers to life guard Schleicher as 'Slichert' and to widow Von Bysing as 'von Byzen'. Kerstin Lindblom, Secretary of the *Linné's Råshult Trust*, first identified them correctly (1998), and has researched the case with the help of Stig Rudberg of the Dingtuna Local History Society, Västmanland.

Life guardsman Isaac Schleicher (1706–1740) came of a German family which had moved to Sweden during the closing phases of the Thirty Years' War. His grandfather Abraham Schleicher (1618–1696) was an accountant, and first settled in Stockholm. It is perhaps of some general interest to note that he was great-grandfather to the mother of Carl Michael Bellman (1740–1795). About 1660 he remarried into a Västmanland family from Ramnäs, a village some seventeen miles north-west of the cathedral city of Västerås, found employment as manager of the brass-foundry in the nearby village of Skultuna, and in 1675 acquired the freehold of the estate of Hummelsta, in the parish of Haraker just north of Skultuna. Much of the accountancy work in the area evidently consisted of looking after the financial affairs of the bishop, the bishop's wider family, and the local landed gentry and mining magnates. Abraham Schleicher's son Isaac (1672–1736) also took up the profession of accountant, married a Sarah Prytz (1682–1722), and between 1701 and 1722 had ten children by her, including the life guardsman Isaac. In 1721 Isaac the elder obtained official confirmation of his unencumbered possession of the freehold of the estate of Hummelsta. In 1735, just prior to his father's death, Isaac the younger, with the help of his wife's money, managed to buy out his siblings' interests in the estate.

In 1725 life guardsman Isaac, then only nineteen years old, married Catharina Eleonora Fägersköld (1696–1738), widow of Bogislav Johan von Greiff (1679–1719), an officer of Silesian extraction who had served with the Uppland regiment, and by whom she had two sons – Gustav Johan (1718–1787) and Fromhold Christer (1719–1783). His wife was the eldest child of Gustav Fägersköld, a nobleman and life guard killed at Kliszów in Poland on July 9th 1702, during one of the skirmishes preceding Charles XII's victory over Augustus and the crown army. She had two younger sisters – Mette Charlotta (1698–1743), who in 1717 married Crispin Jernfeltz (1694–1757), captain in the West Gothland cavalry, and grandson of Olof Pehrsson (1627–1693), the nefarious mayor of Kristinehamn, whose shady doings may well have been brought to Linnaeus' notice by Nils Reuterholm when he visited him in Arboga in the June of 1746 (IV.ii.2.9), and Christina Gustaviana (1699–1740), who in 1724 married Conrad Gustav von Siegroth (1694–1762), knight, colonel and factory-owner.

Through this marriage Isaac Schleicher gained a certain interest in the family assets of the Fägerskölds. These included the estate of Stockumla in the parish of Dingtuna some seven miles west of Västerås, where the three sisters had grown up, and where the widowed Catharina Eleonora lived with Schleicher until they moved to his family-home in 1737, the estate of Gäddeholm in the parish of Irsta just to the east of the cathedral city, where the three sisters lived as newly-weds, the Von Greiff, Jernfeltz and Von Siegroth children all being born there, and the estate of Stora Apelnäs, together with the farms belonging to it, in the parish of Roasjö, West Gothland. The sale of Gäddeholm in 1733 evidently gave rise to a certain amount of friction between the families of the three sisters, and it seems to have been the way in which Schleicher handled his interest in Apelnäs after his wife's death at Hummelsta on July 25th 1738, which brought this friction to boiling-point and gave rise to the frightful assassination recorded by Linnaeus.

The widow with whom Schleicher fell in love so soon after his wife's decease was Charlotta Catharina von Bysing (1703–1784), granddaughter of bishop Carl Carlsson (1642–1707) of Västerås, daughter of district judge Peter Johan von Bysing (1665–1713), former wife of life guardsman Carl Ludwig von Poll (1683–1739). In 1707 most of the bishop's assets and interests had passed to this woman's mother, Catharina Cederström (1682–1739). Since the death of her mother at almost the same time as that of her husband left her an extremely wealthy widow, it is by no means improbable that Schleicher's love was heightened by his awareness of the material advantages due to flow from the unification of the two families. The couple were in any case well-acquainted – although none of the Schleicher–Fägersköld children survived infancy, there were several christenings, and it is evident from the Stockumla church-books, the lists of those who stood as witnesses and godparents, that the efficient accountancy of the Schleichers had given them a certain standing in episcopal circles. In 1728 Helena Cederström (1690–1753), wife of bishop Sven Caméen (1667–1729), took Christina Eleonora (d. 1729) under her care, in 1731 Christina Elisabeth von Bysing (1708–1780), the widow's younger sister, did the same for Catharina Gustaviana (d. 1735), as did bishop Andreas Kalsenius (1688–1750) and the widow herself for Gustav Adolph (d. 1737) in 1736.

Whatever the background to the murder recorded by Linnaeus, the final outcome as recorded in the Haraker Parish Register on December 21st 1740, is stark enough: "Isaac Schleicher, corporal and life guardsman, in his home on the estate of Hummelsta, at nine o'clock in the evening on December 6th, as he was preparing to go to bed, was without warning wickedly and murderously shot dead. Three buckshot were fired through the east window, and all three went straight through him, one passing out through the west window. A frightful and terrible murder".

Although Kerstin Lindblom, in 1998, was the first to identify the persons mentioned here by Linnaeus, this atrocious outcome of social tensions and personal jealousies seems to have been long-remembered, for some twelve years ago (1987) the present owner of Hummelsta, who then knew little of the earlier history of the house, was visited by three elderly ladies from Stockholm, who asked him whether he was aware that a murder had been committed there some two hundred and fifty years before. It is quite clearly no longer possible to reconstruct a fully rounded picture of what gave rise to the murder. It could be a matter of some significance, for example, that three buckshot were fired, no more and no less, for in 1727 local jealousies may well have been roused when the widow's brother-in-law Natanael Franckenheim (1680–1742), veteran of Charles XII's campaigns, was ennobled and granted a coat-of-arms with three bullets on it, commemorating the injuries he had sustained in 1708, at Poltava, and during the Rügen campaign of 1715.

Linnaeus may have known who the murderer was, but if the civil authorities of the time failed to bring the person to justice, attempting to identify the culprit at this distance of time might appear to be a pretty hopeless task. Given what we know of the circumstances, however, it is reasonable to rule out a number of leading possibilities:

1) The person is not likely to have been a member of Schleicher's own family – the husband of one of his sisters – since the ownership of Hummelsta had

been settled in 1735, and in any case Linnaeus informs us that the widow was presented with “an” estate, not that she was granted what the guardsman had inherited from his father. Nor can it have been either of his step-sons, Gustav Johan (1718–1787) or Fromhold Christer (1719–1783) von Greiff, since both were still alive when Linnaeus died.

2) There is little possibility of the murderer’s having been a member of the Von Bysing family, since they were not Schleicher’s “in-laws” and the widow was the beneficiary of the love-gift. What is more, on July 5th 1741, only seven months after Schleicher’s death, his step-son Gustav Johan married the widow’s daughter Catharina Lovisa von Poll, which may account for the Cederströms’ having paid the tax on the Hummelsta estate in 1741.

3) The husband of Schleicher’s youngest sister-in-law was very prominent socially and very comfortably situated in his own right, and both his sons, born in 1725 and 1726, lived on into the nineteenth century.

There is a very distinct probability, therefore, that the person Linnaeus had in mind was Gustav Crispin Jernfeltz (1718–1767), eldest son of Mette Charlotta (1698–1743), the second of the three Fägersköld sisters, who was twenty-two at the time of the murder, and just the sort of person to react rashly and violently to what he regarded as Schleicher’s social pretensions and financial machinations. He had enlisted as a cavalryman at the age of sixteen, and only three months after the murder became involved in the military mobilization which preceded the declaration of the war on Russia at the end of July 1741. The social and political climate of the time probably accounts for the laxity of the legal enquiries which must have taken place after Schleicher’s death. Jernfeltz seems to have served in the war with some distinction, since he was promoted to the rank of corporal in 1742 and to that of lieutenant in 1744.

In 1745 Jernfeltz married Maria Elisabeth Insenstjerna (1721–1789), cousin to Anna Charlotta Adlerberg (1736–1767), the wife of Linnaeus’ close friend and colleague Abraham Bäck (1713–1795). Linnaeus may well have first heard of the murder from bishop Kalsenius and Nils Reuterholm during the early stages of his tour of West Gothland in the summer of 1746, but he almost certainly obtained his inside information on it through Bäck. He probably took a particular interest in it on account of Jernfeltz’s being descended from Olof Pehrsson (IV.ii.2.9), and in this connection it may be of some interest to note that although Jernfeltz’s family died out with his children, neither of whom had issue, the children of his younger brother developed distinct criminal tendencies, one being dismissed from the army for disorderly conduct in 1805, the other being sentenced for forgery in 1818.

Since Jernfeltz did not die until September 25th 1767, it is significant that this case-history does not occur in the London manuscript. Linnaeus quite evidently added it to his collection once he had written out the greater part of what we now know as the Uppsala manuscript during the autumn of 1765

(*Introduction* I.i.2). In the burial book of Lojo, a parish on the eastern shore of Lake Lohjanjärvi, some thirty miles west-north-west of Helsinki, we find it recorded that colonel Jernfeltz was buried on September 29th, after having died of "internal inflammation".

Cf. Elgenstierna, X: 694 (Von Bysing); Anrep I: 432 (Cederström); I: 902–903 (Fägersköld); II: 4–5 (Von Greiff); II: 386 (Jernfeltz); III: 717–718 (Von Siegroth); SBL 17: 256–257 (Von Greiff); 'Inventorium efter Fru Catharina Eleonora Fägersköld', *Svea Hovrätts Arkiv* E IX b: vol. 14; Lojo församlings kyrkoarkiv I C: 2, p. 191, *Åbo Landsarkiv*, Aningaisgatan 11, Åbo, Finland.

III.iii.3.1:

Uppsala ms. 61b. Latin.

III.iii.3.2:

Uppsala ms. 125. Swedish.

Hospitals in Stockholm had been supported out of public funds since 1527, children's homes since an ordinance of Gustav II Adolf dating from 1624.

Johan Bréant (1725–1766) was secretary to the parliamentary body concerned with this public service. Fredrik Georg Olander (d. 1768) was his notary. During the Cap-dominated parliament of 1765/1766, Jakob Serenius (1700–1776) recently appointed bishop of Strängnäs, was one of the most active members of the governing party: IV.ii.2.13.

III.iii.3.3:

London ms. 15, no. 29; Uppsala ms. 84. Swedish.

It was the task of a university bailiff to collect the rents from the farmers on the university estates. Since irregularities were liable to occur not only in the initial payments but also in the way in which the bailiff managed them, it was difficult for the university to keep a check on all the transactions involved. A bailiff had opportunities enough for cooking the books, and since the university remunerated him with nothing more than his accommodation and a hundred silver dollars a year, reason enough for doing so. What is more, in 1725 the office had been brought into line with that of the corresponding crown agents – a bailiff was no longer required to put up caution-money, but was employed simply on the basis of a promissory note: Annerstedt, C., III-2: 568–569.

In 1751 Westrin was suspended on account of complaints received from Östervåla, an administrative district some thirty miles to the north-east of Uppsala, concerning his conduct as bailiff. During the subsequent court-case, evidence against him was also provided by Linnaeus' friend Olof Celsius (1716–1794), professor of history, subsequently bishop of Lund, public prosecutor Per Grizell (d.c. 1773), like Celsius married to one of Kyronius' daughters (III.iii.2.3; III.iii.7.3), and Peter Julinschöld (1709–1768), the university treasurer (III.iv.3.7). The proceedings dragged on for twelve years, but on April 30th 1763 it was eventually decided that Westrin was guilty of embezzlement on two accounts and that he should be removed from office.

III.iii.3.4:

London ms. 16, no. 40; Uppsala ms. 186. Swedish.

Cf. III.ii.4.2; the case has also been placed here on account of Sperling's attitude to Krabbe.

III.iii.4.1:

Uppsala ms. 60, 69. Swedish and Latin. The last three sentences are entered separately at the foot of the page containing the case-history of Cicero (III.iii.4.2).

All the material relating to the defeat of Antonius at Mutina on April 15th 43 B.C., the appointing of the triumvirate on November 27th, and the horse-trading which followed, is taken from Plutarch's life of Cicero, see §46: "Cæsar is said to have contended for Cicero the first two days; but the third he gave him up. The sacrifices on each side were these: Cæsar was to abandon Cicero to his fate; Lepidus, his brother Paulus, and Antony, Lucius Cæsar, his uncle by his mother's side. Thus rage and rancour entirely stifled in them all sentiments of humanity; or, more properly speaking, they showed that no beast is more savage than man, when he is possessed of power equal to his passion".

Julia Domna (d. 217), born at Emesa in Syria, was the second wife of the emperor Severus Septimius (145–211), and the real mother of both his sons – Bassianus (186–217) and Geta (190–212). Gibbon observes that "if we may credit the scandal of ancient history, chastity was very far from being the most conspicuous of her virtues". She was, in fact, reputed to have committed incest with Bassianus, and publicly married him.

Cf. the description in Dio's *Roman History* (bk. 78, c. 2) of Bassianus' having arranged for his brother to be murdered by the centurion Antoninus: "he struck down Geta, who at sight of them had run to his mother, hung about her neck and clung to her bosom and breasts, lamenting and crying: "Mother that didst bear me, mother that didst bear me, help! I am being murdered". And so she, tricked in this way, saw her son perishing in the most impious fashion in her arms, and received him at his death into the very womb, as it were, whence he had been born".

III.iii.4.2:

London ms. 11, no. 5; Uppsala ms. 69. Swedish.

Popilius, a tribune of the people, born at Picenum, when accused of parricide, was saved from disgrace and death by the eloquence of Cicero. On December 7th 43 B.C., when the orator was sixty-three, he was slain by Popilius near Formiae. As in the case of III.iii.4.1, all the material here is taken from Plutarch's life of Cicero, see §47: "His assassins came to the villa, Herennius a centurion, and Popilius a tribune, who had once been prosecuted and defended by Cicero; and they had helpers".

The head and hands were sent to Rome and nailed to the rostrum, after Fulvia, widow of Clodius and wife of Antony, had bored through the tongue with her hairpin.

III.iii.4.3:

London ms. 12, no. 13, where Hallman's surname is given; Uppsala ms. 196, where it is replaced by four dots, inserted above, and then blotted out. Swedish.

Johan Gustav Hallman (1726–1797) was the son of the vicar of the Hedvig Eleonora church in Stockholm. He matriculated at Uppsala in 1742, graduated in medicine under Linnaeus, and on his recommendation then went abroad to take his doctorate at Padua (1750) and study the cultivation of the white mulberry. The attempt to introduce the tree into the Stockholm area with a view to furthering the production of silk (1754/1759) was not successful. Hallman was appointed court physician in 1756, and subsequently became member of the college of medicine (1773) and of the Commission for watering-places or spas.

In 1755 he married Charlotta Fredrika Kryger, daughter of the Stockholm wine-merchant Samuel Kryger (d. 1735), who brought with her the very substantial dowry of 120,000 copper plate, the equivalent of 80,000 silver rix-dollars. He was evidently engaged to her before he left for Italy in July 1749, and it is possible that what Linnaeus saw as his "hasty and unexpected departure" from Uppsala just before Christmas 1748, had something to do with his relationship with the "extremely beautiful gentlewoman". *Letters* II: 125, no. 255, Christmas 1748; IV: 253, no. 823, January 1754; 303, no. 869, September 1754.

III.iii.4.4:

London ms. 13, no. 14; Uppsala ms. 138. Swedish.

Sweden declared war on Russia on July 28th 1741, and within a month Russian troops had taken and plundered the important fortified town of Willmanstrand in south-east Finland. On September 3rd Henrik Magnus von

Buddenbrock (1685–1743), supreme commander of the Swedish land forces in Finland, was replaced by Charles Emil Lewenhaupt (1691–1743). Reinforcements were shipped over, a counter-offensive was planned, and by November 23rd Lewenhaupt had advanced his troops to within twenty-five miles of the Russian fortress of Viborg – a move which precipitated the palace revolution in St. Petersburg two days later. Instead of taking advantage of the changed political situation and continuing his advance, Lewenhaupt went into winter quarters, and when the Russians renewed their offensive during the spring of 1742, simply withdrew to Helsingfors. He was relieved of his command on July 19th, recalled to Stockholm, put on trial with Buddenbrock, and condemned to death: II.ii.3; II.iv.2.7; IV.i.4.3.

Buddenbrock was beheaded on July 16th 1743, but on account of a wave of popular opinion and an appeal lodged by his son, Lewenhaupt's execution was postponed for a fortnight. With the help of Eric Engberg (d. 1765) he escaped from prison on July 29th, and was recaptured on board a ship bound for Danzig on August 2nd. He was fettered, transported back to Stockholm, and executed on August 4th.

The reward offered by the crown for information leading to his recapture was never claimed. The two naval officers who actually arrested him were Sebald Hartman von Graman (1699–1766) and Carl Tersmeden (1715–1797): Malmström, C.G., III: 204–205; Tersmeden, C., *Memoirs* 1912/1919.

Bohus fortress just north of Gothenburg, first erected by king Håkon of Norway in 1308, was acquired by Sweden through the treaty of Roskilde in 1658. It was so thoroughly fortified that in spite of undergoing fourteen sieges it was never taken.

III.iii.5.1:

Uppsala ms. 56, 56b, the paragraphing being partly that of Linnaeus and partly editorial. Swedish and Latin.

The second sentence in the third paragraph is in Latin – a quotation from Juvenal, *Satires* VIII: 121–122.

The last sentence in the third paragraph is also in Latin – a reference to *Genesis* 4: 10; 18: 20; *James* 5: 4; perhaps also to *Ecclesiasticus* 35: 17.

The fourth paragraph is entirely in Latin; on Linnaeus' interpretation of the Lisbon earthquake, see IV.ii.1.2.

The final paragraph, also in Latin, is taken from the following passage in Seneca's *De Clementia* (I.18.2): "Even slaves have the right of refuge at the statue of a god; and although the law allows anything in dealing with a slave, yet in dealing with a human being there is an extreme which the right common to all living creatures refuses to allow. Who did not hate Pollio Vedius even more than his own slaves did, because he would fatten his lampreys on human blood, and order those who for some reason incurred his displeasure to be thrown into his fish-pond – or why not say his snake-preserve? The monster! He deserved to die a thousand deaths, whether he threw his slaves as food to lampreys he meant to eat, or whether he kept lampreys only to feed them on such food!"

Pollio Vedius (d. 15 B.C.) was a freedman's son and a friend of the emperor Augustus. The fish-ponds were on his estate near Naples which he named Pausilypon or 'Heartsease', still extant in the name of the modern town Posillipo: Pliny *Natural History* IX: 78; Dio *Roman History* bk. 54, c. 23.

III.iii.5.2:

Uppsala ms. 134. Swedish and Latin.

Abraham Söderberg (1728–1803), evidently shortly after having served in the Pomeranian war (1757–1762) as surgeon attached to the Uppland regiment, settled in Uppsala and set about getting himself appointed surgeon to the university, eventually succeeding, with the help of the king, in 1774: III.ii.1.2. It could be that his entertaining of Mrs. Dahlman, wife of Lars Dahlman (1701–1764), professor of moral philosophy, and of the von Linnés, had something to do with this furthering of his professional ambitions. The Dahlmans and von Linnés were in any case on familiar terms with one another: II.iii.2.5.

The meeting must have taken place on or about August 16th 1764 – the day on which Nils Wallerius (1706–1764), professor of theological controversy, died of a sudden attack of fever. On the character and subsequent misfortunes of his wife Anna Margaretha Boy (1712–1801), see III.ii.2.4.

Well acquainted as Nemesis is with the secrets of the heart, with our solitary thoughts and private utterances, it is only natural that she should judge and condemn from what she hears rather than from what she sees. Man is "obliged to learn of the three kingdoms of nature through the five external senses" (*System of Nature*, 1735, observations), but his awareness of things spiritual tends to be by means of hearing rather than sight, smell, taste or touch: II.iii.1.1; II.iii.2.1–II.iii.2.5.

God, the creator of nature, man and Nemesis, is immanent within the senses: "He is wholly and completely *Sense*, wholly and completely *Sight*, wholly and completely *Hearing*; and although both *Soul* and sense, he is also solely *Himself*' (*System of Nature*, 1758: 10; *Introduction* I.ii.2; III.iii.1). One of the constant themes of the *Nemesis Divina* is therefore that God sees and hears everything: II.ii.1; III.iii.8.1; IV.iii.2.1; IV.iv.3.2.

It is usually the case that omniscience is attributed to God on naturalistic grounds: Pettazzoni, R., 1956, 12. "The attribute of omniscience is not originally implicit in the idea of deity generally, but organically connected with the peculiar nature of all-knowing gods, who are all-knowing because they are all-seeing and all-seeing because they are luminous, as being in the first place sky- and astral-gods".

In the case of Linnaeus, however, the attribution derives from the conception of God as the creator of the world of nature known to us through the senses, and as the lawgiver and supervisor of our inner or ethical world: note II.iii.1.6.

III.iii.5.3:

London ms. 2, no. 48, 29, no. 71, more extended; Uppsala ms. 174. Swedish.

Linnaeus may well have noted this down from a newspaper report; the fact that the case is also included in the London papers indicates that he must have done so prior to the late summer of 1765; it is to be presumed, therefore, that the events took place around the middle of the century. The records of the Göta High Court at Jönköping are extant for this period, but they have not yet been worked through and analyzed: letter, March 24th 1997, from the Archivist, Göta Hovrätt, Lantmätargränd 4, 550 02 Jönköping.

Courts of this kind originated in an ordinance of 1614, by which the "king's judgement" or "court law" was invested in a body consisting of the chancellor, the executor of the king's justice, aided by a panel of assistants divided into three classes: four from the privy council, five from the nobility, and four "learned and experienced in the law". In 1615 the status and procedures of such courts were decided upon: they were to function as courts of appeal preliminary to final appeal to the king in council, they were to reach their judgements solely on the basis of the written evidence supplied by the lower courts, they were to have jurisdiction over a specific area – hence the establishment of such a body at Åbo for Finland in 1623, at Dorpat for Estonia in 1630, at Jönköping for Småland and Öland in 1634.

The seventeenth-century records of the court at Jönköping have been worked through, the analysis showing that the body made the final decision on ninety-four per cent of the death sentences passed by the lower courts and revoked seventy-seven per cent of them, with the result that between 1635 and 1699 the imposition of the death penalty in the area fell by seventy per cent. This constituted a very significant humanization of the principles implicit in Swedish common law and in the Mosaic code adopted by the state in 1608: Thunander, R., 1993; *Introduction* III.i.3.

It is to be presumed that this general tendency continued throughout the eighteenth century. The case recorded by Linnaeus, in that it involved the imposition of the death penalty, was therefore to a certain extent untypical of the functioning of the court during this period. Since he evidently regarded the fatal retribution for the thrashing as justified, one wonders whether he had any reservations about the docking of the salaries.

III.iii.5.4:

London ms. 27; Uppsala ms. 144. Swedish. In both manuscripts, Linnaeus notes that he had come across the case in Friess, F.C., 1763, and that Friess had taken it from Pontoppidan's *Annals*. In the London manuscript there are two words for "the lucarne windows in the church tower"; when Linnaeus copied the case out, however, he first wrote "kyrkiogluggarne", and then inserted "torns" between the two elements.

Friess (1758 ed., 77–79) presents the case as an example of talion working not by the ordinary means of evident laws, but by the extraordinary means of incomprehensible connections, there being no foreseeable link between lying and not prospering, having a father such as Grubbe and being reduced to begging, publishing an untruthful and scurrilous pamphlet and falling from a church tower. The case has been classified here simply as an example of the extent to which malice is at odds with the enforcement through an effective legal system of the respect due to others.

The bishop of Århus in question, Morten Madsen (1596–1643), married Kirstine Hansdotter (1602–1672) in 1623; it seems to have been on account of his mother-in-law's maiden name that he was sometimes referred to as Lælius. Their daughter Ellen (d. 1679) married the clergyman Christen Nielsen Bundtz (d. 1679) in Århus on July 2nd 1643. At about the same time, and just prior to the bishop's death on October 17th, the rumour got about that Bundtz had had a child by his mother-in-law which had been done away with. Erik Grubbe (1605–1692), the local justice of the peace, was appointed to investigate the matter. Evidently motivated in the way that Linnaeus maintains he was, he engaged the author and printer Hans Hansen Skonning (1579–1651), whom the bishop had sacked from his position as cathedral sexton in 1641, to give false evidence and to compose *Kattenis Rættergang med Hundene* (Aarhus, 1650), a satirical animal allegory in the manner of *Reynard the fox*, written in doggerel rhyming-couplets, which despite its distasteful background and homely style has its place in literature as the earliest original epic poem composed in Danish. In 1647 the court presided over by Grubbe decided that the bishop's widow was guilty and that Bundtz should be removed from office, a decision which was reversed on appeal two years later.

In 1690 Grubbe's property at Tjele near Viborg in Jutland was taken over by his son-in-law, who in January 1692 had him declared completely incapable of governing his own affairs. The singular careers of his three daughters have captured the imagination of various writers and given rise to a considerable literature. Marie (c. 1643–1718) entered on public life by marrying the future governor of Norway Ulrik Fredrik Gyldenløve (1638–1704) in Copenhagen in 1660, took the Jutish nobleman Palle Dyre (d. 1707) as her second husband in 1673, and after marrying a storeman cum organ-grinder in 1691, ended her days running an inn on the island of Falster: Holberg, L. (1684–

1754) *Epistler* no. 89; Pontoppidan, E. (1698–1764) *Annales* III: 170–173; Zwergius, D.G. (1699–1757) *Det Siellandske Clerisie* I: 552–558; Blicher, S.S. (1782–1848) *En Landsbydegns Dagbog*; Andersen, H.C. (1805–1875) *Hönse-Grethes Familie*; Jacobsen, J.P. (1847–1885) *Fru Marie Grubbe* (Eng. tr. New York, 1927); Kjær, S., *Erik Grubbe og hans tre Døttre, Anne Marie Grubbe, Marie Grubbe, Anne Grubbe* (København, 1904).

Skonning did in fact die by falling from the window about which he had lied; he was buried in Århus on April 16th 1651: Ehrencron-Müller, H., 1924/1939, VII: 367–370.

DBL III: 75 (Bundtz); V: 309 (Grubbe, Erik); V: 310 (Grubbe, Marie); IX: 325 (Madsen); XIII: 449–450 (Skonning).

III.iii.5.5:

Uppsala ms. 133. Swedish. In the last paragraph Linnaeus wrote "werld" or "world" instead of "wärd" or "host".

The absence of the case from the London ms. is understandable given the date of Dahlberg's death. Linnaeus may have heard of it through his pupil Nils Dahlberg (1736–1819), who studied under him at Uppsala from 1752 until 1763, and went on to become physician to crown prince Gustav (1768) and vice-president of the college of medicine (1773): *Letters V*: 319–526; his correspondence is preserved in the town library at Linköping.

Arvid Magnus Dahlberg (1718–1763), son of the famous warrior-autobiographer Alexander Magnus Dahlberg (1685–1772) and his wife Anna Margareta Holst (1699–1761), was born at Gävle: SBL 9: 601–603. He joined the Västerbotten regiment at the age of thirteen, was drafted to Finland, and captured by the Russians when they took Willmanstrand in August 1741: II.iv.2.7; III.iii.4.4. In 1744 he transferred to the Västmanland regiment as a second lieutenant, became staff lieutenant in 1754 and full lieutenant in 1757.

On February 12th 1757 he married Helena Elisabet Gyllengahm (1737–1817) at Romfartuna, which is some forty-five miles south-south-east of Hedemora, probably a few months prior to the encounter with the skittle-players. Later that year he was sent on active service to Pomerania, where he was captured by the Prussians at Boitzenburg on November 18th 1758, being released as part of the peace settlement on May 10th 1762: Sparre, S.A., 1930, pt. IV.

Since Dahlberg's parents retired to live in Karlshamn in 1751, and it was there that he died on February 20th 1763, it looks as though Linnaeus was misinformed concerning the identity of the host and hostess.

The case has been classified here on account of Dahlberg's attitude to his fellow soldiers – the legal system failed to provide proper recompense for the fatal outburst of pique. It is also of interest as yet another instance of the role horses play in the fulfilment of Nemesis: II.iii.3.6; II.iii.3.7; II.iv.2.4;

II.iv.2.5; III.iii.1.6.

III.iii.5.6:

Uppsala ms. 68. Swedish. Cf. Friess, F.C. 1758, 86; IV.iv.3.1.

Christian II (1481–1559), king of Denmark, Norway and Sweden, the last of the monarchs of the Kalmar Union (1397–1520), succeeded his father as king of Denmark and Norway in 1513. In Copenhagen castle on August 15th 1515 he married Elisabeth (1501–1526) granddaughter of the emperor Maximilian, sister of the future Charles V (1500–1558).

His succession in Sweden was less straightforward. At a council meeting held in Stockholm in 1512, Sten Sture the younger (c. 1492–1520), backed by Lübeck, had been elected regent. Opposition to this election was strengthened in 1514, when Gustav Trolle (c. 1488–1535), backed by Christian II and by Rome, was elected archbishop of Uppsala. Support for Sture was confirmed by the council meetings held at Arboga and Stockholm in 1517 and he was encouraged to resort to arms. In the course of the hostilities the archbishop was captured and forced to resign. Christian sent a fleet to the Stockholm area and landed troops, and after a number of serious setbacks the regent was defeated. Sture struggled back towards Stockholm, mortally wounded, and in the February of 1520 was buried there in the church of the grey friars, the Riddarholmskyrkan. Christian promised to govern according to the laws and usages of the country and not to exact vengeance for past events. On November 4th 1520 he was finally crowned as hereditary monarch of Sweden by Gustav Trolle in the Great Church in Stockholm, and the next three days were given up to congenial celebrations and banqueting. On the 8th and 9th of November, however, the king set about the extirpating of heresy – Sten Sture's remains, together with those of his little child, were exhumed and burnt, and in the ensuing bloodbath some eighty or ninety persons were executed.

Christian's having "gnawed Sture's bones in order to demonstrate how furious he was with the dead", is certainly in character, and is recorded by Olof von Dalin (1708–1763) in *Svea Rikes Historia* (2 vols., Stockholm, 1760/1761) II, ch. 20, §28, who gives as his source Johannes Messenius (1579–1636) *Scondia illustrata* (ed. J. Peringskiold, 13 vols., Stockholm, 1700/1705) IV: 89.

In 1523 Gustav Vasa established himself as king of Sweden and the duke of Holstein became Frederick I (1523–1533) of Denmark and Norway. Christian fled to the Netherlands, recouped his resources, and after having spent eight years in exile, returned to Scandinavia with a fleet and with the support of the emperor Charles V, in order to recover his kingdoms: he was defeated and captured, and spent the rest of his life in prison – sixteen years in the castle at Sønderborg, eleven years in the keep at Kalundborg. The groove made by his thumb in the table round which he walked while in solitary confinement at

Sønderborg, is still to be seen there.

The case has been classified here, on account of Christian's attitude to Sture and his political opponents in Sweden, as a particularly shocking instance of malice. Linnaeus does seem to imply, however, that it also has a wider significance – just as Christian would not allow the dead to rest, so was he not allowed to find rest in death.

III.iii.6.1:

Uppsala ms. 20. Latin and Swedish. On the opening sentence, see Ovid *Tristia* I.ix.5.6; Walther, H., 1963/1986, no. 6535; on the Latin concerning dogs and being bitten, Håkanson, L., 1982/1983, 95; IV.iv.2.1.

On Gustav Kierman (1702–1766), mayor of Stockholm, see IV.iii.2.6.

Linnaeus originally concluded the section by quoting *Proverbs* 17: 15, subsequently deleting the reference.

III.iii.6.2:

Uppsala ms. 69. Swedish. These three sentences are entered separately at the foot of the page containing the case-history of Cicero; cf. III.iii.4.1; III.iii.4.2.

The basic tone of Linnaeus' treatment of friendship certainly calls to mind this interesting variation on being hoist with one's own petard.

III.iii.6.3:

Uppsala ms. 121, 121b. Swedish. The quotations from *Ecclesiasticus* were noted down separately on 121b, and subsequently deleted.

Lars Sedlin (1719–1782) was curate in the parish in which Linnaeus acquired the two properties of Sävja and Hammarby in 1758 – hence his detailed knowledge of his family life: cf. II.i.1.8; III.i.3.13; III.ii.1.6; III.iv.2.4. Since he went on to become rector of the parish of Huddunge, some twenty-five miles to the north-west of Uppsala, the paralysis brought on by the stroke cannot have been too serious.

Although the mother-in-law had shown kindness and consideration, Sedlin and his new wife had not, and the retribution for this came through death in childbirth and through the stroke.

III.iii.6.4:

London ms. 15, no. 37, 20, no. 52b, Latin; Uppsala ms. 79, Swedish. In the Uppsala version Böstling's name is spelt incorrectly.

It looks as though Linnaeus must have taken note of the Uggla–Böstling involvement while he was working in Stockholm 1738/1741. Gottfried Böstling, merchant in Stockholm in 1727, was still alive in 1745. Samuel Uggla (1712–1782), a Stockholm lawyer, had as his first wife Charlotta Aurora Folcker (d. 1742), and as his second Maria Kristina Herkepaeus (1707–1772), born Nordenflycht, widow of the appeal court judge Christian Johannes Herkepaeus (d. 1736), whom he married on June 27th 1744; Anrep III: 49.

Although it is by no means clear here what is being retributed and with what degree of justification, the friendship shown by Uggla to the revenue official in Vasa is evident enough, and has determined the placing of the case.

III.iii.6.5:

Uppsala ms. 192. Swedish, without the heading, which has been taken from the first three words of the text.

Johan Gyllenborg (1682–1752) joined Stenbock's dragoon regiment at the age of eighteen and by 1706 had risen to the rank of captain. He was serving under general Adam Lewenhaupt (1659–1719) in Riga when news came in the March of 1708 that the king was planning to march east from the winter quarters at Smorgoni, and that the army in Livonia should prepare to march south and join up with him. Lewenhaupt did not actually receive marching orders until June 3rd, and was not able to get his 12,500 men, 16 cannon and huge baggage-train on the move before the end of the month. Half the men and most of the supplies were lost on the way south, the remnant of Lewenhaupt's forces not joining the main army until October 11th.

It was evidently during this march south that Gyllenborg arranged for the Russian captain to be spared. He himself fell ill and was lying incapacitated outside his tent two days after the battle of Poltava, when the retreating Swedish army, cornered between the Dnieper and its tributary the Vorskla near the town of Perevolochna, finally surrendered to the Czar. After his capture he was taken to Kasan, Solikamsk and Tobolsk, not returning to Sweden until 1719.

On June 8th 1720, on his estate of Skenäs near Vingåker, he married Margaretha Eleonora von Beijer (1692–1732). His son Gustav Fredrik Gyllenborg (1731–1808), after studying at Uppsala, had a distinguished career as a civil servant, writer, dramatist and poet. He himself became member of the privy council in 1741, and in 1742 chancellor of the university of Lund; IV.iii.1.2; Anrep II: 69–70.

Linnaeus knew the family well: Carl Gyllenborg (1679–1746), Johan's brother, was chancellor of the university of Uppsala from 1739 until 1746, and founder of the zoological museum there; Henning Adolph Gyllenborg (1713–1775), Johan's nephew, took part in the famous botanizing excursions: Fries, Th.M., I: 263; II: 9, 208; *Letters* VI: 229–237.

III.iii.7.1:

Uppsala ms. 61. Swedish and Latin.

The advice concerning the fulfilling of one's office is in Latin: cf. Luther, M., *Tischreden*, 1912/1921, III: 498, no. 336; Wallenberg, J., 1928/1941, II: 493; IV.iv.2.1; IV.iv.3.2.

Elias Frondin (1686–1761) took his degree at Uppsala in 1710 and became treasurer of the university in 1721. In 1729 he was considered for the professorship of poetry, and in 1730 appointed professor of history. He retired in 1747. In his teaching he was not quite such a nonentity as Linnaeus implies, since he went out of his way to encourage an interest in Swedish as opposed to general or classical history. His son Birger Frondin (1718–1783), also an historian, became university librarian in 1747.

From the very beginning of his academic career, Linnaeus was always very aware of the way in which his energy and enthusiasm for the subjects he was teaching set him somewhat apart from the general life-style of many of his colleagues: *Introduction* I.ii.3.

III.iii.7.2:

Uppsala ms. 182. Swedish.

On Maja Hierpe (d. 1773), her background, and the outbreak of fever, see II.iii.1.3.

Jonas Sidrén (1723–1799) joined the medical faculty at Uppsala as an assistant in 1752. In 1758/1759 his application for permission to give public lectures in surgery, obstetrics and medical law was rejected. In 1767 he was assessed as being "one of the most accomplished physiologists in the country", and appointed to a chair. In 1772 he drew up the list of duties Carl von Linné the younger (1741–1783) was expected to fulfil in the faculty: Fries, Th.M., II: 108, 141–146, 165; *Letters* I: 172–173; V: 3.

The case has been placed here not on account of Sidrén but on account of Maja Hierpe. Just as Linnaeus was rewarded for his industry and diligence with obloquy and plagiarism (III.iii.7.1), so the servant girl from Småland was rewarded for her dedication, love and loyalty with disease, death and violation.

III.iii.7.3:

Uppsala ms. 193. Swedish.

Per Grizell (d.c. 1773), as crown district commissioner for Uppland, was responsible for the general order and security of the area, and for supervising the collecting of taxes. The duties associated with the office had been regulated by an ordinance of 1687: cf. III.iii.3.3.

Anders Johan Blomberg (1752–1788), as district tax-collector, was technically under Grizell's supervision, actually able to call him to account for irregularities.

Anders Hamrén, as inland revenue clerk, was responsible for the final check on the tax-accounts.

On the financial collapse of the university treasurer Peter Julinschöld (1709–1768) in 1768, see III.iv.3.7.

The distilling of brandy in Sweden is first recorded towards the close of the 1460s. During the next century or so the harmful social and economic effects of the practice were such, that in 1550 Gustav Vasa issued a decree making it illegal. In 1638 authorized production was introduced, and taxed. In 1718 private production was strictly forbidden, and the proceeds from the taxation of the authorized production were used in order to finance the social services. In 1731 home production for domestic use was allowed, and in 1747 anyone was allowed to distil the liquor, as long as the production was simply for home consumption and duty was paid.

It seems possible, therefore, that Grizell began his career and built up his reputation as a tax-collector prior to 1731.

III.iii.8.1:

Uppsala ms. 58. Swedish and Latin.

Linnaeus' conception of the law can be seen as arising directly out of the case-histories and categories dealt with in the preceding section of the *Nemesis Divina*: III.iii.1.1.

His presentation of it as a decalogue is evidently meant to call to mind *Exodus* 20: 1–17 and *Deuteronomy* 5: 6–21: it is only the fourth of the Mosaic commandments, that concerning the keeping holy of the sabbath day, which has no equivalent in Linnaeus, and only the tenth of the Linnean injunctions, that concerning the avoidance of intrigue, which has no equivalent in the Old Testament texts.

In a manuscript concerning faith, morality and social life, now in the keeping of the Linnean Society of London and probably dating from the early 1750s (Malmeström, E., 1939: 63; *Introduction* I.i.1, I.i.2), one finds what is evidently a preliminary version of this attempt at reformulating the general principles of the law:

1. Acknowledge the God of the universe. 2. Beware of using abusive language. 3. Know that you are created to the glory of God. 4. Be grateful for benefits received. 5. Harm no one. 6. Avoid living polygamously. 7. Do not take what belongs to another: leave to each what is his own. 8. Do not call God to witness: He is not blind. 9. Do not appropriate under false pretences.

As one would expect from the whole tone of the *Nemesis Divina*, there is no reference in either of these contexts to *Matthew* 22: 36–40 or *Mark* 12: 29–31.

One noteworthy feature of Linnaeus' opening triad is the injunction concerning perjury, which is evidently the equivalent of the commandment concerning graven images: IV.iii.1.1–IV.iii.3.5; *Introduction* II.iii.2.

III.iii.8.2:

Uppsala ms. 40. Swedish.

These three lines are noted down separately, and would appear to have no particular connection with the material around them.

They have been placed here, in close proximity to the re-statement of the Mosaic decalogue, in order to emphasize the general obliqueness of Linnaeus' approach to these central issues: III.iii.1.1; III.iii.5.1; III.iii.5.2; III.iii.6.1; III.iii.7.2.

III.iv:

Cf. III.iv.1.2.

III.iv.1.1:

Uppsala ms. 2. Latin.

An analogue of this story, evidently based on *Matthew* 20: 1–16, is to be found in the thirteenth-century *Gesta Romanorum*. "A king proclaimed that whoever came to him would be granted all they requested. The noble and the rich requested dukedoms, counties or knighthoods, treasures of silver and of gold. Yet whatever they requested they were granted. Then came the poor and the simple, and they solicited a similar boon. 'You have come rather late', said the king, 'the noble and the rich have already been, and they have taken away all that I have'. Hearing this disturbed them exceedingly, and the king was moved. 'My friends', he said, 'though I have given away all my temporal possessions, I am still in possession of my omnipotence, since no one requested it of me. I therefore appoint you to be their judges and their masters'." Swan, C., 1905, no. CXXXI, 278–279.

Fortune was usually conceived of as closely associated with Plutus, and only in the second instance as the mother of mankind and the daughter of Jupiter: Patch, H.R., 1927, 61–65; Boethius, *Consolation of Philosophy* II.ii.5.

Cicero makes mention of her as nursing Jupiter, however, and Menander regarded her as "a divine breath or understanding which guides and preserves all things": *De Divinatione* II.85; Stobaeus I.vi.1.

It is this second conception which seems to have appealed to Linnaeus, and formed his high opinion of Fortune's timorous darlings. Linné, C. von, 1907, 240; Wissowa, G., 1894/1972, 2nd series, XIII: 1671–1673.

III.iv.1.2:

Uppsala ms. 2. Latin.

Aristophanes *Plutus* (388 B.C.): why is it so often the case that the ungodly prosper while the righteous remain poor and needy? The reason is that Plutus is blind: the tables will be turned once he regains his sight. How is it then, that he became blind? "Jupiter, jealous of mankind, brought it upon me. When I was a little chap, I used to boast that I would only benefit the wise, the good and the orderly. He blinded me so that I would be unable to pick them out. He is incurably envious of the good" (87–92).

III.iv.1.3:

Uppsala ms. 2. Latin and Swedish.

It is evident from the manuscript that III.iv.1.2 was inserted later, between this observation and III.iv.1.1. The original sequence of Linnaeus' thinking was, therefore, from Jupiter to the coronation of kings: cf. IV.i.5.1.

The basic thinking here is in fact a continuation of that in III.iv.1.1: it is certainly not literally the case that "pushers get nothing", or that "those who rush get very little".

The first Swedish king to be crowned was Erik Knutson in 1210. Christopher of Bavaria was crowned in Uppsala cathedral in 1441, as were all monarchs up to Ulrika Eleonora in 1719, with the exception of Christian II (1520), Christina (1644) and Charles XII (1697), who like Frederick I (1720) and Adolph Frederick (1751), were crowned in Stockholm.

III.iv.2.1:

Uppsala ms. 45. Swedish and Latin.

The penultimate sentence is in Latin: Juvenal *Satires* VIII: 121–122; cf. III.iii.5.1.

III.iv.2.2:

Uppsala ms. 56. Latin.

A reference to *Genesis* 4: 10; 18: 20; *James* 5: 4; perhaps also to *Ecclesiasticus* 35: 17; cf. III.iii.5.1.

III.iv.2.3:

London ms. 20, no. 49; Uppsala ms. 67. Swedish and Latin.

Erik Sohlberg (1660–1739) was known to Linnaeus from his visits to the mining-district of Falun in Dalecarlia during the Christmas vacations of

1733/1734 and 1734/1735. He was impressed by the drinking-culture of the area; "The foundry owners put a firkin on the table: the cups then race over the table as briskly as the piss-pot under it until the firkin is empty. I saw one of them incapable of lying down as the beer gushed out of his full stomach, and Erik Sohlberg squeeze his belly so that the beer spurted onto the wall opposite". He also found the manner of doing business worthy of note. Sohlberg proposed that Linnaeus should be tutor to his son on a study tour to Holland, with a salary of three hundred copper dollars a year. When the time for departure came, however: "No further mention was made of the salary. All I got was twelve copper plate. I could not now draw back from the journey, nor could I say anything to the old man who had housed and fed me for six months. The Sohlbergs already owed me an agreed sum of thirty plate for my assaying. I put everything into the hands of God, who had wonderfully advanced me into the world, continuing faithfully to serve my travelling companion, knowing that God always rewards us according to our deserts". Blunt, W., 1984: 77, 85.

The son, Claes Sohlberg (1711–1773), fellow student of Linnaeus at Uppsala, took his doctorate at Leiden in 1735, and later became a general practitioner and mine physician at Filipstad. Since he subsequently had a crown appointment as resident physician at the Loka mud-baths, a fashionable resort some twenty miles north of Karlskoga, it looks as though Linnaeus' concluding observation here may be something of an exaggeration: cf. Walther, H., 1963/1986, no. 5081; III.iv.3.4; III.iv.3.8.

III.iv.2.4:

Uppsala ms. 143. Swedish.

Linnaeus quite evidently knew of the family and the financial transactions of Thure Jacob Sylvander (1739–1772) on account of the interest he took in the parish of Danmark once he had acquired the properties there in 1758: cf. II.i.1.8; III.i.3.13; III.ii.1.6; III.iii.6.3.

Johan Funck (1703–1773) published on legal matters and had sympathies with the Caps politically. After beginning his legal career in the Stockholm court of appeal in 1724, he was appointed president of the court of appeal in Uppsala in 1743, and in 1762 lord lieutenant of the Uppsala area. He took as his second wife Anna Catharina (1738–1823), daughter of Fredric von Friesendorff (1707–1770), and died in Uppsala on February 24th 1773: Anrep I: 881, 893, 894; Annerstedt, C., III-1: 404. Cf. *Luke* 6: 31: And as ye would that men should do to you, do ye also to them likewise.

III.iv.2.5:

Uppsala ms. 41. Swedish.

Cf. II.i.2.1.

III.iv.3.1:

Uppsala ms. 17, 17b. Swedish, the last sentence being in Latin. The paragraphing is editorial.

III.iv.3.2:

London ms. 11, no. 4; Uppsala ms. 67. Swedish.

Though separated from III.iv.2.3 in the London manuscript, this case was noted down together with it in the Uppsala version.

Anders Jöranson (1678–1744) was a friend of Linnaeus' wife's family in Falun. he took part in the celebrations associated with Linnaeus' engagement to Sara Elisabeth Moræa (1716–1806) at the beginning of February 1735, and at the farewell party two weeks later, when Linnaeus was leaving for Holland with Claes Sohlberg, Jöranson gave him two gold ducats, a coin roughly the equivalent of the half-sovereign: Fries, Th.M., I: 196.

In the London ms. Linnaeus also mentions that Jöranson was a magistrate. This is omitted from the Uppsala version, evidently in order to emphasize the character of the self-made man.

III.iv.3.3:

Uppsala ms. 169. Swedish.

Olof Håkansson (1695–1769), in 1710, inherited from his father the farm at Lösen in Blekinge which the family held as tenants of the crown. He was first elected to parliament for the local area in 1726, and for the whole of Blekinge in 1731. He was vice-speaker for the peasantry in the parliament of 1734; speaker from 1738 until 1762, and then again in 1769.

The Russian ambassador, who had much to do with Håkansson under conditions of the strictest secrecy, reported back to St. Petersburg: "I can certainly say that I have never before come across a farmer as ingenious and intelligent as this, with such an insight into and knowledge of affairs in general, who speaks in such an effective and natural manner". His contemporary biographer the archivist Sigfrid Lorentz Gahm Persson (1725–1794), comments on his ability for making: "the most complicated subjects intelligible to those of his colleagues who had not had the opportunity to acquaint themselves with the business in question": *Svenska galeriet* 2, 1783.

He was typical of so many of the politicians of the period, in that although he remained generally loyal to his class, in this case the peasantry, he was open to bribes and influence from all quarters, taking money from the Russian, French, British and Danish ambassadors, and regularly switching allegiance between the Caps and the Hats.

The money he made from the way in which he functioned as speaker in parliament was well invested in Blekinge: he made credit available against security and at interest, he put money into industry, he purchased estates, and after his first wife died in 1742 he entered into a socially advantageous marriage.

His sudden death in Stockholm on November 18th 1769, soon after he had once again (98 votes to 52) been elected speaker of his house, probably accounts for Linnaeus' rhetorical moralizing. His sons by his second marriage lived well from his acquisitions, one of them Anders af Håkanson (1749–1813), who married in Stockholm in 1771, having a distinguished career in the law and politics: SBL 19: 573–580.

III.iv.3.4:

Uppsala ms. 123. Swedish and Latin. Cf. IV.iv.2.1.

Erland Carlsson Broman (1704–1757) came of a legal family and studied at Lund. He entered court and legal life 1722/1724, played a prominent part in parliamentary and privy council affairs 1734/1752, was appointed marshal of the court in 1741.

His first wife, Eva Johanna Drakenhielm (1704–1747), whom he married in 1726, died of drink. His second wife, Vilhelmina Magdalena Taube (1720–1757), whom he married in 1748, was the younger sister of Hedvig Taube (1714–1744), whom he had inveigled into becoming the mistress of king Frederick I (1676–1751). In a certain sense, therefore, his second marriage brought him into a family relationship with the crown.

After Adolf Frederick (1710–1771) had been accepted as heir to the throne and had married Louise Ulrica (1720–1782) the sister of Frederick the Great (1743/1744; IV.i.5.10), Broman looked to the future and set about meeting their pecuniary requirements by juggling with the state finances, a procedure facilitated by the introduction of non-convertible banknotes. In 1745 he persuaded Katarina Ebba Horn (1720–1781) to become the ageing king's mistress, and was rewarded by being given control of the Hessian budget.

Towards the end of his life his financial affairs got completely out of hand; just before his final bankruptcy he was reduced to borrowing at 12% and lending at half that rate. Between 1769 and 1774, his heirs were still struggling with his debts.

Carl Gustav Tessin (1695–1770) once observed that the key to Broman's career was that since he had been born without the resources necessary for supporting the sort of life he led, he had learnt from his earliest youth to improvise his way out of difficult situations. It is hardly surprising, therefore, that his schemes and policies were often as shady as they were ingenious. Tessin added that Broman was "always buying and selling, and could make more money out of a chicken than most people could make out of a Spanish mare". He never read anything, and all he ever wrote consisted of receipts, promissory notes and bank bills or cheques. Nevertheless, his financial dealings could hardly have avoided ending in the complete disaster which eventually overtook them. Moralists condemned him as "a person driven by baseness and self-interest, devoid of decency, completely lacking in honour". But he was also widely appreciated as jovial and easy-going, as simply too weak to deny himself or others the good things of life, as the dupe of his own resourcefulness and ingenuity: Fredrik Vilhelm von Ehrenheim (1753–1828) *Tessin och Tessiniana* (1819); SBL VI: 364–372.

On the final observation, which is in Latin, see Walther, H., 1963/1986, no. 5081; cf. III.iv.2.3; III.iv.3.8.

III.iv.3.5:

Uppsala ms. 192. Swedish.

Charles XI was born in Stockholm Castle on the feast day of St. Chrysogonus, November 24th 1655, and died there of stomach cancer after a long and painful illness, on Easter day, April 5th 1697. He was a dedicated, conscientious and God-fearing monarch, and by his foresight and reforms laid the foundations of the military achievements of his son Charles XII: cf. Roberts, M., 1967; 1973; 1979.

In many respects his "reductions" were the economic and social cornerstone of these reforms. Throughout the sixteenth and seventeenth centuries the crown had granted land and rents for services rendered, thereby creating a well-endowed nobility and diminishing the resources of the central government. Parliament made a serious attempt to curb this process as early as 1604, and in 1655 set up a body for the properly organized "reduction" of the power of a nobility which was far too often more mindful of its privileges than it was of its public duties.

The disastrous military outcome of the Scanian war of 1674/1679 made it clear to Charles XI that if the country was going to survive as a great power, radical reform was necessary. In the parliament of 1680, a coalition between the king's supporters, the sub-nobility and the commonalty pushed through the necessary legislation, instituting a very considerable strengthening of the power of the crown, and an equally significant reduction of the power of the upper nobility, all estates yielding more than a certain level of rent a year reverted to the crown. The process initiated was regulated by further enactments in 1682, 1686 and 1687, and by the end of Charles XI's reign the measures had increased the annual income of the central government by some two million silver dollars.

In 1690 Nicodemus Tessin (1654–1728) was commissioned by the king to redesign Stockholm castle. After the fire of May 7th 1697, as a result of which much of the old building was destroyed, a programme of radical reconstruction was embarked upon.

III.iv.3.6:

Uppsala ms. 194. Swedish and Latin.

Cf. III.i.3.4; the case has also been placed here on account of Lars Molin's "money-grubbing" and his son's debts.

III.iv.3.7:

Uppsala ms. 164, 164b. Swedish. *Job* 16: 12 is referred to but not quoted.

Peter Julin (1709–1768) succeeded Nils Rommel (d. 1754) as university treasurer in 1740. He was the son of one of the sisters of Johannes Steuch (1676–1742), archbishop of Uppsala, vice chancellor of the university, and took the name Julinschöld in 1756, when he was ennobled for the excellent work he had done in getting the university's finances in order. He had inherited and augmented a very considerable private fortune, and he dealt with the university's assets in much the same manner – the annual accounts were meticulously presented, salaries were paid on the dot, the investments made proved to be immensely rewarding. His general character inspired confidence and commanded respect, his personal life-style was anything but extravagant; he was, moreover, an intellectual and a scholar in his own right, at one time being in the running for the professorship of poetry (*Letters* I: 120; 8.10.1745). In 1755 the university salary he commanded was three times that of an ordinary professor; his influence on more purely academic matters such as appointments was by no means negligible, and to some extent reflected his political affiliation with the Hats.

His bankruptcy came as a complete surprise to nearly everyone. The university consistory was informed of it on January 29th 1768, and little more than a fortnight later Julinschöld was dead. Linnaeus, as member of the consistory, was directly involved in the protracted wheeling and dealing which followed, the final outcome of which was that on June 23rd 1769 the university settled for a total loss of some 20,000 silver dollars: Annerstedt, C., III-1: 422–425, III-2: 60–65, 548–559.

Linnaeus must have written to his friend privy councillor Anders Johan von Höpken (1712–1789) concerning the bankruptcy, since in the February and the March he replied as follows: "I thought at first that his misfortune, like that of so many others, had had its origin in the general credit-crisis then prevailing, and in the rapid and freakish fall in the exchange rate. I now realize with great dismay that the heart of the man was by no means pure .... It goes against the grain to speak ill of the dead, but comparing the list of his debts with his assets does not put him in a good light personally. I am sorry for those involved, since they have the prospect of a long wait and little return. Your reflection on the real cause of his fall is very reasonable and edifying. Were the truth of it to be generally accepted, no one could be blaming anyone else and each would find personal fulfilment in compassion, equity and being helpful". *Letters* VII: 173, 174, nos. 1569, 1570; cf. II.iv.2.7.

Here in the *Nemesis Divina* the essence of Linnaeus' reasoning concerning Julinschöld is as follows: he had relied upon family connections and intrigue in unjustly ousting Rommel from the treasurership, cf. III.iii.1.9; when treasurer he had in all likelihood misbehaved sexually and he had certainly continued to scheme underhandedly, cf. III.iii.3.3, IV.ii.2.6; in spite of his competence as a financier and his services to the university he had therefore been punished with bankruptcy, cf. III.iii.7.3, III.iv.3.11.

The actual historical context of the case is, however, less clear-cut: Rommel retired on a full salary in 1740, and continued to draw it at Julinschöld's expense for the next seven years, Annerstedt, C., III-1: 62–63; since like Julinschöld, Nils Rosén von Rosenstein's wife Anna Christina von Hermansson (1718–1782) was the child of one of archbishop Steuch's sisters, it was only natural that he should have been invited to take his meals with her family, Fries, Th.M., I: 181 note; although Julinschöld certainly used university assets to underpin his private financial dealings, von Höpken was right to regard the bankruptcy as part of a wider financial crisis brought on by Cap policies and a collapse in the price of corn, Annerstedt, C., III-1: 425.

As a result of the settlement the university made with Rommel, it acquired possession of his house, which it put at the disposal of Anders Berch (1711–1774), professor of commercial law, who took on responsibility for the university's finances after Julinschöld's collapse, Annerstedt, C., III-2: 62. It was Carl Fredrik Georgii (1715–1795), professor of history, who lightened the last fortnight of Julinschöld's life by mediating with the consistory: *Letters* II: 260, no. 358.

III.iv.3.8:

Uppsala ms. 188. Swedish and Latin.

Charles De Geer (1669–1730), son of the Amsterdam merchant and military man Louis De Geer (1622–1695), inherited the estate of Lövstad near Norrköping in 1692 from his unmarried uncle Emanuel De Geer (1624–1692). He renovated not only the estate but also the local church, both of which were burnt by the Russians in 1719, and endowed the church at Hammarby, near Linnaeus' home, with an altar-piece. He died at Lövstad on September 15th 1730: Anrep II: 550.

Carl Niklas Hultman (1696–1770) was of German extraction, his family having emigrated from Mansfeld to Sweden in 1647; his wife Christina died in 1782: *Svensk Slägt-Kalender för år 1888* (Stockholm, 1887): 315–316. He must have bribed his way into becoming barber-surgeon under the medical professors Olof Rudbeck (1660–1740) and Lars Roberg (1664–1742) at about the same time as Linnaeus was attempting to get himself established at the university: Annerstedt, C., III-1: 514.

Evald Ziervogel (1728–1765) was the university printer; he had the honorary title of professor and was also an expert numismatist: Annerstedt, C., III-1: 319; III-2: 356, 531, 542.

On the final observation, which is in Latin, see Walther, H., 1963/1986, no. 5081; cf. III.iv.2.3; III.iv.3.4.

III.iv.3.9:

London ms. 27; Uppsala ms. 179. Swedish. In both manuscripts Linnaeus notes that he came across the case in Friess, F.C., 1763.

Friess (1758 ed., 124–129) informs his readers that he had gathered it "by word of mouth from a trusted friend living in the Svendborg area on the island of Funen", and presents it as an example of talion working other than by means of evident laws, there being no predictable connection between not settling an account and having a man hanged, and unwittingly purchasing stolen goods and being hanged oneself. Linnaeus heightens this aspect of its significance by altering Friess's statement that it was "shortly afterwards" that the sheriff bought the two horses from the stranger.

Friess also presents the case as an example of "God's visiting the same suffering upon a sinner as he has caused another to undergo", and has a long note indicating its similarity to a famous case recorded in Pontoppidan, E., 1741/1752, IV: 478–480: As in the first case, a sheriff buys some horses from a farmer without settling the account and then looks as though he is going to lose his job; the farmer retrieves the horses and the re-instated sheriff has him hanged for it. In the second case, however, Kaj Lykke (c. 1625–1699) of Rantzausholm, the great landowner and womanizer, then buys the same horses from a sheriff by the name of Hans Lauridsen without settling the account, just prior to being arrested for conduct injurious to the dignity of the crown by public prosecutor Søren Kornerup (1624–1674); in 1661, while Lykke's estate is in the process of being confiscated, Lauridsen retrieves the horses and the public prosecutor has him hanged: DBL 8: 189–190; 9: 228–230.

The case has been classified here on account of its centering principally upon buying and selling.

III.iv.3.10:

Uppsala ms. 97. Swedish.

Anders Norrelius (1679–1750), after a basic training in oriental languages at Uppsala, travelled abroad for three years studying the same subject in Amsterdam, Leiden, Utrecht and possibly also Oxford, eventually returning home in 1719 when his funds ran out. Although he concerned himself extensively with rabbinical literature, even promoting the idea of establishing a university synogogue, he was not by nature suited for the ordinary academic work of teaching students and organizing projects. He first found his niche in the university administration and then in the library. In 1740 he put up the backs of the professorial body by not allowing them to borrow books and take them home, a measure which Linnaeus in particular found immensely annoying, and which he did not manage to get lifted until 1745.

Norrelius' disastrous first marriage (1726–1753) with Margareta Benzelstierna (1708–1772), daughter of archbishop Erik Benzelius (1675–1743), seems to have been entered into not so much in order to found a family, as to enlist the support and influence which might have enabled him to secure an academic position for himself: cf. IV.ii.2.4; IV.ii.2.14.

By the time he entered into his second marriage – with Anna Christina von Friesendorff (1726–1745) in 1744 – his view of the usefulness and purpose of the institution had quite evidently changed. She was the daughter of the military veteran Carl Magnus von Friesendorff (1701–1758), the nephew of the two maiden ladies who at one time inhabited Linnaeus' house at Hammarby (II.i.1.8), who was clearly incapable of furthering Norrelius' proposed academic career: Anrep I: 883; SBL 27: 589–592.

As Linnaeus indicates by means of his final observation, it is Norrelius' parsimony and his monetary relationship with his second father-in-law which constitute the central significance of the case.

III.iv.3.11:

Uppsala ms. 128. Swedish and Latin.

Johan Brauner (1712–1773) the agriculturalist was the brother-in-law of Linnaeus' colleague Johan Ihre (1707–1780), whose first wife Sara Charlotta Brauner (1714–1758) applied herself so devotedly to the development of the estate of Sandbro in the parish of Björklinge to the north-west of Uppsala: III.ii.2.2. The financial resources of Anna Brita von Kock (1727–1801) daughter of the mill-owner lieutenant Anton von Kock of Håkansbols in Värmland, widow of sergeant Lorens Möller, whom Brauner married on September 11th 1763, enabled him to develop the similar estate of Grenome in the parish of Stavby, some fifteen miles to the north-east of Uppsala. Brauner was an expert in both the basic techniques and the economics of estate management, publishing on such diverse topics as meadowland, deep ploughing and arable, clover, food crops and woodland, cattle, poultry and game.

He came of a well-known Småland family, began to study at Uppsala at about the same time as Linnaeus, and went on to become member of the chamber of commerce in 1739, to attend the parliaments of 1742/1743 and 1751/1752, and in 1760 to be awarded the title of councillor: Samuel Gustaf Hermelin (1744–1820) *Åminnelse-tal öfver ... friherre Johan Brauner* in the proceedings of the Royal Academy of Science October 16th 1776; Anrep I:

304; SBL 6: 118–123.

Linnaeus probably learnt of Brauner's financial involvement with Julinschöld through having to deal with the bankruptcy proceedings as member of the university consistory: III.iv.3.7. He so presents the case that one can hardly avoid concluding that the motivation behind Brauner's marrying found its proper punishment in penury and social degradation. The fact is, however, that the marriage took place after Julinschöld's collapse, and that Brauner's wife did not die until 1801.

III.iv.3.12:

Uppsala ms. 150. Swedish.

Johan Gustaf Gyllenkrok (1704–1764), gentleman of the court, master of the hunt in Västmanland, crown forester, married Johanna Elisabeth Cederström (1721–1795), daughter of marshal of the court Sven Cederström (1691–1767), on September 6th 1744. He lived on his estate at Frötuna in the parish of Rasbo, some eight miles east-north-east of Uppsala. His son Johan Georg Gyllenkrok was born on September 2nd 1745, studied at Uppsala, took a commission in the horse guards, and died of small-pox at Frötuna on November 6th 1766. His widow married Gregorius von Kothen (1723–1776), whose wife had died in 1764 and who had four children: Anrep II: 106, 491.

Linnaeus treats the ending of the family connection with Frötuna as retribution for family trouble-making visited upon the Gyllenkroks. When taken together with Gyllenkrok's treatment of his wife, however, it can also be seen as a fitting punishment for similar trouble-making among the Cederströms. Johanna's mother Charlotta Elisabeth Blixenstjerna (1702–1776) held considerable property in her own right. When her granddaughter Ulla Cederström had spurned her wishes and refused to marry Anders Cederström (1729–1793), she had simply transferred the whole of her possessions to her chosen heir, disinheriting her own descendants: Anrep I: 438.

III.iv.3.13:

London ms. 29, no. 72; Uppsala ms. 100, 100b. Swedish. The last sentence is written in the hand of Linnaeus' son Carl von Linné the younger (1741–1783): cf. I.i.

Anders Nordenflycht the elder (1674–1743) married Christina Rosin (1678–1756), daughter of a clergyman, and after a competent career as an accountant and administrator was ennobled in 1727. The family employed Linnaeus' services when he was working as a physician in Stockholm 1738/1741, and the daughter Hedvig Charlotta (1718–1763), renowned as a poetess, published verses in praise of his *West-Gothland Journey* (1747): Fries,

Th.M., I: 258, 362. His son, also Anders Nordenflycht (1710–1762), became a member of the chamber of commerce in 1728, and in 1736 married Jacobina Lohe (1696–1738), a twin, one of the eighteen children of the immensely wealthy sugar manufacturer and shipping magnate Johan Lohe (1643–1704), cf. III.iii.2.1, and widow of Baltzar Bunge (1692–1736), director of the Stockholm auction-room. After the death of his wife on April 28th 1738, he took up the position of director of a mining concern in Courland and married again, with Fredrika Juliana Auerbach von Quedlinburg (d. 1760): Anrep III: 50.

The unmarried brother who lived with the Nordenflychts, stole wood and was enticed forth on holiday, was Johan Fredric Lohe (1680–1741), who died at Wernigerode on July 2nd and was buried in the churchyard there.

Another brother, Adolphe Tobias Lohe (1683–1759) and a sister Johanna Lohe (1690–1759), lived together with the mother (d. 1731) in Stockholm at Lilla Nygatan 5. When this house was renovated in 1937, eighteen thousand silver coins dating from 1626–1741, together with other silver objects, were found under the floorboards.

The sole male heir to the family fortune was Conrad Anton Lohe (1685–1763), who married Anna Beata Skeckta (1710–1763), and whose son Adolph, as the younger Linné notes, died in 1759: SBL 24: 91–94.

Although financial preoccupations are central throughout, this is in fact a double case-history: Nordenflycht's unprincipled self-centredness has its reward in his being murdered; the eccentric self-centredness of the Lohes results in the extinction of the family.

III.iv.3.14:

Uppsala ms. 116. Swedish.

Eric Råfelt (1662–1724), born into the peasantry, rose on account of his ability in handling the finances of Charles XII's campaigning, and was eventually ennobled for his services.

His elder son Johan Råfelt (1712–1763), the main subject of this case-history, studied law at Uppsala and in 1731 started work in the Stockholm court of appeal. In 1747 he was appointed president of the court of appeal in Nerike, and in 1756 lord lieutenant for the Örebro area. Early in the 1740s he married a widow, by whom he had one daughter (b. 1745). His younger brother Anders Råfelt (1714–1740) served in the life guards, died unmarried, and was buried in the Clara church in Stockholm: Anrep III: 571. Linnaeus, while working as a physician in Stockholm 1738/41, may well have attended him in his final stages.

Antoinetta Maria Stjerncrona (1718–1773) was the only child of Peter Stjerncrona (c. 1690–1726) and his wife Elisabeth Amyn (d. 1719), daughter of a Gothenburg merchant. Her father had studied economics and commerce in England, France and Spain, and served in a diplomatic capacity in Madrid before returning to Sweden and improving his capital resources by lending money to the crown. Orphaned at the age of eight, she was put under the guardianship of Peter Galle (d. 1747), warden of Stockholm castle. She lived at Söderby in the parish of Österhaninge just south of Stockholm, and in 1735 married Carl Broman (1703–1784) – marshal of the court in 1748, appointed governor of the county of Stockholm in 1751: Anrep I: 321–322; IV: 170.

This case-history involves a new element: instead of financial retribution being brought about by a predominantly impersonal network of events, individual decision-making is now the crucial factor.

III.iv.3.15:

London ms. 12, no. 13b; Uppsala ms. 160. Swedish.

The Cederhielm family had close connections with Linnaeus' home parish of Stenbrohult in Småland. Germund Cederhielm (1635–1722) of Möckelnäs studied at Växjö School and Uppsala, and first made his mark in dealing with the Danish freebooters, 'snapphanarna' active in Scania after the province became Swedish in 1660: cf. III.ii.4.2. He was appointed lord lieutenant of the Skara area in 1712 when he was in fact too old for the job, and retired in 1716.

His son Germund Cederhielm (1661–1741) was a member of the law commission 1702/1716, and in 1719 lord lieutenant of Södermanland, when he was severely criticized for the inadequacy of the defences against the Russians. He was appointed president of the Göta court of appeal in 1723. His nephew Carl Vilhelm Cederhielm (1705–1769), together with Linnaeus, was one of the founding members of the Academy of Science: Fries, Th.M., I, Bil. vi.

The two bright sons were Carl Gustav Cederhielm (1693–1740) and Mauritz Bleckert Cederhielm (1695–1728). Although they began their careers promisingly enough in a legal and diplomatic capacity, high-life in Paris proved to be their undoing. The elder seems to have spent a comfortable time in prison, however, and in the course of time became something of a celebrity: it became customary for Swedish travellers of the better sort, notably the courtier, historian and poet Olof von Dalin (1708–1763), to pay him a visit, and he conducted an extensive correspondence with a whole range of experts on the Hobbesian issue of natural law and our duties to God and our neighbours: Anrep I: 418–420.

In this case-history it is the collective decision of the court in Paris which finally determines the fate of the family.

III.iv.3.16:

London ms. 25; Uppsala ms. 85, 85b. Swedish.

Cf. III.ii.2.4; the case has also been placed here on account of the financial issues involved, the distinct possibility of arson, and the attempt made by the university to ameliorate the situation.

III.iv.3.17:

Uppsala ms. 74, 74b. Swedish.

Since no mention is made of this case in the London ms., it looks as though Linnaeus must first have taken note of it after 1765, probably from a newspaper report. Stade is situated near the left bank of the river Elbe some twenty miles west-north-west of the centre of Hamburg. It came into the possession of Sweden by the Treaty of Westphalia and from 1654 until 1719 was the capital of the Swedish province of Bremen-Verden, a status which is still very much in evidence in its main municipal buildings. The surviving records of eighteenth-century legal proceedings in the area are very patchy, and despite a careful search have yielded no further information concerning the people Linnaeus is referring to.

The case-study has been placed here largely on account of the concluding observations: the ill-gotten gains of the father and the criminality of the daughter are set aright through the routine procedures of the established legal system.

III.iv.3.18:

London ms. 21, no. 73, nine lines, name spelt 'Blackwell', 'Blackwall', Swedish and Latin; London ms. 25, nineteen lines, name spelt 'Blackwall', Latin; Uppsala ms. 131, 131b, fifty-four lines, name spelt 'Blatckwell', 'Blackwel', 'Blackwell', 'Blackwähl', the first seven paragraphs mainly Latin, the last three mainly Swedish.

Alexander Blackwell (c. 1700–1747) came of an outstanding Scottish presbyterian family, his father Thomas Blackwell (c. 1660–1728) being professor of divinity at Marischal college in the university of Aberdeen, his brother Thomas Blackwell (1701–1757) being professor of Greek and later principal at the same institution, his sister being the mother of James Fordyce (1720–1796) the poet and popular preacher and of Sir William Fordyce (1724–1792) the physician: DNB 2: 142–143; 147–149; Jackson, B.D. in *Journal of Botany* 48 (1910) 193.

He received his earliest education from his father, and by the age of fifteen had acquired remarkable proficiency in both Greek and Latin. He was matriculated at the age of sixteen, and soon distinguished himself on account of his understanding of the classics and his mastery of "other useful learning". Although French was not taught at the college, he acquired a good working knowledge of it in his spare time, evidently with an eye to leaving academic life as soon as possible, entering business and seeing the world, since he quitted his studies without taking a degree and travelled down to London, where he found employment with a well-known printer by the name of Wilkins.

He turned out to be an excellent copy-editor and proof-reader, and took an interest in many other aspects of the trade. Once he had got on his feet as assistant to Wilkins, he married Elizabeth (c. 1700–1758), daughter of an Aberdeen stocking merchant, "a virtuous gentlewoman with whom he received a handsome portion", hired a house in the Strand and set up as a printer in his own right. Since he had not served his apprenticeship, however, he was sued and bankrupted by a syndicate of fellow traders, who forced him to spend two years in prison. His wife managed to pay off the debts and gain his release by applying her talent for painting to the delineation of medicinal plants. This work brought her into contact with Sir Hans Sloane (1660–1753), Dr. Richard Mead (1673–1754) and Isaac Rand (d. 1743), curator of the botanic garden at Chelsea, who recommended that she should take a house close to the garden, so that she could be easily supplied with fresh specimens: "She not only made th drawings with her own hands, but engraved them on copper plates, and colour'd them. Her husband explained their uses in several different languages, in order to make them acceptable abroad; and, from the produce, they maintain'd their family very well". *Gentleman's Magazine* 17: 424–426, September 1747; DNB 2: 144; Blackwell, E., *A Curious Herbal containing five hundred cuts of the most useful plants, which are now used in the practice of Physick* (2 vols., London, 1737/1739; German ed. 6 pts. Nuremberg, 1750/1773.

The couple became London celebrities as a result of this work. Mrs. Blackwell was permitted to present the publication in person to the college of physicians. They were visited at home by a whole series of eminent physicians, surgeons and apothecaries. Blackwell himself was encouraged by the success to devote himself not only to the study of medicine but also to estate management, the outcome being, *A New Method of Improving Cold, Wet, and Barren Lands: Particularly Clayey-Grounds. With The Manner of burning Clay, Turf, and Mole-hills; as practised in North-Britain. To which is added The Method of cultivating and raising Fruit Trees in such Soils* (London, 1741) 121 pp., Bodleian catalogue Vet. A4 e.1145. The work was dedicated to Cockin Sole (d. 1796), member of an ancient Kentish landowning family, and as well as giving evidence of Blackwell's knowledge of classical authors and Scottish farming methods, contains much sound advice on drainage, burning, planting, fish-ponds, compost, ploughing and orchard management. Evidently as a result of his expertise in such matters, and certainly before he had completed and published this book, he was engaged by James Brydges (1673–1744), first duke of Chandos, patron of Pope and Handel, Purgotti and Paolucci, to supervise the management of the grounds and gardens of his sumptuous residence at Cannons near Edgware: Hasted, E., 1797/1801 VI: 202, 575; Baker, C.H.C. and M.I., 1949, ch. 7: 152–162.

Jonas Alströmer (1685–1761) first settled in London in 1707 as a merchant's accountant. By 1710 he had established himself as a ship's broker with both English and Swedish rights, and the general experience thus gained led to his being appointed Swedish consul in 1722. In 1724 he returned to Sweden in order to devote himself to applying what he had learnt in England for the improvement of his country's agriculture and commerce. His main undertaking was an establishment at Alingsås some twenty-five miles north-east of Gothenburg devoted to the breeding and rearing of sheep and the manufacture of woollen goods. The enterprise was recognized by parliament in 1726, and also attracted a grant from the crown: SBL 1: 556–563; Anrep I: 38–39.

When the Hats eventually gained the upper hand in the parliament of 1738/1739 Carl Gustav Tessin (1695–1770) was elected marshal, and backing Alströmer's initiatives became a major aspect of national policy. Tessin himself took a great interest in the work of Alströmer and Linnaeus, thus preparing the ground for the foundation of the Academy of Science on June 2nd 1739. Alströmer was encouraged to contact K.M. Wasenberg, the Swedish minister in London, concerning the possibility of finding someone capable of improving still further the work being done on the experimental farms. Since Elizabeth Blackwell's *Herbal* was already gaining international recognition, since her husband's *New Method* had just been published and Chandos was known to be cutting down on his expenses and laying people off, it is hardly surprising that the choice fell on Blackwell. He took leave of his wife and child, promising that if things turned out well he would send for them, and travelling via Denmark arrived in Sweden during the May of 1742: Fries, Th.M., I: 260; SBH 2: 600–601; Dixelius, O., 1984/1985, 89–107.

Blackwell first travelled to Gothenburg to visit Nicolas Sahlgren (1701–1776), friend of Alströmer and director of the Swedish East India Company. He then stayed with Alströmer for a while in Alingsås, before travelling through West Gothland, Scania and Blekinge with his host's brother-in-law Johan Clason (1704–1790), making a survey of agricultural resources and methods. On December 6th 1742 the Chamber of Commerce requested that consideration should be given to providing him with financial support, and during the following year he submitted proposals for the improvement of agriculture and stock-rearing. On May 28th 1744 the king decreed that the cost of the journey to Sweden should be met. The ministry of trade and industry put him on an annual salary of two thousand silver dollars, granted him travelling money, freed him from paying tax and postage, and added a generous allowance for his family. He was then given the lease of certain crown lands at Ållestad in West Gothland on the understanding that he would make use of them for giving instruction in agricultural methods: SBL 4: 730–733.

This preferential treatment aroused the same reactions in Sweden as his activities as a printer did in London. During the course of 1744/1745 he published four tracts, in Swedish, on agricultural subjects, all of which were given hostile reviews. During the summer of 1744 Alströmer introduced him to Stockholm society, and as Linnaeus notes, the success of his practising as a physician there was somewhat mixed. The broker with whom he was lodging died under his treatment and he inherited his wife. On August 2nd Anders von Drake (1682–1744), an ardent Hat, president of the ministry of trade and industry, husband of Sophia Lovisa Psilanderhielm (1710–1758), relative of close friends of the Linnaeus family in Stenbrobult, also succumbed to what he prescribed: III.i.3.12; Selling, O.H., 1960: 85, note 37; Anrep I: 602; SBH 1: 250.

Blackwell's successful treatment of the king resulted in his being appointed a physician in ordinary on January 7th 1745, but this only served to increase the pretty general aggravation and suspicion. On February 8th 1745 Linnaeus wrote to his close friend the physician Abraham Bäck (1713–1795) that "a barber-surgeon's assistant, a charlatan Englishman (sic) by the name of Blackwell (sic), has been elevated to the position of physician in ordinary": *Letters* IV: 39, no. 639. Linnaeus' colleague Nils Rosén von Rosenstein (1706–1773) expressed the opinion that Blackwell "knew very little about medicine", Linnaeus' pupil Peter Jonas Bergius (1730–1790) maintained that he did no more than prescribe old wives' remedies, and often in an ignorant manner which he attempted to cover up by never expressing a firm opinion of his own: SBL 4: 731–732; Strandell, B., 1940.

When Linnaeus visited the experimental farm at Ållestad on July 3rd 1745, he was therefore strongly predisposed to giving it a bad write-up in his journal: inadvisable alterations had been made to the house and the fencing, the kitchen garden, the hop-fields and the potato and bean patches had been mismanaged, the fields had been ploughed too deeply and for the wrong reasons, the drainage system had been messed up: Linnaeus, C., 1747, 105–106.

Blackwell undoubtedly had a certain gift for putting people's backs up. It looks as though he must have got under Alströmer's skin by the summer of 1745, perhaps, as Linnaeus indicates, on account of what he had written in his letters home. By the autumn of 1746 he had certainly realized that it would be advisable to look around for alternative employment, since it was evidently to this end that on October 15th he first contacted Walter Titley (1700–1768) the British envoy-extraordinary in Copenhagen. Titley replied on November 5th and wrote again on December 24th, Blackwell receiving the letter about the beginning of January and replying on February 7th. During February he was summoned to Stockholm, ostensibly in order to give an account of the progress being made in the teaching of farming methods at Ållestad. He arrived on March 6th and was given eight days in which to prepare his report. The day after his arrival he received an unsigned letter from Copenhagen, written in the hand of Titley's secretary and dated February 10th. Two days later he made the "rash move" mentioned by Linnaeus. During an audience with the king arranged on account of his appointment as a physician in ordinary, he evidently drew his majesty's attention to the fact that a subsidy of

£100,000 was available if he was prepared to further the improvement of the country's relationship with Denmark and Great Britain. The king seems to have responded by pointing out the risk he was running by presenting such a proposition, by indicating the implications it had in respect of the constitution and the succession, and by referring him to the marshal of the court Erland Carlsson Broman (1704–1757). Broman informed Tessin, and on the following day, March 10th, Tessin summoned Blackwell for questioning and then had him arrested by Carl Gustav Löwenhielm (1701–1768) the minister of justice: Remgård, A., 1968; III.iv.3.4; SBH II: 111–112.

The constitutional issues raised by the proposed subsidy were indeed radical and far-reaching. Since the death of Charles XII effective constitutional and political power had been vested in the parliament and the privy council rather than the crown. The king held his position through his marriage with Ulrika Eleonora (1688–1741), sister of Charles XII and queen in her own right 1718/1720, and election by parliament. To acquire "sovereignty" by mean of an alien power, to be rendered financially independent of parliament and council by a foreign subsidy, would be to enter into an entirely new constitutional position.

Nevertheless, given the conflict of interests within the country, the question of the succession and the international situation, entertaining the possibility of such a position was by no means out of the question. Only nine years after the Blackwell episode there was a serious attempt by the court party to override Hat-Cap differences and get more power vested in the crown: III.i.1.4; IV.i.2.5. The king had no legitimate offspring. At the conclusion of the disastrous war of 1741/1743 and at the bidding of the empress of Russia, the Hats had settled the succession on her cousin Adolf Frederick (1710–1771), very distant descendant of Charles IX, prince-bishop of Lübeck, administrator of the duchy of Holstein-Gottorp. They had done so in the face of militant opposition from the Dalecarlians, who had almost stormed the capital during the midsummer of 1743, and from Denmark, where it was hoped that the far better hereditary claims of crown prince Fredrik (1723–1766) would be recognized. In fact until the July of 1744, acceptance of the settlement had had to be enforced by the quartering of a Russian army division of 12,000 men in the heart of the Swedish mainland.

On December 11th 1743 crown prince Fredrik of Denmark married Louise (1724–1751), daughter of George II of England, and on the death of his father Christian VI on August 6th 1746, succeeded to the throne. During the early months of 1745 there had been some Anglo-Danish diplomatic activity centred on the idea that Fredrik's sister Louise (1726–1756) might marry George II's youngest son William Augustus (1721–1765), the victor of Culloden. It was certainly the case, therefore, that while Blackwell and Titley were corresponding, both Denmark and Great Britain had a certain interest in improving relationships with the Swedish court: DBL 4: 540–542. Doing so by means of such a subsidy was, moreover, commnon practice at the time. The attack on Russia in July 1741 had only become a real possibility once the French subsidy had been secured in the February, II.iv.2.7; prominent politicians such as Olof Håkansson (1695–1769) throve on the pecuniary advantages accruing from their foreign contacts, III.iv.3.3; between 1764 and 1771 the British Civil List disbursed not far short of £100,000 on Swedish corruption: Roberts, M., 1967, 291.

Tessin was chairman of the chancery committee which prosecuted Blackwell. The defendant maintained that he had done no more than put to the king the proposal of improving the country's relationship with Denmark and Great Britain. Tessin was intent on uncovering a plot, on eliciting from Blackwell the admission that he and not the king had first raised the issue of the constitution and the succession, that he had been purposely pursuing a policy hostile to the government. Although the statements Blackwell made on the three occasions when he was put under torture (April 1st, April 15th, May 6th) appeared to confirm Tessin's contentions, they were too incoherent to be taken seriously, and on May 21st it was decided that there was no point in further interrogation. On June 5th the committee found him guilty of "activity prejudicial to the good order of society" – plotting to employ foreign money for the setting aside of the established government and succession – and the death-sentence was passed five days later: Remgård, A., 1968, 191–236.

On the eve of his execution, August 8th (N.S.), he made his confession to Erik Tolstadius (1693–1759), the well-known pietist preacher. In fact there is no evidence at all for Linnaeus' assertion that Blackwell was an "atheist", in any ordinary sense of the word, or even that he ever considered questioning the basic tenets of religion inculcated in him by his father. It would, indeed, be an extremely instructive exercise to compare the basic principles of Linnaeus' own theology (*Introduction* III.iii.3) with those put forward by Thomas Blackwell the elder in his *Ratio Sacra* (Edinburgh, 1710) and *Schema Sacrum* (Edinburgh, 1720), especially in respect of atheism, deism, predestination, and the complementarity of natural and revealed religion.

Blackwell's execution took place in the middle of Stockholm, with much publicity and in the presence of a huge crowd. He met his fate with remarkable fortitude. When the executioner indicated that he was kneeling on the wrong side of the block, he remarked without the least emotion that it was scarcely to be wondered at that he should need a little instruction, since it was his first experiment in being beheaded: Anon., *A Genuine Copy of a Letter from a Merchant in Stockholm* (London, 1747): 17.

It looks very much as though Tessin exploited Blackwell's dabbling in high politics in order to further his own political objectives. He was then preoccupied with the process of strengthening his position in the Hat administration and creating some sort of cross-party consensus by finding common ground with the moderate Caps. In December 1746 he had failed in his attempt to become president of the chancery; by the end of 1747 he had reversed this failure, and largely as a result of having strengthened his position by the successful prosecution of Blackwell. The well-publicized outcome of the trial also served as a salutary warning to his political opponents of what might happen to them if they refused to co-operate with him. It therefore helped to ensure that when parliament dispersed in December 1747, Tessin and his council were left with a clear mandate for shaping national policy, which they continued to act on for the next four years. In foreign affairs this policy involved attempting to free the country from swinging between a French alliance hostile to Russia and a Russian alliance hostile to Denmark, and developing a more independent position for itself by cultivating an alliance with Prussia. In 1744 Tessin had played an important part in arranging the marriage between Adolf Frederick the heir to the throne and Lovisa Ulrika (1720–1782) the sister of Frederick the Great. Highlighting the Blackwell "conspiracy" helped to bring out the soundness of this policy, and during the trial Tessin was constantly hammering the point home among his colleagues.

Thure Gustaf Rudbeck (1714–1786), Cap military man and politician, firm supporter of Adolf Frederick (IV.i.2.11), passed on to Linnaeus' friend the surgeon Abraham Söderberg (1728–1803) the information concerning the corpse discovered in Tessin's house (III.ii.1.2; III.iii.5.2). The precise construction put upon the discovery is by no means self-evident. Presumably Rudbeck was of the opinion that Tessin had had the courier murdered because he might have provided evidence that Blackwell was not as guilty as he was finally deemed to be. Carl Nyrén (1726–1789), who had close contacts with Alströmer, was deeply involved in the Alingsås project and studied under Linnaeus at Uppsala, published anonymously a short account of Blackwell's Swedish career: *Alexander Blackwells Lefvernesbeskrifning* (Norrköping, 1763).

The case-history has been placed here on account of the central importance of the subsidy on offer to Fredrik I, and the fact that in this particular instance monetary considerations failed to play a determining role in national decision-making.

III.iv.3.19:

Uppsala ms. 12. Latin.

Seneca *De Consolatione ad Marciam* XII.4.1-2. Linnaeus reverses the sequence of the two sentences, which he notes down on a separate sheet, apparently out of any particular context.

They have been included here in order to round off the first three main sections of the *Nemesis Divina* – at this point the biological, psychological and sociological aspects of individuals give way to national, ecclesiastical and religious involvement. As in the case of Everyman, what goes "with us to the end" is not beauty, strength, five wits or discretion, nor is it fellowship, kindred and goods: it is the knowledge and good deeds of our higher ethical and religious life: *Everyman* (c. 1500); Christopher John Wortham, 'An Existentialist approach to Everyman', *Journal of the Australasian Universities Language and Literature Association* 19 (1978): 333–340.

IV.i:

Cf. IV.i.1.1.

IV.i.1.1:

Uppsala ms. 31. Swedish.

Collective responsibility, God's judging not simply individuals but also nations, one of the central themes of the Old Testament, is given due emphasis both in the Pauline epistles and in most seventeenth- and eighteenth-century works on Divine retribution. Erik Pontoppidan (1698–1764), for example, in his widely-read *Mendoza* (3 pts. Copenhagen, 1742/1743; Germ. tr. 1747, 1759^4; Dutch tr. 1749, 1776^4; Swedish tr. 1749, 1771^2), which recounts in a series of letters how the Asiatic prince of that name travels the world in search of Christians and comes across very few of them, raises the issue in epistle fifty-one. Friess, F.C., 1758, does so under the general heading of retributive justice working other than by means of evident laws (§7: 58–64): "I call God's *Retributive Judgement* national when it is passed on whole countries and peoples, when God punishes general national sins as it were, and does so by visiting upon them the most requisite and therefore the most apposite national plagues. By national sins I mean those common to a whole people and country – general lack of respect for God's word, general blood-guilt, general pride, licentiousness, lasciviousness, and other such vices".

By and large, Linnaeus' treatment of the topic, like the rest of the *Nemesis Divina*, is based upon the case-histories of individuals. The central theme comes through by means of the administrative and national contexts in which these individuals are acting – parliament (IV.2), ministerial responsibility (IV.5), military leadership (IV.4), kingly and imperial power (IV.5). The *Nemesis Divina* as a whole might be regarded as a philosophical assessment of that distinct phase of Swedish history which began with the accession of Gustavus Vasa and closed with the death of Charles XII.

IV.i.1.2:

Uppsala ms. 40. Latin.

Virgil *Eclogues* I: 71–72; *Aeneid* X: 472; Seneca *De Consolatione ad Marciam* XXI.6.4; cf. II.i.3.2, no. 34.

After the second line Linnaeus wrote, in Swedish: "Cato can turn Caesar's son into Brutus", and then deleted it.

IV.i.1.3:

London ms. 29, no. 70; Uppsala ms. 157. Swedish and Latin. No heading.

It is evident from the London ms. that Linnaeus had this case by word of mouth from Edvard Carleson (1704–1767), who had visited Paris during his period abroad 1726/1730, when he had travelled through Denmark, Holland, England and France, and made an extensive study of the agriculture, industry and commerce of these countries: SBH 1: 164–165; SBL 7: 428–441; see especially his *Journal* (1726/1727) in the Royal Library in Stockholm, sign. M.229; cf. II.iii.1.2.

Carleson was in Constantinopel in a diplomatic capacity from 1734 until 1745, when he corresponded with Linnaeus concerning cotton and tobacco: *Letters* II: 287; Fries, Th.M., II: 98, 241. He had Hat affiliations, became a member of the Academy of Science in 1746, and was one of the committee which interrogated Blackwell in 1747. He became president of the Chamber of Commerce in 1762.

The case has been placed here as an instance of the ineffectiveness of state power, the evident inefficiency of an institutionalized legal system.

IV.i.1.4:

London ms. 15, no. 36, Latin; Uppsala ms. 72, 72b, Swedish, duplicated material deleted from the first sentence of the penultimate paragraph. Cf. IV.iv.1.4.

Lorentz Christoffer Stobée (1676–1756) was the son of Andreas Stobaeus (1642–1714) professor of poetry at Lund, and the elder brother of Kilian Stobaeus (1690–1742), Linnaeus' landlord and tutor 1727/1728, while he was studying at the Scanian university: II.iii.2.3. It seems very likely, therefore, that Linnaeus gathered this information verbally from his tutor, and that the inaccuracies in it are due to his not having noted it down until long after it was communicated: II.ii.3.

Lorentz Stobée excelled in mathematics at Lund, and at the age of twenty-three was sent to Finland as a lieutenant responsible for fortifications. In 1702 he was promoted to the rank of captain and made responsible for the fortifications at Keksholm and Viborg. In 1706 he was promoted to the rank of lieutenant-general, and the year after ennobled.

Stobée and his wife Christina Åkerhielm (1691–1716) were in Viborg in 1710 when the place was besieged by the Russians for thirteen weeks. When it was eventually taken, the Russians violated the terms of the capitulation and sent Stobée off to work in the Czar's parks and gardens. He proved to be so effective in this employment, that he was offered the rank of major if he would take a commission in the Russian service. When he refused he was banished to Arkhangelsk.

He escaped in 1715, and after a trek of some two thousand miles got back to Sweden. Johan Kock von Gyllenstein (d. 1724) had told his wife that he was dead. The Holstein party came into political prominence as soon as Charles XII was dead. Since the king was childless there was no direct heir to the throne. The elder of his sisters Hedvig Sophia (1681–1708), who had married Frederick IV (1671–1702) of Holstein-Gottorp in 1698 and given birth to a son Charles Frederick (1700–1739), was the focal-point of the Holstein party programme. In 1725 her son married Anne (1708–1728), a daughter of Peter the Great (1672–1725) and his second wife Catherine I (1684–1727). The younger of his sisters Ulrika Eleonora (1688–1741), who had married Frederick of Hesse (1676–1751) in 1715, became queen in her own right for two years, and was then succeeded by her husband, who reigned in his own right as Frederick I until his death. Such marital affiliations and complicated succession-rights heightened the national and international tensions inherent in the Swedish scene in the immediate post-war years, and gave rise to the exercise in impersonation which cost Anders Gyllenroth, executed in Stockholm April 23rd 1722, his life: Anrep II: 120.

In 1723 Stobée was put in charge of fortification works in Scania and in 1730 made responsible for those at Stralsund. In 1739 he was appointed director general of fortifications, and soon afterwards became a member of the Academy of Science. In 1749 he retired to his estate of Ågård near Lidköping in West Gothland. He wrote both Latin and Swedish poetry: SBH II: 538.

Since this case is not headed 'Gyllenstein' or 'Gyllenroth', there would appear to be good grounds for placing it here as an instance of the relative effectiveness of state power, the evident efficiency of the institutionalized legal system.

IV.i.2.1:

London ms. 20, no. 52; Uppsala ms. 25. Swedish.

In Linné, C. von, 1968: 175, this is included not as a general heading but as a case-history. Evidence that Linnaeus thought of parliamentary affairs as yielding a distinct category of case-histories is to be found in the account given by Johann Beckmann (1739–1811) of his visit to Uppsala in the winter of 1765: "Court Physician Linnaeus told me that it was not only in the things of nature that he had found so many proofs of God's Providence .... He assured me that there were cases of this kind in the present Parliament": *Introduction* I.i.3.

Josias Carl Cederhielm (1734–1795) had experience in commerce and in the diplomatic service before entering parliament as a Cap in 1760. Carl

Fredrik Pechlin (1720–1796) distinguished himself in the Pomeranian war before entering the same parliament as leader of the Land party, a group of disgruntled Hats who had found common ground with the Caps. In January 1761, however, he transferred his allegiance back to the Hats, a move which led to retaliation from Cederhielm, who on August 24th 1761 eventually managed to get him excluded from the house of nobility: Anrep I: 420; SBL 8: 51–59; 28: 772–775; IV.i.2.8.

Carl Estenberg (1728–1815), a lawyer and a man of letters, first made his mark by capitalizing on the general disillusionment with the Pomeranian war (1757/1762) – by attacking those who had started it without parliament's consent. On June 30th 1765, after Cederhielm had courted unpopularity by criticizing parliament, Estenberg and his supporters had him excluded from parliament for expressing views detrimental to the dignity and interests of the institution: SBL 14: 622.

Rudman Bergenstråhle (1724–1813) managed to get Estenberg excluded in 1766, and on August 31st 1771 was himself excluded for servility to the crown, those engineering the move being the Caps Fredrik Ulrik von Essen (1721–1781) and his friend Claes de Frietzcky (1727–1803), together with Claës Wilhelm Grönhagen (1732–1777). SBL 3: 468; 14: 571–576; 10: 439–452; 17: 381–383.

The name of Daniel Tilas (1712–1772) noted down at the foot of the entry would seem to indicate that it was he who called Linnaeus' attention to this sequence of tit for tats. It may well have been one of the few things on whicn they were in agreement. Tilas was a founding member of the Academy of Science, a mineralogist of some repute, and in 1762, when Linnaeus had to have his coat of arms drawn up, the herald in charge of the procedure. By and large, they found themselves at odds whenever a joint opinion was required of them: *Letters* II: 5, 255, 327, nos. 179, 355, 419; Fries, Th.M., I: 325; II: 351.

IV.i.2.2:

London ms. 26, Swedish and Latin; Uppsala ms. 190, Swedish.

Per Herkepaeus (1691–1764) came of a Finnish clerical family, one of his forebears, Erik Herkepä (d. 1578) having been bishop of Viborg. His father, Andreas Herkepaeus (d. 1705) was a clergyman with a living at Pälkärvi, and shortly before his father's death the whole family was captured by the Russians. He was not released until 1722, when he took up employment as a clerk attached to a dragoon regiment stationed at Åbo. In 1726 he became mayor of Enköping, and in 1737 judicial mayor of Uppsala. He was elected to the parliaments of 1731 and 1740/1741.

In Uppsala his abrasiveness was constantly bringing him into conflict with the magistracy and the university. Recent archival research has led to his being characterized as "monomaniac in his hankering after prestige, insatiable in his thirst for power, quarrelsome and obstinate": Petré, T., 1958: 255. It was Nils Kyronius (c. 1700–1784) who in 1741 accused him of "using offensive language in the public council chamber with respect to the Hat party commission": cf. III.iii.2.3. It was Carl Wilhelm Cederhielm (1705–1769) who was put in charge of the investigation.

As Linnaeus notes, the investigator's father Josias Cederhielm (1673–1729) had come into contact with Herkepaeus in Russia. Captured at Poltava, he was used by the Czar in a diplomatic capacity, travelling backwards and forwards between St. Petersburg and Stockholm, helping to broker the peace, arranging the release of prisoners, and not returning to Sweden permanently until 1721. He was appointed ambassador to the Russian court in 1724, made knight of the Russian Order of St. Alexander in 1726, and retired in 1727: Anrep I: 420.

Cederhielm had had Herkepaeus sent to prison for three weeks and put on bread and water. He had left it to parliament to decide whether he was eligible for re-election. Parliament pardoned the offence and declared him electable, and he represented Uppsala in all the parliaments between 1746 and 1762. In 1752 he was awarded four thousand silver dollars in damages for what he had had to undergo eleven years earlier: Annerstedt, C., III-1: 286f.

The point of the case-history must be, therefore, that the destruction of Cederhielm's house and belongings in 1751 was a higher confirmation of the decision already reached by the estates of the realm.

IV.i.2.3:

London ms. 13, no. 18b; Uppsala ms. 119. Swedish. Linnaeus was uncertain of the date of the accusation, writing "174.." in both versions.

Börje Filip Skeckta (1699–1770) was evidently known to Linnaeus on account of his having his residence at Wilunda in the parish of Hammarby, Uppland: cf. II.i.1.8; III.ii.1.6; III.iii.6.3; III.iv.2.4. He volunteered for the Life Guards in 1718, was made knight of the Order of the Sword, and eventually retired from the service having attained the rank of captain. Sometime after 1740 he married Margaretha Silfversparre (1715–1774) of Wilunda, whose father had fought with Charles XII and been captured at Poltava. Anrep III: 746.

Abraham Hedman (1699–1766) was a factory-owner with a whole range of further commercial interests who owned a large estate at Haraker just north of Västerås. Although after 1739 he profited from the backing the Hats were giving to commerce and manufacturing, he had strong Cap sympathies. In 1742 he entered into a financial arrangement with the Danish envoy, agitating openly for the acceptance of the Danish crown-prince as heir to the Swedish throne and supporting the unrest which eventually gave rise to the Dalecarlian march on Stockholm. On March 29th 1743 he was arrested for plotting concerning the succession, but in the September the case was dropped with inconclusive results. Hedman subsequently attempted to improve his financial position by also taking money from the Russians for raising the question of the succession. He was re-arrested on February 13th 1747, and the court-case dragged on until the end of 1748, once again without any very conclusive results. After this second case, his manufacturing and commercial interests went into what eventually turned out to be a permanent decline: SBL 18: 496–497; cf. III.iv.3.18.

In 1746/1747 Skeckta lost a lot of money on fruitless ore-prospecting projects in the Nyköping area. In October 1756 he was arrested for the abusive manner in which he had attacked Olof Hakånsson (1695–1769) the speaker of the peasantry, and imprisoned in the fortress at Varberg, some sixty miles SSE of Marstrand, from which he was not released until 1766: Anrep III: 792; III.iv.3.3.

IV.i.2.4:

London ms. 15, no. 32, Swedish and Latin; Uppsala ms. 65, Swedish.

Sven Lagerberg (1672–1746) was one of the most renowned of those who had fought with Charles XII. He was shot through the body at Poltava and trampled over by the enemy troops during their advance; a badly-wounded dragoon had saved his life by picking him up and slinging him across the back of his horse. He was with the king in Turkey, and in May 1710 had been sent to negotiate with the Tartars in the Crimea. In the political world of post-war Sweden he was respected by all on account of his firmness of character, soundness of judgement and courtesy of manner – as close personal adviser to the king's sister when she became queen, as land marshal of the 1723 parliament, as privy councillor, as president of the court of appeal. He was ennobled in 1731. His daughter Anna Christina Lagerberg (1728–1803) corresponded with Linnaeus 1758/1759, and it may have been through her that he gathered some of the details recounted in this case-history: Anrep II: 564; *Letters* VIII: 158. The royal treasurer David Lagersparre (1685–1749) was certainly putting himself in a precarious position by falling foul of Lagerberg, since it was well-known that if one had a mind to bring to light irregularities in treasury affairs, one could always do so.

Christopher Springer (1704–1776) must have furthered his career by giving evidence against Lagersparre, since ten years later he was a Cap member of parliament energetically engaged in distributing Russian money. Under §47 of the 1720 constitution members of parliament were required to "give expression to the views of those they represent". Springer maintained that in respect of the succession question the Levant merchant, mayor of Stockholm and member of the house of burghers Thomas Plomgren (1702–1754) was not doing so. The Hats took the matter up with a vengeance in the parliament of

1746/1747, and the Hat judge Johan Georg Lillienberg (1713–1798) saw to it that both Springer and the Cap leader Samuel Åkerhielm (1684–1768) lost the case and with it their political credibility: Anrep II: 729; Malmström, C.G., III: 219f.; SBH II: 60–61; 202; 504–505; Lagerroth, F., 1915; Höjer, T., 1938.

Springer was imprisoned in Marstrand in January 1748. He escaped in June 1752 disguised as an old woman with a brandy-cask under his arm. After spending three days at sea in a small boat he landed in Jutland, and passed on from there to Russia where he found employment in the Chamber of Commerce. When Sweden pressed for his extradition he emigrated to London, where he ended his days as a greatly respected member of the Swedish church community: *Letters* III: 269.

IV.i.2.5:

London ms. 16, Fate no. 44, Latin, cf. 30, no. 65; Stockholm ms. 2, no. 3, Swedish; Uppsala ms. 127, 127b, Swedish, by far the fullest account.

Cf. III.i.1.4; the case has also been placed here as an instance of the tensions subsisting between the judicial system, parliament and the crown.

IV.i.2.6:

London ms. 13, no. 15; Uppsala ms. 115. Swedish.

Although Samuel Rogberg (1698–1760) was born in Jönköping, he grew up in Unnaryd, Småland, where his father was vicar and rural dean. After attending Växjö School he went on to study at Uppsala, and in 1726 returned to Växjö, where he taught Linnaeus history, politics and ethics during his last year at school. He took holy orders in 1730, and subsequently held clerical appointments as vicar of Fröderyd some thirty miles north of Växjö and as rural dean of Västra, as well as playing a prominent part in diocesan politics.

He was elected to the parliaments of 1738, 1741 and 1755/1756. During the last of these he sat as ecclesiastical representative together with Engelbert Halenius (1700–1767) bishop of Skara, on the commission which tried and condemned those who had conspired to procure more constitutional power for the crown: III.i.1.4. As Linnaeus notes, his conduct on the commission did him no credit, and helped to raise the general issue of the advisability of requiring churchmen to play a part in such decision-making: BL XII: 128–130.

Not long afterwards he ran into trouble in his own parish over the acquisition of accommodation for one of his chaplains. He became involved in a highly vituperative dispute with the church council as a result of entering into negotiations for the property without first consulting them. The whole project had to be abandoned, and Rogberg was so disturbed by what had transpired that his health was affected. He went off to take the waters at Växjö and died there on August 1st 1760: Anrep II: 158; Elgenstierna, G.M., 1912, I:

573–576.

Rogberg worked for years on a general history of Småland, to which Linnaeus contributed a chapter on the flora and fauna (*Introduction* III.iii.3). The work was published in a badly-edited form ten years after his death.

IV.i.2.7:

Uppsala ms. 112. Swedish. This sheet is now missing, although it was present when the manuscript was purchased by the University of Uppsala in 1845 (*Introduction* I.i.3). The text here translated is that published by Elias Fries in 1848, who notes that he has had to base it on a transcription – which has now also been lost: Linné, C. von, 1968: 220.

Olof Bidenius Renhorn (1706–1764) was born at Härnösand in Ångermanland, son of a clergyman. In 1735 he gained a position in the customs service at Helsingfors, soon rising to become mayor of the city and its parliamentary representative (1738/1741). Since he was a warm supporter of the 1741/1743 war against Russia, the disastrous outcome of it forced him to leave Helsingfors for Stockholm. Once he had settled in the capital permanently, he soon made his mark as a forceful and effective member of the Hat party, and was co-opted onto their secret committee.

Due to Hat influence he was elected mayor of Arboga in 1747 and represented the town in the parliament of 1751/1752. In the 1755/1756 parliament he came out very strongly indeed against the conspiracy to procure more constitutional power for the crown, and as Linnaeus notes, was well rewarded by his party for the fanatical manner in which he pressed for the conviction and execution of the plotters: III.i.1.4. The party organization provided him with permanent employment, and in 1761 he retired to Arboga on full salary. He published several works, including a translation of *Gulliver's Travels* (1744/1745; 1772[2]) and an exposition of the Platonic background to Rudbeck's *Atlantis* (1751).

The incident at the spa involved captain Magnus Ahlström (1714–1790). It is possible that Linnaeus got to hear of it from his friend Jonas Alströmer (1685–1751), who had various connections, family and otherwise, with the Swedish merchant navy: III.iv.3.18.

Renhorn died in Stockholm on July 24th 1764: SBH II: 328; SBL 29: 782–786.

IV.i.2.8:

Uppsala ms. 110. Swedish.

Carl Fredrik Pechlin (1720–1796) was born in Holstein and first came to Sweden when his father was appointed Holstein envoy to the Swedish court in 1726. He studied for a while in Lund, and in 1737 joined Adlerfelt's regiment

(III.iii.1.7) as a second lieutenant. He served in the 1741/1743 war against Russia, and after switching regiments many times was appointed lieutenant colonel of the Kalmar regiment in 1751, during which year he was also naturalized and ennobled. He served with distinction in the Pomeranian war of 1757/1762, and in 1770 reached the rank of major general.

He made a very distinct impression on all who knew him, his courage and resolution, his resourcefulness and capacity for careful calculation, being as evident in all he did as his selfishness, his cunning and his utter unreliability. Naturally enough, he got on very well indeed in the Swedish political scene of the time. He entered parliament in 1742/1743, and in 1749 married Anna Kristina Plomgren (1727–1788), daughter of the Thomas Plomgren (1702–1754) who had just seen off an aggressive onslaught from the Caps on the representation issue (IV.i.2.4), a move which bound him pretty firmly to the Hats. During the 1755/1756 parliament he was as busy distributing French money among the nobility as he was advocating the confining of the queen to Gripsholm. After returning from the Pomeranian war he got on the wrong side of Linnaeus' close friend the Hat leader and privy councillor Anders Johan von Höpken (1712–1789), and in the 1760/1762 parliament gathered other disgruntled Hats around him and set up the Land party, which in the period prior to January 1761 found much common ground with the Caps: II.iv.2.7; IV.i.2.1. In the 1769/1770 and 1771/1772 parliaments he played much the same sort of game, switching his allegiance from the Hats to the Caps and then attempting to find common ground between the two parties.

When Gustav III started the revolution in Stockholm on August 21st 1772, the governor of the city Thure Gustaf Rudbeck (1714–1786) happened to be away, having left Pechlin in charge: IV.i.2.11. Pechlin, declaring his opposition to what was happening, marched his regiment south and two days later was captured at Jönköping. He was imprisoned in Gripsholm, then in Stockholm and court-martialled, but the evidence against him made conviction difficult, and when after five months he agreed to take the oath of allegiance to the new government, he was released.

He retired to his estate at Ålhult in the county of Kalmar, and in 1775 so got under the skin of the locals as a result of the regulations concerning brandy-distilling which he introduced, that there was an attempt to assassinate him: SBH II: 273–274; SBL 28: 772–775.

IV.i.2.9:

London ms. 11, no. 6; Uppsala ms. 149. Swedish.

Eric Gylling (1709–1773) could only have held the lectureship at the grammar school in Karlstad for a very short period, since the Hat Reinhold Antonsson (1717–1800) was elected member of parliament for the town in the October of 1760. The information concerning the way in which Antonsson helped to further the career of his brother dean Bengt Antonsson (1720–1799) could have come to Linnaeus' notice either through the Alströmer family or through his contacts with Eric von Stockenström (1703–1790), a fellow member of the Academy of Science: III.iv.3.1; IV.i.2.10. Although Linnaeus visited Karlstad on July 30th 1746, there is no evidence that this resulted in his making any lasting contacts in the town: Linnaeus, C., 1747.

IV.i.2.10:

Uppsala ms. 159. Swedish. The final phrase was evidently added at a later date.

Reinhold Antonsson (1717–1800) began his career as a lawyer before switching to trade, especially the export of timber, and becoming the wealthiest man in Vermland. Much of his success as a businessman resulted from his contacts with Jonas Alströmer (1685–1761) and his son Patrick Alströmer (1733–1804), who enabled him to develop international trading-projects through Gothenburg: III.iv.3.1. He quite evidently revelled in exercising influence throughout the local area, constructing a wooden road along the falls at Trollhätten, acquiring various farms and factories in his home province, donating various buildings, including a new town-hall, to Karlstad. In 1760 he was elected to represent the burghers of Karlstad in parliament, only one vote being cast against him, and took the lead for the Hats in many of the ensuing debates: Dalgren, L., 1939, II: 48, 59, 84, 157; Brolin, P.-E., 1953, 168–169.

Erik von Stockenström (1703–1790) studied chemistry at Uppsala and became a member of the Mining College in 1725. After travelling abroad and studying mining methods in Norway, Germany, Holland, England and France, he was appointed surveyor of mines in 1738, and nine years later assessor to the Mining College. He co-operated with Linnaeus in the setting up of the Academy of Science, was ennobled, and in 1758 appointed chancellor of the justiciary. Politically, he was strongly affiliated with the Hats. If the issues at stake between him and Antonsson concerned policy-making in mining matters, it was therefore the height of presumption for the latter to "frighten the life out of him": BL 16: 75–86.

Antonsson's reward came when he appeared to be at the height of his power. He already had the title of deputy judge, and in 1762 he was granted the power of deputizing for the president of the court of appeal. In 1764 he was again elected to represent the burghers of Karlstad, this time unanimously, and when parliament assembled on January 15th 1765 he was put forward by his party as candidate for the position of speaker. The Cap candidate was Carl Fredrik Sebaldt (1713–1792), one of the Stockholm representatives. The Hats called in question Sebaldt's eligibility, and when their objection was disallowed left the chamber in protest. Taking their chance, the Caps put the matter to the vote in their absence and got their candidate in by 57 votes to

2. The Caps continued to dominate the parliament throughout and toward the end of the session passed a resolution making persons who held legal positions such as Antonsson's ineligible for re-election. Antonsson fell out with the magistracy at Karlstad, and in 1769 the mayor of the town Anders Kruskopf was elected to represent the burghers: Brolin, P.-E., 1953, 15–16; BL XIV: 185–188.

IV.i.2.11:

Uppsala ms. 117. Swedish. Since the handwriting in the final paragraph is somewhat different, it may well have been added to the main text at a somewhat later date.

Thure Gustaf Rudbeck (1714–1786) was the grandson of Olof Rudbeck (1660–1740), Linnaeus' teacher at Uppsala: II.i.2.2. His father Olof Rudbeck (1690–1716) was a captain in the Uppland infantry, and Thure followed in his footsteps, joining the army at the age of seventeen. During the 1741/1743 war against Russia he saw active service on the Åland islands, distinguishing himself by taking a position and then holding out against the enemy with great resourcefulness for three and a half months. In 1743 he was promoted to the rank of lieutenant, for services rendered to the defence of Stockholm during the Dalecarlian uprising. He became a lieutenant colonel in 1749 and a major general in 1766.

In 1745 he married Magdalena (1726–1809), daughter of Johan von Mentzer (1670–1747), owner of a large estate near Tidaholm and governor of the district of Jönköping. It was through the wealth thus acquired that in 1757 he was able to purchase Edsberg estate near Sollentuna, formerly in the possession of the Oxenstierna family: Anrep II: 905; III: 536–539.

He first entered politics as a Cap strongly opposed to Hat policies in 1742/1743. He supported the court party in the crisis of 1756, but seems not to have been directly implicated in the plot to secure more constitutional power for the crown: III.i.1.4; IV.i.2.5. After serving for a while in the Pomeranian War (1757/1762) he returned to politics, but in an attempt at getting an appointment as privy councillor lost out to Johan Diedric Duwall (1723–1801), the son-in-law and protégé of the Stockholm businessman Gustaf Kierman (1702–1766), a setback he was able to avenge three years later once the Caps had elected him speaker of the house of lords: IV.iii.2.6.

Linnaeus may originally have intended to conclude this case-study with the tribute to Rudbeck's piety – in which case it was the dramatic events of 1772/1773 which gave rise to his adding the final paragraph. Rudbeck was appointed governor of Stockholm by parliament in February 1772; when Gustav III's coup took place during the following August he was inspecting the construction of the docks at Karlskrona; he hurried back to the capital to oppose the king, was arrested, pardoned, and subsequently became one of the monarch's most ardent supporters: BL XII: 340f.; SBH II: 382–383.

This summary of Rudbeck's career provides a fitting conclusion to the series of parliamentary case-studies: cf. IV.i.2.1, since it not only rounds off what has been presented as a chronological sequence, but also introduces the new element of parliamentary infighting being resolved by the intervention of the crown.

IV.i.3.1:

London ms. 27; Uppsala ms. 68. Swedish. In both manuscripts Linnaeus notes that he had come across the case in Friess, F.C., 1763. Friess (1758 ed., 86–88) presents it as an example of talion working not by the ordinary means of evident laws, but by the extraordinary means of incomprehensible connections – there being no foreseeable link between instigating a bloodbath and being beheaded. He does, however, indicate the steps by which Christian II took Slagheck into his confidence, allowed himself to be influenced by him, and then attempted to wash his hands of responsibility for the massacre by having him executed: as his source for the historical facts he cites Erik Ludvigsen Pontoppidan (1698–1764) *Annales ecclesiæ Danicæ diplomatici* (4 vols. Copenhagen, 1741/1752) II: 404.

Didrik Slagheck (d. 1522) was a Westphalian who got his training as a barber-surgeon in Holland, went on to take a master's degree in Italy, and arrived in Scandinavia in 1517 with the papal legate Johannes Angelus Arcimboldus (d. 1555), expert in canon law and later archbishop of Milan, the object of whose visit was to help settle the civil and ecclesiastical unrest in Sweden. The legation proved to be eminently unsuccessful, and in 1519 Slagheck transferred his allegiance to Christian II, advising his new master in a manner which four days after his coronation on November 4th 1520 gave rise to the Stockholm bloodbath – the summary execution by the Danish authorities in the city of some eighty to one hundred persons, ostensibly for "blatant heresy", actually for having frustrated the king's political objectives. Slagheck was rewarded by being given civil administrative powers and by being appointed bishop of Skara. In the autumn of 1521 he was promoted to the position of archbishop of Lund, then a Danish see, entering the cathedral at his installation on November 25th with great pageantry and to the sound of martial music. During the same autumn, however, a papal legation in Copenhagen had been investigating what had taken place in Stockholm after the coronation, and five days after his installation as archbishop Slagheck was summoned to give an account of the advice he had given and the manner in which he had acted. Christian II decided to abandon him in an attempt to clear his own reputation, and Slagheck was executed and burnt in Copenhagen on January 24th 1522: III.iii.5.6; DBL 13: 504–505.

IV.i.3.2:

London ms. 27; Uppsala ms. 109. Swedish. In both manuscripts Linnaeus notes that he came across the case in Friess, F.C., 1763. Friess (1758 ed., 88–90) presents it as an example of talion working not by the ordinary means of evident laws, but by the extraordinary means of incomprehensible connections, there being no immediately foreseeable link between inciting a king to murder and being beheaded. Linnaeus follows Friess, just as Friess follows his source Holberg, L., 1733/1735, II: 461, 462, 475, and, indeed, all previous interpretations of the events put forward since the sixteenth century, in presenting Jöran Persson (c. 1530–1568) as the evil genius of Erik XIV (1533–1577), as a heartless and ruthless instigator of wickedness who thoroughly deserved the frightful pain and humiliation of his final torture and execution.

In both manuscripts Linnaeus gives 1658 as the date of the king's deposition. All he adds to Friess' narrative is the fact that the murder took place at Uppsala. He omits those parts of Friess' text which have a bearing on the psychology of the actors – Persson's "cunning", Sture's "submissiveness", the considerations which led to the cutting short of the final torture. He takes over from both Friess and Holberg the somewhat misleading information that Jöran Persson was "of Sahlberg": he was born in the town of Sala, some forty miles west of Uppsala; the got his *name*, of Sahlberg, from his father, also known as "Salemontanus".

Jöran Persson was the son of a clergyman. He was educated at the cathedral school in Strängnäs and at the university of Wittenberg, where he acquired a proficiency in civil law, came under the influence of Melanchthon, and developed a great interest in astrology. When he first returned to Sweden in 1555 he was employed in an administrative capacity by Gustavus Vasa, but after three years he switched his employment and joined the court of Erik, then heir apparent, in Kalmar. By the time Erik succeeded his father in 1560, Persson was very much in charge of the legal and administrative aspect of his affairs. Anyone petitioning the king had to do so through him, and he was also in charge of a newly-instituted Star Chamber court.

Although Persson seems to have had no influence on Erik XIV's foreign policy, he played a large part in determining the way in which the king reacted to the opposition and scheming of his two brothers, Johan duke of Finland (1537–1592), the later John III, and Karl duke of Västergötland (1550–1611), the later Charles IX, and of the higher nobility in general, prominent among whom were several members of the Sture family. There is therefore some justification for regarding him as having been partly responsible for Erik's having stabbed Nils Svantesson Sture (1543–1567) in Uppsala on May 24th 1567, and for the subsequent assassination of Sture and his supposed confederates by the royal bodyguard. The precipitation of these murders was also due in part to the king's incipient mental instability, however, and there is every likelihood that those who subsequently prosecuted and condemned Persson for what had happened were well aware of this. Persson was sentenced to death by the council of the realm on September 5th 1567, but not executed until September 22nd 1568, by which time the royal dukes had gained military control of Stockholm and the country at large, and were preparing the deposition of their brother: SBL 20: 553–555; IV.i.5.8.

In the life of Erik XIV published in 1774 by Olof Celsius (1716–1794), Linnaeus' colleague at Uppsala, the traditional view of these events still predominates. The modern assessment of them, the shifting of the emphasis from Persson the evil genius to Erik's own psychology, first comes into its own in the classic history of the Swedish people published by Erik Gustaf Geijer (1783–1849) in 1832/1833 (c. XI).

IV.i.3.3:

London ms. 11, Latin and Swedish; Uppsala ms. 145, Latin.

Peder Schumacher, count Griffenfeldt (1635–1699), the Danish statesman, secretary to Frederick III 1665/1670, privy councillor and prime minister to Christian V 1670/1676, was the son of a Copenhagen wine merchant. He was educated at Queen's College Oxford 1657/1660, studying constitutional law, Arabic and Persian under Thomas Barlow (1607–1691), subsequently bishop of Lincoln, and before returning to Copenhagen in 1662 travelled widely in France, Spain and Italy.

As a result of what has been described as an "almost disquieting diligence", he rose rapidly in the civil service, and by the early 1670s was playing a prominent role in nearly every aspect of the country's central administration. His most lasting contribution to its political life was the establishing of an absolutist monarchical constitution which remained in effect until 1849, the drawing up in November 1665 of the famous *King's Law*. In this truly remarkable document he spells out succinctly, and in immediately understandable terms, the full and complete theory of an all-powerful Lutheran kingship: the Augsburg Confession has validity to the thousandth generation; the Estates having voluntarily yielded up their authority, the king is to be acknowledged as supreme on earth, as having no superior but God, as not to be bound by any man-made covenants or oaths; the moment the king dies, the next prince of the blood is "actually, and without any further ceremony" absolute ruler, with the same powers as his predecessor, alone responsible for maintaining the peace, declaring war, sustaining the church: Fabricius, K., 1920; *English Historical Review* 36: 622.

When Griffenfeldt was knighted in 1671, he took "the pen guides the sword" as his motto. By the August of 1675, however, Christian V was beginning to make use of the power bestowed upon him by Griffenfeldt's pen in order to bring what he regarded as an overweening busybody to heel. On

March 11th 1676 he had the author of the *King's Law* arrested and put on trial for treason, not as was usually the case before the Supreme Court, but before an extraordinary tribunal of ten dignitaries, none of whom was particularly well-disposed towards the accused. The chief prosecutor was a quite evidently partial German lawyer Otto Mauritius (1637–1682), who soon after the trial was himself imprisoned for allegedly traitorous activities. Despite the blatant irregularity and injustice of the proceedings, on May 26th Griffenfeldt was declared guilty and condemned to "loss of life, honour and goods". Only one of the ten dignitaries, the Danish nobleman Christian Skeel (1623–1688), dissented, on the ground that since the accused had not been proved to be guilty of treason, there was no justification for having him put to death. The execution was fixed for June 6th, and the king waited until Griffenfeldt was actually on the scaffold and the axe was about to descend, before granting pardon and commuting the sentence to life imprisonment: Rothe, C.P., 1745; DBL 15: 290–295; 9: 481.

On the surface of it, Linnaeus appears to be instancing a form of talion also illustrated in the subsequent fate of those who sat in judgement on Görtz and assassinated Caesar (IV.i.3.4; IV.i.5.6). It is by no means unlikely, however, that his evident interest in Griffenfeldt (II.iii.3.5) had rather more to do with the contrast between the smooth and efficient conduct of Danish affairs throughout the eighteenth century, and the chaotic state of Swedish politics prior to the coup of 1772 (IV.i.2.8; IV.i.2.11).

IV.i.3.4:

London ms. 13, no. 18, Swedish; Uppsala ms. 148, Swedish and Latin. Linnaeus misses out the name of Ribbing and mis-spells the names of Polhem and Fehman.

Georg Heinrich Görtz (1668–1719) came of a noble family established at Schlitz near Fulda in Hesse since the mid-thirteenth century. His grandfather built the Ottoburg, the baroque residence which still dominates the town, placing in the cellar of it the stone inscription which is still to be seen there: "The grace of the Lord be with us and ours for ever". Görtz planned a career for himself in administration and diplomacy, and studied at the university of Jena, where he lost an eye in a duel. Soon after entering the service of duke Frederick IV of Holstein Gottorp in 1698, he accompanied him to Stockholm for the celebration of his marriage with Hedvig Sophia (1681–1708), elder sister of Charles XII. In 1711 he sold the Ottoburg to his uncle for sixty thousand guilders: Sippel, H., 1987: 117–118, 166–168.

For well over a decade after the death of duke Frederick IV in 1702, Görtz was fully occupied with Holstein Gottorp diplomacy. This became particularly complicated and demanding after the resumption of hostilities between Denmark and Sweden in 1709, since the young duke Karl Frederick (1700–

1739) was potential heir to the Swedish throne, and the duchy was under constant threat of being invaded by the Danish army. The duke was sent to Sweden when his mother died in 1708, and in 1713 the Danes did in fact occupy the duchy.

There were many in Sweden who took offence at the ways in which Görtz attempted to pursue the interests of the duchy during these extremely difficult years, notably Ulrika Eleonora (1688–1741), the younger sister of Charles XII, who was to succeed her brother as monarch in her own right, and Arvid Horn (1664–1742), who was appointed governor to the orphaned duke once he had settled in Stockholm. Görtz was therefore unpopular in certain circles long before his meeting with Charles XII in Stralsund in the December of 1714, and these circles were not at all pleased when the king proceeded, against all expectations, to take the Holstein diplomat into his confidence. From the very beginning of their well-known co-operation in economic and diplomatic affairs, there was a general feeling that Charles was being duped by a man who had other than Swedish interests at heart.

As Linnaeus indicates, there can be no doubt that this was not the case. Görtz might be regarded as having entered the service of Charles XII early in 1715. For most of the period from the middle of 1716 until the autumn of 1718 he was abroad, attempting in an ultimately unsuccessful but nevertheless intelligent, resourceful and constructive manner to sort out the country's hopeless financial plight on the Amsterdam stock exchange, and resolve its equally hopeless diplomatic predicament by negotiating with the Russians on the Åland islands. The decision to debase the coinage, to issue the notorious "Görtz money", was made before he had begun to concern himself with such matters, very possibly on the advice of Christopher Polhem (1661–1751): SBL 17: 168–177; 29: 388–393; Polhem, C., 1951.

Görtz was on the way to a meeting with Charles XII when the king was killed (II.iv.3.5). Only two days later, while he was staying overnight at the rectory in Tanum, some twenty-seven miles south of where the king had fallen, he was arrested on the orders of Frederick of Hesse (1676–1751), the future Fredrik I, husband of Ulrika Eleonora, and despatched under a strong escort to Stockholm. The question of diplomatic immunity was ignored, and on December 18th the Council decided that he should be tried by the Court of Appeal. On the following day, however, Arvid Horn had this decision quashed: the prisoner was to be brought before a special commission, and it was made quite clear that the main qualification for sitting on it, was to be that one had a grievance against the way in which the country had been run. Pehr Ribbing (1670–1719), confidant of Ulrika Eleonora, was appointed president, Thomas Fehman (1665–1733), brisk and aggressive, chief prosecutor, and seven further nobles, two clergy and three burghers, were also co-opted. As Linnaeus indicates, the trial may have answered to the widely-felt public need, but it was a complete travesty of justice. Görtz admitted sins as a Christian but denied offences as a minister. Little or no attempt was made to sift ostensibly conflicting evidence, and on February 19th the prisoner was summarily beheaded: SBH I: 328–329; II: 339.

The significance of the trial and execution remained a living issue throughout Linnaeus' lifetime, mainly on account of their financial consequences. In 1723 parliament decided that Görtz's activities had cost the Swedish state one and a half million silver dollars, and claimed the amount from his heirs. The family lodged an even higher counter-claim, and sold the responsibility for negotiation to a consortium. As a result of Russian diplomatic pressure the case was finally settled during the 1770s, some one million two hundred thousand silver dollars being paid to the consortium. In 1776 Gustav III wrote a celebrated letter to the last-surviving of Görtz's daughters, Georgina von Eyben, paying generous tribute to her father's integrity and services to the crown, and making her a personal gift of sixty thousand silver dollars, in order that "the realm might be cleansed of blood-guilt". Moser, F.C. von, 1776.

IV.i.3.5:

London ms. 15, no. 26; Uppsala ms. 102. Swedish.

Ernst Johann Biron (1690–1772) was the grandson of a groom in the service of the duke of Courland. After studying at the university of Königsberg, from which he was expelled for rioting, he got a position in Mittau at the court of Anne (1693–1740), the widowed duchess of Courland, niece of Peter the Great. She became infatuated with him, and when she became empress of Russia in 1730 created him her grand-chamberlain, effectively prime minister, showered further honours and riches upon him, including having him elected duke of Courland in 1737, and on her deathbed arranged for him to be declared regent of the country during the minority of her successor, her great-nephew Ivan VI (1740–1764) – at the time only ten weeks old.

Biron had made himself immensely unpopular on account of the shameless and rapacious manner in which he had exploited this imperial benevolence. His regency lasted for no more than three weeks. Anne died on October 28th 1740, and an November 19th he was seized in his bedroom and put under arrest. On April 11th 1741 the commission appointed to try him condemned him to death by quartering, but this was commuted by the clemency of the new regent, the czar's mother, to life imprisonment at Pelym on the Ob in Siberia.

As Linnaeus notes, the driving-force behind Biron's downfall was Burkhard Christoph Münnich (1683–1767), a German from Oldenburg who had begun his career by serving successively in the armies of France, Hesse-Darmstadt and Saxony. After being taken into service as an engineer by Peter the Great, he built up a reputation for ruthless efficiency in the construction of the Ladoga canal and the development of various mining projects. Having been put in charge of the ministry of war and appointed field-marshal, he carried out a comprehensive reform of the Russian army, which bore fruit in the military successes of the War of Polish Succession (1733/1735) and the major campaigns against Turkey (1736/1739). On Münnich's involvement in the assassination of the Swedish officer Malcom Sinclair (1690–1739) at Grünberg near Naumburg in Silesia on June 17th 1739, see IV.iii.2.8.

It was effectively on Münnich's authority that Biron was arrested, tried and condemned. On November 25th 1741, however, when Elizabeth (1709–1762), Peter the Great's daughter, with Sweden's connivance, seized the throne, the authority of the regency council was swept away (II.ii.4; II.iv.2.7). Münnich fled for the frontier, was captured, tried and condemned to death. It is indeed curious that when he had already mounted the scaffold, the new empress should have commuted his sentence to one of life imprisonment, and that he should have been sent to Pelym, Biron being moved on to a slightly more salubrious gaol at Jaroslav on the Volga. Both of them remained in prison until 1762.

Linnaeus' interest in and knowledge of these cases is probably to be accounted for by the fact that during the build-up to the war against Russia, that is, between September 1738 and the autumn of 1741, he was living in Stockholm, and that it was during this period that he first got to know privy councillor Anders Johan von Höpken (1712–1789), who remained a life-long friend and had inside information on the diplomatic dealings between Sweden and Russia: II.iv.2.7.

IV.i.3.6:

London ms. 7, listed; Stockholm ms. 2, no. 6; Uppsala ms. 135. Swedish. It is apparent from the Stockholm ms. that Linnaeus was uncertain of the dates here: in the Uppsala ms. he gave 1758 as that for the vote to send the expedition against Prussia.

During the early summer of 1757, Carl Johann von der Osten-Sacken (1721–1794), Saxon/Polish ambassador to the Swedish court, arrived in Stockholm together with his twenty-four year old wife Henriette Erdmuthe von Brühl (1733–1762), daughter of August III's minister count Heinrich von Brühl (1700–1763). Claes Ekeblad (1708–1771), married since 1741 to Eva de la Gardie (1724–1786), founder with Linnaeus of the Academy of Science, one time Swedish envoy to Paris, prominent Hat politician, was then member of the council of state. On June 22nd the council decided, by one vote, on its own initiative and without consulting parliament, to uphold the treaty of Westphalia and send troops into Pomerania. It was in this way that Sweden entered the Seven Years' War (1756/1763) in which France, Saxony/Poland and Austria were ranged against Prussia, Hanover and Britain. A French subsidy was forthcoming as from the September of 1757.

Henriette von der Osten-Sacken died in childbirth in 1762, at about the same time as the Swedish war-effort in Pomerania petered out into a stalemate and the peace treaty was signed in Hamburg. In November 1764, as a result of Ekeblad's casting vote as chancery president, the council of state decided to resume acceptance of the French subsidy. The decision was an anathema to the Caps, and when they got the upper hand in parliament in January 1765, they forced Ekeblad to resign: Malmström, C.G., 1893/1901, IV, chs. 23, 24; *Deutschbaltisches Biographisches Lexikon* 1970, 563–564.

Ekeblad's wife is a matter of interest in her own right, since she was elected member of the Academy of Science in 1748, nearly two centuries before the next woman, for having shown that schnapps can be prepared from potatoes. Ekeblad had an experimental estate in Västergötland, and put a lot of energy and money into demonstrating the food-value of potatoes: Roberts, M., 1986, 141–142.

The king himself objected to the decision of November 1764: since Ekeblad had allowed a personal attachment to affect his decision-making on a public issue, it was only right that the outcome of his policies should have forced him to relinquish his office.

IV.i.3.7:

London ms. 12, no. 8; Uppsala ms. 189. Swedish. The dangers of the situation in 1756 (III.i.1.4) are emphasized by inserting "dire consequences" above "involvement".

Arvid Bernhard Horn (1684–1742) played a leading role in the establishment of the elective monarchy after the death of Charles XII. He was himself elected speaker of the house of lords in 1720, and for the following eighteen years controlled both the domestic and the foreign affairs of the country, increasing the influence of parliament and its committees, pursuing a broad policy of reconciliation with Russia, and so helping to overcome many of the potentially harmful legacies left by Caroline absolutism and the unfruitful military exploits of the Great Northern War. The overwhelming success of this policy led to its being taken for granted and then called in question. In 1734 there was strong opposition in parliament to his decision not to make a stand against Russian policy in the War of Polish Succession (1733/1735). Four years later he was eventually forced to resign by the impetuous onslaught of the triumphant young Hat party. He retired to his country estate at Ekebyholm, in the parish of Rimbo, some twenty-five miles west-south-west of Uppsala.

It was during these final years in office that Horn was most frequently accused of lack of patriotism, and even of having been in the pay of Russia. The truth of the matter is, however, that he had begun to invest in land as early as 1710, had inherited extensive estates in 1720, and between 1724 and 1736 had received large sums from the crown as repayment for war loans. He purchased the estate at Ekebyholm in 1724.

Arvid's son, Adam Horn, earl of Ekebyholm (1717–1778), had a reasonably successful career in the military, seeing service in the Hessian army and in the 1741/1743 war against Russiat before being appointed colonel of the East Goth cavalry regiment in 1747 and rising to the rank of lieutenant general in 1759.

One might have expected this career to be enhanced by his marriage, on July 7th 1741, to his cousin Anna Catharina Meijerfeldt (1722–1779), since her father had fought at Poltava and just prior to the capitulation at Perevolochna been appointed by Charles XII to negotiate with the czar. His father-in-law's policy of gaining the release of prisoners by exchange had been resented by many of the high-ranking Swedish oflicers in Russian captivity, however, and association with his family does not appear to have brought Horn any particular advantage in his military career. What is more, as Linnaeus notes, his wife became insane, and Horn was unsuccessful in his attempts to get a divorce; Anrep II: 292; 888–889.

The younger Horn's political career also turned out to be much less successful than it might have been. He began it by following in his father's footsteps and speaking up for the eminently ineffective Cap party in the parliament of 1740. When the Hat administration threatened to eliminate the crown's prerogatives in 1756, Horn, then one of the royal family's closest confidants, was sent to St. Petersburg with a secret appeal for help. The court party's scheming (III.i.1.4; IV.i.2.5) came to light some three months after Brahe had left Stockholm, and on July 23rd the public executions took place. Fearing for his life, Horn requested royal permission to stay abroad for at least two years, and this was granted on November 25th. He redeemed his national standing by serving successfully in the Pomeranian War (1757/1762).

In 1760 he was recognized as leader of the Caps, and was put forward by them as candidate for the position of speaker of the house of lords. Since the Pomeranian war had turned out to be no more than a stalemate and the Hats were in a state of disarray, it looked as though Horn would be successful. As things turned out, however, he was defeated by Fredrik Axel von Fersen (1719–1794) the Hat candidate, who had also seen active service in the war. In 1761 Horn was appointed privy councillor and then lord chamberlain, positions which simply reflected his longstanding affiliation with the court. When the Caps eventually came to power in 1765 he was offered the position of president of the privy council but declined. Although he was made a member of Gustav III's cabinet in 1772, he played little part in determining its policy.

Linnaeus quite evidently regarded the younger Horn's frustrations and missed chances as flowing from the fundamentally unwholesome policies of his father. This summary of his career provides a fitting conclusion to the series of prime-ministerial case-studies, since it not only rounds off what has been presented as a chronological sequence, but also introduces the new element of a continuity over the generations – which finds its focal-point in the principle of monarchy (IV.i.5.4).

IV.i.4.1:

Linné, C. von, 1907, 241; Uppsala ms. 105. Swedish.

Although the general drift of Linnaeus' argument is accurate and cogent enough, there is a certain amount of historical confusion here.

John Churchill duke of Marlborough (1650–1722) did indeed establish his European reputation as a military commander during the war of Spanish Succession (1702/1713), by defeating the armies of Louis XIV at the battles of Blenheim (1704), Ramillies (1706), Oudenarde (1708) and Malplaquet (1709). His career impinged directly on that of Charles XII during the early summer of 1707, when the Swedish army was encamped at Altranstädt near Leipzig, and there was a chance that the French might persuade Charles to invade Austria. Marlborough found it worth his while to travel from The Hague and discuss the general European situation with the Swedish high command. He concluded correctly from the interview that Charles' overriding objective was to march east and attempt to settle his account with Russia: Thomson, G.M., 1979: 194–195.

Churchill married Sarah Jennings (1660-1744) in 1677, and they were "at the height of their fortune and favour" with Anne (1665–1714) before she became queen in 1702. On her accession she did indeed make Sarah "mistress of the court", appointing her groom of the stole, mistress of the robes, keeper of the privy purse: Rowse, A.L., 1956: 280. By 1704, however, the personal friendship between the two women was quite clearly on the wane, mainly because Anne had realized the dangers of allowing Sarah to capitalize on their relationship by exercizing political influence: Harris, F., 1991: 91.

Although there was no very clear-cut party system at the time and Marlborough could hardly be regarded as a firm party man, it was the Whigs not the Tories who backed him most consistently in his military campaigning, the Tories not the Whigs who brought about his dismissal at the end of 1711 and the conclusion of peace with France two years later. All four of the general elections which took place during Anne's reign reflected the ministries already appointed – the Tory majority of July 1702, the heavy Whig gains of June 1705, the clear Whig majority of May 1708, the crushing Tory victory of October 1710: Holmes, G., 1987. It could be argued, moreover, that the quarrel between the duchess and the queen which blew up during August 1708 was the beginning of the end of Marlborough's political influence: Sarah took offence when Anne declined to wear the jewels she had chosen for the thanksgiving service in St. Paul's for the victory at Oudenarde, Anne decided that the time had come to make quite clear that the personal friendship between them was over: Harris, F., 1991: 148. It was certainly not the case, however, that the ending of Marlborough's military career was brought about simply on the whim and fancy of the monarch.

The alleged incident concerning the pair of gloves dates from before Anne's accession. It is reputed to have taken place shortly after the death, on July 30th 1700, of her eldest son William duke of Gloucester. In 1697 Sarah had managed to get Abigail Hill (d. 1734), one of her many impecunious cousins, appointed woman of the bedchamber to Anne. While ladies of the bedchamber, such as Sarah herself, were women of noble birth or marriage who served as social companions to the princess, the woman of the bedchamber performed the menial tasks of a personal maid, which included "pulling on the princess's gloves when she was unable to do it herself": Croker, J.W., 1824, I: 292–293; Gregg, E., 1980: 110–112. On this particular occasion Abigail is reputed to have asked Sarah for a pair of Anne's gloves, which Sarah had put on by mistake. Not realizing that the door was open and that Anne was in the next room, Sarah is said to have expressed disgust at having donned "anything which has touched that disagreeable person", and to have asked Abigail to take them away. The story is almost certainly apocryphal, and if there is any substance in it, the incident has to be dated much later than 1700, since Anne was incapable of dissimulating personal animosity, had no possible motive for doing so, and made no attempt to conceal the cooling off of the friendship after 1704: Brown, B.C., 1935: 104–105.

It is not difficult to find an answer to Linnaeus' final query. When Marlborough accompanied his sovereign into the west country in the November of 1688 in order to deal with the Orange invasion, he deserted him and went over to the enemy. In the May of 1692 he was quite rightly arrested for high treason when he objected to the way in which his sovereign was surrounding himself with foreigners and entertained the idea of removing him from power. It was only fitting, therefore, that he should have had no support from his sovereign when the political tide began to turn strongly against him during the autumn of 1710.

IV.i.4.2:

London ms. 2, 3; Uppsala ms. 129. Swedish and Latin.

Voltaire had made use of the execution of admiral Byng in order to discredit still further Leibniz's conception of this being the best of all possible worlds: "In this country, however, it is a good thing to execute an admiral now and again, since it encourages the others": *Candide* (1759) ch. 23. It looks as though Linnaeus may have attempted to counter the famous quip by providing this analysis of the background to the event: cf. Rousseau, A.-M., 1960; IV.ii.1.2. He may have been justified in his final objective, but in this particular instance he is on rather weak ground historically.

There is no evidence that admiral George Byng (1663–1733), viscount

Torrington, ever "rampaged" over anyone. His biographer, who knew him well, summed up his general character as follows: "His great Diligence, joined with excellent natural Parts, and a just Sense of Honour, made him capable of conducting difficult Negotiations and Commissions with proper Dignity and Address". His principal contact with Minorca took place during a period of two days in July 1718, when he landed four regiments of foot at Port Mahon, before sailing on to Sicily: Corbett, T., 1739, 16, 197–199; Payne, J., 1793, III: 408–412.

When British troops first took Minorca in September 1708, during the war of Spanish Succession, they were keenly supported by the local militia, the population at large having many grievances against the regime of Philip V. Later they strongly supported Charles III. Britain was mainly interested in the island on account of the excellent natural facilities for a naval base provided by the harbour at Port Mahon. No attempt was made to interfere with local affairs or the administration of justice. What is more, under the governorship of Sir Richard Kane (1666–1736), who made a particular point of furthering trade and industry, the island flourished as never before – a road was constructed from the former capital Ciudadela to Port Mahon, the island's economy was stimulated by the lavish expenditure of the garrison, encouragement was given to the growing of fruit and vegetables and the rearing of poultry, sheep and cattle. As is apparent from a debate on the matter in the House of Lords in January 1742, the British administration became unpopular not on account of oppression, but on account of the opposition of the Roman Catholic church: "The Lieutenant-Governor of the island has told you, that, tho' the inhabitants of the island be now much richer, than they ever were, whilst under the dominion of the Spaniards; yet, in his opinion, they would certainly join the Spaniards in case they should invade the island. As this, my Lords, is a very extraordinary circumstance, it must be owing, either to the people's thinking themselves oppressed by us, or to their continuing bigotted to the Popish religion". Tunstall, B., 1928, 89–110; *The Scots Magazine* V: 153–167 (Edinburgh, April 1743).

Byng's son John Byng (1704–1757) entered the navy in 1718, and to a certain extent his subsequent promotion was undoubtedly a reflection of his father's reputation rather than his own abilities: "he chose to spend his time for the most part as senior or sole officer at Port Mahon. This may have been very pleasant, but it was not exercising him in the duties of his rank, or training him for high command": DNB 3: 118–121. He was promoted to be rear-admiral in 1745, vice-admiral in 1747, admiral of the blue in 1756.

In the opening moves of the Seven Years' war there was a build up of French shipping, troops and armaments at Toulon. On April 5th 1756 Byng sailed from England with a small squadron and orders to keep an eye on the Mediterranean, and if Minorca was attacked, "to use all possible means in his power for its relief". The French landed at the western end of the island on April 18th, and when he arrived at Gibraltar Byng learnt that they had overrun it and were besieging the fort at Port Mahon. On May 8th he sailed from Gibraltar with thirteen ships of the line and three frigates but without the extra troops he had asked for, and eleven days later appeared off Minorca. On May 20th, after bearing down on the defending French squadron and getting the worst of the encounter, he withdrew and spent four days refitting. He then summoned the council of war mentioned by Linnaeus. His officers quite evidently had no conception of the strategic value of their fleet as a menace to the French army, which contained hundreds of men incapacitated by dysentry and wounds and was already despairing of success. They chose to ignore the fact that troops for the relief of the garrison could have been safely landed in St. Stephen's cove. Almost unanimously, they therefore decided: firstly, that an attack on the French fleet provided them with no prospect of relieving Minorca; secondly, that even if there were no French fleet, it would be impossible for them to raise the siege. Byng therefore left Minorca to its fate and sailed for Gibraltar, arriving back at Spithead on July 26th.

At the instigation of captain Thomas Frederick Cornwall (1706–1788), the propriety of whose conduct during the campaign had been questioned, Byng was tried by court-martial for neglect of duty. It seems to have been Cornwall who made the suggestion that "any order has to be obeyed, be it to send the men to hell". Tunstall, B., 1928, 168; Valentine, A., 1970, 204–205; *The Commissioned Sea Officers of the Royal Navy 1660–1815* (3 vols., n.p. 1954) I: 199, Bodleian Library sign. B.3.251.

On January 27th the court pronounced Byng guilty of neglect of duty, an offence which under the twelfth of the articles of war carried the death penalty. The matter was referred to the king, who passed it on to a panel of twelve judges, who confirmed the legality of the sentence Byng was therefore shot, on the quarter-deck of the Monarque, in Portsmouth harbour, on March 14th 1757.

Linnaeus is certainly justified in emphasizing the effect of public opinion on this course of events, and on Byng's posthumous reputation. The ministry needed a scapegoat for the loss of a naval base of such immense strategic importance, the opponents of the ministry were only too ready to urge the point that Byng had been executed as a cloak to ministerial neglect. Byng conducted himself in an exemplary manner throughout, and concluded his valedictory epistle with an observation of universal significance: "The supreme judge sees all hearts and motives, and to him I must submit the justice of my cause".

There are no grounds for following Linnaeus and regarding the imagined unrest of the Minorcan population under the oppressive regime of the elder Byng as having its counterpart in the popular clamour which accompanied the execution of his son. It is a fact, however, that when on March 13th 1708 the elder Byng intercepted the French fleet supporting the old Pretender at the entrance to the Firth of Forth, he failed to take full advantage of the situation.

At the time there was a demand for a parliamentary enquiry. The public discontent soon subsided, however, and the matter was allowed to pass. Linnaeus could have argued with some plausibility that it was the son who faced the final retribution for this neglect of duty.

On the final observation, which is in Latin, see Walther, H., 1963/1986, no. 26060; cf. III.ii.4.1; IV.ii.2.14; IV.iv.3.2; IV.iv.3.3.

IV.i.4.3:

London ms. 7, listed; Uppsala ms. 122. Swedish.

Otto Arnold Paykull (c. 1662–1707) was born in Livonia – which Sweden took from Poland in 1625/1626, which was tacitly recognized as Swedish by the truce of Altmark in 1629, and which the Polish crown confirmed as a Swedish possession by the treaty of Oliva in 1660. He joined the Saxon army, and on July 9th 1702 was wounded when Charles XII defeated the Saxons at the battle of Kliszów. In April 1700 Charles had issued a declaration absolving his Livonian subjects from prosecution if they served in a foreign army. In December 1702 he ordered the Swedish high court to annul it, and to condemn Paykull, in his absence, to loss of life, honour and goods. Paykull was captured by the Swedes during the Polish campaign of 1705, and claimed that he had never taken the oath of allegiance to the Swedish crown. Charles allowed the case to be tried by the high court in Stockholm, and it was evidently at this stage in the proceedings that Henric Gotthard von Buddenbrock (1648–1727), a member of the Livonian gentry in Swedish service, came out for having Paykull executed.

The case aroused great public interest, and as Linnaeus notes, the general opinion was that Paykull ought to have been let off. The widowed queen mother and both Charles' sisters pleaded with him to intervene and quash the proceedings. Paykull, who was a highly respected alchemist, offered to reveal to the king the art of turning base metals into gold. The court therefore decided to refer the matter back to the crown, and on advice given by his secretary Olof Hermelin (1658–1709), the king eventually decided that Paykull should be executed: "even if he is capable of turning the whole of the Brunkeberg into gold". The decision was carried out in Stockholm on February 4th 1707, the spot in Norrmalm on the Brunkeberg where the execution took place being long known as Paykull's pightle: SBL 28: 768–770.

Like his father, Magnus Henric Buddenbrock (1683–1743) was born in Livonia and joined the Swedish army, serving under Lewenhaupt in Courland in 1708, fighting at Poltava and following Charles XII over the Dnieper into Turkey, and also accompanying him in the final Norwegian campaign. After becoming naturalized in 1729, he was appointed major-general. Toward the end of the 1730s he joined forces with the Hats in urging war with Russia, and was rewarded for his enthusiasm by being put in charge of the troops in Finland. After the defeat at Willmanstrand on August 23rd 1741 he was relieved of his command. When Helsingfors capitulated on August 24th 1742 he was recalled to Stockholm and put on trial for: 1) failing to control his troops; 2) failing to defend Willmanstrand; 3) failing to co-operate effectively with the commander who succeeded him. Since the government was in desperate need of a scapegoat for the disastrous outcome of the war, he was found guilty on May 21st 1743, the sentence was confirmed by parliament, and he was beheaded on July 16th – on the very spot where Paykull had been executed thirty-five years before: SBL 6: 651–658; II.ii.3.

IV.i.4.4:

London ms. 26; Uppsala ms. 202. Swedish. Linnaeus inserted "reaching out his hands" after having completed the main text.

The rising of the peasantry in Dalecarlia was the direct result of the disastrous course of events in the 1741/1743 war against Russia. Their province had suffered economically from the closing of the frontier with Norway, they deplored the military set-backs in Finland, and they were determined that those in Stockholm who had brought the country to this sorry state should themselves be brought to book. Their main positive proposal was that Fredrik (1723–1766), the crown prince of Denmark, should be accepted as heir to the Swedish throne. While the rising was getting under way, the empress of Russia let it be known that she was prepared to restore nearly all the conquests she had made in Finland if Adolf Frederick (IV.i.5.10), the guardian of her nephew Karl Peter Ulrich (1728–1762) of Holstein Gottorp, were elected heir apparent in Stockholm.

It was this move which sealed the fate of the Dalecarlians, who by the midsummer of 1743 had marched down and occupied Norrmalm, just beyond the water within sight of the royal palace. On June 22nd troops loyal to the government attacked them successfully in Norrmalmstorg, the present Gustav Adolfs Torg by the Royal Opera, and on the following day the four estates of the realm elected Adolf Frederick successor to Fredrik I: Beckman, B., 1930.

Christian Joachim Klingspor (1714–1778) joined the horse guards in 1727, became a lieutenant in 1743, but was not promoted to the rank of captain until 1747. In 1758 he married Helena Christina De Besche (1730–1766) of Nävekvarn, just south of Nyköping, with whom he had four children, his eldest son Gerdt Adolph (1753–1814) also becoming a captain in the cavalry. He was "frightfully cut about the head and face" at the battle of Taschenberg in 1760, his father Stephan Klingspor (1690–1766), curiously enough, having suffered in much the same way at the battle of Hälsingborg in 1710: SBL 21: 324; Anrep II: 452–453.

Better than any of the case-studies which precede it, this sketch of Klingspor brings out the central paradox of the military – the bane and guardian of human, civil and divine authority. Cf. IV.i.2.5.

IV.i.5.1:

Uppsala ms. 2. Latin and Swedish.

Cf. III.iv.1.3; III.iv.1.1. The passage was included in the former context on account of its bearing on the distribution of wealth. It is included here on account of its bearing on regal and imperial authority.

IV.i.5.2:

Uppsala ms. 38. Swedish.

Linnaeus notes this down at the bottom of the page, with IV.i.5.5 at the top, the context here evidently being of no particular significance. Cf. Friess, F.C., 1758, 124.

The conquest of Canaan, begun by Joshua, was completed when the Judahites slaughtered ten thousand of those who were defending their land at Bezek. Adoni-bezek, the king of the Canaanites fled, but the Judahites pursued, captured and mutilated him. It is certainly implied in the Biblical narrative that this was justified, in that the king had formerly treated his seventy captives in the same manner.

It was then a commonplace to trace royal authority back to the historical narratives of the Old Testament, and there is every likelihood that Linnaeus noted down this instance of talion with such a general thematic context in mind: Normann, C.-E., 1948.

IV.i.5.3:

Uppsala ms. 8. Swedish.

Linnaeus notes this down at the bottom of the page, with II.iii.3.6 at the top, the context here evidently being of no particular significance.

When Gideon had fought off the Midianites, the Israelites asked him to rule over them: “you, your son and your grandson”. Gideon replied that neither he nor his son but the Lord would do so. The immediate outcome was, however, that unity in the leadership was only maintained by Abimelech’s “going to his father’s house in Ophrah and butchering his seventy brothers, the sons of Jerubbaal, on a single stone block”. It is stated explicitly in the Biblical narrative, that when the woman mortally wounded Abimelech by dropping the millstone on his head: “God repaid the crime which Abimelech had committed against his father”.

The Old Testament may therefore be regarded as having provided Linnaeus with plenty of opportunities for justifying the divinity of kingship on Hobbesian grounds: Lundius, C., 1691.

IV.i.5.4:

Uppsala ms. 23, 23b. Swedish and Latin. The last three anecdotes are noted down separately at the bottom of the second page: on the first of them, see II.iv.2.8.

The chronology of the Roman emperors mentioned here by Linnaeus is as follows: Augustus (63 B.C.–A.D. 14), reigned 27 B.C.–A.D. 14; Tiberius (42 B.C.–A.D. 37), reigned 14–37; Caligula (12–41), reigned 37–41; Claudius 10–54, reigned 41–54; Nero (37–68), reigned 54–68; Vitellius (15–69), reigned 69, 354 days; Vespasian (9–79), reigned 69–79; Titus (40–81), reigned 79–81; Domitian (51–96), reigned 81–96; Nerva (35–98), reigned 96–98; Trajan (53–117), reigned 98–117; Hadrian (76–138), reigned 117–138; Antoninus Pius (86–161), reigned 138–161; Marcus Aurelius (121–180), reigned 161–180; Commodus (161–192), reigned 180–192; Severus Septimius (145–211), reigned 193–211; Caracalla (186–217), reigned 211–217; Macrinus (164–218), reigned 218, 58 days; Maximinus (172–238), reigned 235–238; Gordian III (224–244), reigned 238–244; Philip (c. 204–249), reigned 244–249; Decius (201–251), reigned 249–251; Trebonianus Gallus (c. 206–253), reigned 251–253; Aemilianus (c. 207–253), 253, about 60 days; Valerian (c. 200–260), reigned 253–260; Gallienus (219–268), reigned 260–268; Claudius Marcus Aurelius (214–270), reigned 268–270; Diocletian (243–313), reigned 284–305: Grant, M., 1985; Kienast, D., 1990.

It looks as though this has to be regarded as an attempt to reduce the whole history of the Roman pagan institution to a single formula: the only significant omissions are Heliogabalus and Alexander Severus (218–235), Tacitus, Probus and Carus (275–283). The only significant inaccuracies are the number of years given for the reigns of Vitellius, Gallier and Claudius Marcus Aurelius. The basic information is derived from Tacitus, Suetonius, Dio Cassius, Herodian, etc. One presumes that it was extracted from a history textbook or an encyclopedia: Prideaux, M. and J., 1664, 189–209; Zedler, J.H., 1737, XV: 285–292.

The formula is audaciously simplistic, though hardly more so than the highly successful arithmetic of the sexual system he formulated for the classification of plants. Just as plants are distinguished from the minerals they depend on and the animals they serve, so the pagan emperors are distinguished from the rulers of the former republic and those of the Christian empire which came after them. Just as plants are classified according to the nature of their sexuality by counting stamens and pistils, so Roman emperors are classified by correlating the soundness of their morality with the length of their reign.

This is therefore quite clearly the proper place for this material within the re-structured *Nemesis Divina*, linking parliamentary affairs and ministerial and military policy-making (IV.i.2.1–IV.i.4.4), with the case-studies of Christian dynasties and rulers (IV.i.5.6–IV.i.5.17).

Although such an approach to those who held the imperial office is wholly in tune with the central theme of Linnaeus' theodicy (*Introduction* II.iii.2), he may well have felt that the drastic simplification it involves needed some correction. This would account for the adding of the last three anecdotes, touching as they do upon subconscious awareness, moral decision and theological orientation.

On the first and the second of these anecdotes, see II.iv.2.8 and III.iii.4.1. In the third, Linnaeus wrote 'Aureolus'. He must, however, be referring to the way in which Marcus Aurelius (121–180) reacted to the unsuccessful and fatal revolt of Avidius Cassius (130–175): Dio *Roman History* LXXI 27, 2; *Scriptores Historiæ Augustæ* (Loeb ed. I: 193–195); *The Thoughts* XII, 2, 23.

IV.i.5.5:

London ms. 17, Latin; Uppsala ms. 167, Latin and Swedish; 24, Swedish: cf. II.iv.3.4.

Although historically Caesar's career precedes the establishment of the empire proper, as an individual he embodied a very distinct anticipation of it. This case-study has been placed here as a fuller and more rounded exemplification of the general formula established in IV.i.5.4.

IV.i.5.6:

London ms. 8; Uppsala ms. 38. Latin.

Linnaeus has not got the main course of events entirely clear, probably as a result of an inadequate interpretation of the sources by the professional historians of the time, since even Gibbon's account of Justinian II's character and career differs substantially from that now generally accepted.

Justinian II (668–711), Byzantine emperor, son of Constantine IV, first acceded to the throne in 685.

In 692 he had Leontius, one of his generals, thrown into prison, possibly because he had been in command of the army when it was defeated by the Arabs at the battle of Sebastopolis. Three years later he ordered his release, and put him in command of troops destined for frontier duties. Leontius, thinking that Justinian was sending him to his death, took the opportunity to lead a palace revolution, captured his sovereign, had him mutilated and exiled to Cherson in the Crimea, and was himself proclaimed his successor.

In 698 Carthage fell to the Moslems, and Tiberius Apsimar, one of the officers in the defeated Byzantine fleet, led an uprising in Constantinople which toppled Leontius and installed Tiberius II in his place. Leontius was mutilated in the same way as Justinian had been, and was forced to enter a local monastery.

In 702 Tiberius II banished Philippicus Bardanes (d. 714), one of his officers, to Cephallenia, an island in the Ionian sea, because he suspected him of having designs on the throne.

While in exile, Justinian married Theodora, the daughter of the ruler of Khazaria. In 705, with Bulgar help, he forced the capitulation of Constantinople, and had Tiberius executed. His second reign only lasted until 711, when the banished Philippicus Bardanes took Constantinople. Justinian II was executed on November 14th, and Philippicus proclaimed emperor: Gibbon, E., 1776/1788, ch. 48; Head, C., 1972.

IV.i.5.7:

Uppsala ms. 154, 153. Swedish. There would appear to be no particular significance in the second account of the career of Birger Magnusson (1280–1321), the reference to the sacrament being the only noteworthy variation in it.

The basic theme here is the weakness of the monarchy in Sweden during what is generally known as the Folkungar period (1250–1365), – the uncertainty concerning the precise nature of the constitutional bond linking the various provinces of the country, the absence of any established pecking order between the great families of the realm, the lack of any clear consensus as to whether the kingly office was elective or hereditary. Christianity had been introduced and institutionalized, but the relationship within the country between civil and ecclesiastical authority was as yet ill-defined, as was the status of any appeal to Rome.

The successive generations of the Folkungars, who were partly a party of the higher nobility and partly a family in their own right, reflected the tensions created by this situation. Birger Jarl (c. 1210–1266), the greatest of Sweden's mediaeval statesman, reputed founder of Stockholm and effectively ruler of the country for the last twenty years of his life, had married the sister of Eric Ericsson, monarch from 1222 until 1250 and the last of his dynastic line. Birger's son Valdemar (d. 1302), then only a child, was elected to succeed: just as Eric had been subject to his top magnates, however, so Valdemar was faced with intense rivalry from his brothers, and in 1275 was obliged to cede the throne to Magnus Ladulås (d. 1290), the most forceful of them. Although Magnus managed to secure certain privileges for the crown, and although his son and successor Birger (1280–1321) was well-served during the early years of his reign by his "faithful lord chamberlain" Torgils Knutsson (d. 1306), in 1308 the ruling monarch was forced by his brothers to agree to a power-sharing arrangement, which was to last for more than a decade.

The first concrete chain of events by means of which Linnaeus illustrates this general situation took place in the village of Ramundeboda, now the town of Laxå, some thirty miles south-west of Örebro, during the evening of June 15th 1275. Eighteen miles to the south-west of this village, by Hova, between the great lakes Vänern and Vättern, king Valdemar's troops were doing battle with those of his brother Magnus. The rhyme-chronicle records that while the king lay down and dozed off, "some heard mass and some sat around, some walked about and some dined, and the queen took part in a game of chess". A bloody and bedraggled messenger burst in to say that the battle was lost. Valdemar and his family fled through Värmland to Norway. Magnus was hailed as king at the Mora stone, very near to Linnaeus' Hammarby, on July 22nd, and on May 24th 1276 crowned in Uppsala cathedral.

The second series of events took place between 1306 and 1319. Since the very beginning of Birger's reign, Knutsson had backed him both against his brothers and against the clergy, who as a result of policies promulgated by pope Boniface VIII (1294–1303), were intent on curtailing the monarch's power to tax them. In the December of 1305, however, the king was pursuaded by his brothers to allow them to arrest the lord chamberlain, who was transported to Stockholm bound under the belly of a horse, executed there on February 6th 1306, and laid to rest in unhallowed ground. During the following autumn, while the king was entertaining his brothers on his estate at Håtuna, some seventeen miles due south of Uppsala, they "bent the rules of the game", as they put it, and incarcerated him in the castle at Nyköping, only releasing him two years later, after he had promised them autonomy in their duchies. After waiting eleven years for the opportunity, he managed to turn the tables on them in the manner described by Linnaeus, but their deaths during the summer of 1318 created such an uproar throughout the country that he was deposed by general consent on Midsummer-day 1319, and replaced by his three-year-old nephew Magnus Ericsson (1316–1374), who united the crowns of Sweden and Norway.

In many Scandinavian folk-ballads, as in Shakespeare's *Tempest*, the game of chess is treated as an analogue to the play of affection between two lovers. Chess-games between lovers are frequently represented on wedding-chests. If Linnaeus had such associations in mind, his mentioning of the game at Ramundeboda is certainly ironical, the inaccuracy in his account of it deriving from his having attempted to supplement Dalin's narrative (1750) from memory, without checking on the sources. Dalin had simply recorded that "Valdemar and Sophia kept well in the background, and amused themselves in Ramundeboda". It was well-known that there was no love lost between king Valdemar and queen Sophia (c. 1240–1286). She was the daughter of Eric Ploughpenny, king of Denmark (1216–1250), and since marrying Valdemar in 1260 had borne him several children, mainly daughters. In 1273, however, she was visited by her sister Jutta, a nun from a convent in Roskilde, "as fair as an angel from heaven", so we are told by the rhyme-chronicle. The king fell in love with his sister-in-law, had a son by her, and was forced by public opinion and by the church to atone for the aberration by going on a pilgrimage to Rome, from which he had only recently returned when the trouble blew up with his brother Magnus: Dalin, O. von, 1742/1761, pt. 2, ch. 8, §5.

It seems more likely, therefore, that in the chronicle, as in the text, the game is mentioned on account of its political and fatalistic connotations. The chess-pieces and their moves are seen as the analogues of what happens in social and political life. The idea was popularized in a moralizing work written about 1290 by the Italian Dominican Jacobus de Cessolis. It caught on all over Scandinavia, and remained popular throughout Europe during the middle ages, one of the first books published by Caxton being an English translation of it. Events seemed to bear the idea out. It was said to be the case that in 1028 Canute the Great arranged for his brother-in-law Ulf to be assassinated while he was playing him at chess. The *Knytlingasaga* records that in 1157 Valdemar the Great arranged for his brother Svend to be eliminated in the same manner. And it was thus that queen Sophia's own father had met his end at the hands of his brother Abel. Death was often depicted giving checkmate to a king, and in one of the vaults of Täby church, just north of Stockholm, there is a mural dating from 1490 of a man playing chess with death, the text being: "I say you checkmate": Murray, H.J.R., 1913; Blomqvist, G., 1941; KLNM 2: 223–226; Curman, S. and Roosval, J., 1950, 101.

IV.i.5.8:

London ms. 12, Latin; Uppsala ms. 147, Swedish.

The central theme here is evidently the manner in which the Vasa dynasty managed to unify the country during the course of the sixteenth century. To a great extent this was the direct result of the personal leadership of Gustav Vasa (1496–1560), who established the country's independence by putting an end to the Union of Kalmar (1397–1520), and helped to free the Swedish church from Rome by adopting Lutheranism and the general principles of the Reformation. To some extent, however, it was also the result of developments that had taken place during the union with Denmark and Norway, since once Magnus Ericsson (1316–1374) had established the precedent of being proclaimed king by the general consent of the estates of the realm (IV.i.6.7), once it had become common practice for the newly-elected monarch to undertake the royal progress known as the *Eriksgata*, to visit each of the separate provinces in turn, confirming their various legal codes and privileges, there was much less of a chance that the nobility might find local support against the crown.

Nevertheless, as Linnaeus notes, tensions and differences within the house of Vasa often appeared to be imperilling the general progress. Erik XIV (1533–1577), Gustav Vasa's eldest son by his first wife Katarina (1513–1535), daughter of the duke of Sachsen-Lauenburg, proved to be unstable in character, if not deranged. In 1568 he was captured by his brothers, deposed and imprisoned (IV.i.3.2), and eventually, on February 26th 1577, in Örby castle, put to death by means of poisoned pea-soup. John III (1537–1592), Gustav Vasa's eldest son by his second wife Margareta Leijonhufvud (1516–1551), daughter of one of the king's councillors, was crowned on July 7th 1569. In 1562 he had married Katarina Jagellonica (1526–1583), sister of the king of Poland, in 1577 he was converted to Roman Catholicism, in 1586 his eldest son Sigismund (1566–1632) was elected king of Poland. Charles IX (1550–1611), Gustav Vasa's second eldest son by his second wife, father of Gustavus Adolphus (1594–1632), despite having been excluded from power by his elder brother, inherited the national interest and so became the focal-point for the future development of the country. In 1593 a broadly-based meeting of the clergy at Uppsala re-asserted the protestantism and independence of the Swedish church. In 1599 parliament assembled in Stockholm and deposed Sigismund.

Örby castle is on lake Vendel, some twenty-five miles due north of Uppsala, and the cell where Erik XIV consumed his pea-soup can still be seen there. The heart of it dates from the fifteenth century, and as Linnaeus observes, it was extended by Gustav Vasa, who paid for this from revenues obtained by confiscating church property. It looks as though Linnaeus is implying that this treatment of the established church is to be seen as accounting for certain of the misfortunes of his descendants.

He is certainly implying that John III's treatment of his younger brother is to be seen as accounting for his own subsequent misfortunes, as well as those of his son.

The agreement between the brothers seems to have been entered into during the autumn of 1567, soon after John and his wife had been released from a prolonged imprisonment in Gripsholm castle. The oak in question was not in Östergötland but in Värmland – more precisely, in Knappfors in the parish of Bjurtjärn, which is situated at the northern end of lake Alkvettern some ten miles north-west of Karlskoga. The historian Erik Gustav Geijer (1783–1849), whose uncle purchased the estate of Alkvettern in 1769, informs us that in his day it was still known locally as the King's Oak: 1832/1836, 2: 212.

The first important political result of the alliance was the joint occupation of Vadstena in Östergötland on July 12/13th 1568. The castle there was held by Märta Leijonhufvud, their aunt, who enabled them to pay their troops, many of whom were German or Scottish mercenaries, by making over to them the blood-bullion recently paid to her by Eric XIV in compensation for the murder of her husband Svante Sture (1517–1567). The local mint-master Mikael Hansson was employed to design and produce the five four-cornered silver coins (8 marks, 4 marks, 2 marks, 1 mark, 4 öre) now known to numismatists as "blood-tokens". They were well-executed, and quite evidently designed with the current political situation in mind. On one side they display their value and the three crowns of Sweden, on the other the date and the crowned Vasa emblem of a sheaf, bound by a twisted rope which on either side of the sheaf enfolds the brothers' initials, J and C. The eight-mark piece has around its edge, on both sides, a motif repeated thirty-two times which has been taken to be an acorn set in two oak-leaves. It is in fact the central upper part of the crown, originating in the fleur-de-lis found on some Swedish mediaeval coins: Lagerqvist, L.O., 1995, 152–153; letter from Deputy State Herald Dr. Henrik Klackenberg, 3.11.1998.

The brothers' followers were said to have worn oak-leaves in their caps to symbolize their joint purpose. Elias Brenner (1647–1717), in his book on the coinage of Sweden, seems to have assumed that since the blood-tokens were minted in Östergötland, the oak must also have been located in that province: Brenner, E., 1691, 1731[2], 87; Werwing, J., 1746/1747, I: 18; Dalin, O. von, 1747/1761, pt. 3, bk. 1, ch. 11, §22.

IV.i.5.9:

Uppsala ms. 147. Swedish.

Linnaeus himself appended this to the previous case. Just as Gustav Vasa, by "destroying monasteries and sacred objects" helped to fuel the rivalry between his progeny, put an unprecedented responsibility on the Lutheran clergy, created the prison in which his eldest son was done to death, so Nebuchadnezzar, by "taking the vessels of gold and silver from the temple at Jerusalem" helped his son Belshazzar to seal his fate – drinking out of them together with his concubines and courtesans, requesting that Daniel should interpret the writing, falling to the sword of Darius the Mede.

Although the popular and individualistic assumptions of contract theory were by no means entirely absent from Scandinavian political theory; Johannes Lenæus (1573–1669), for example, archbishop of Uppsala, had suggested by means of his *De jure regio* (Uppsala 1633/1647) that the majesty of the monarch must include his being constrained by the will of the people, there was a very strong tendency to look at things from the other end of the telescope, that is to say, to justify the constitution from a conception of God or an interpretation of the Bible (IV.i.3.3). Hans Wandal (1624–1675), bishop and professor at the University of Copenhagen, in his *Juris regii* (1663/1672), the most authoritative work on the subject in Denmark after the introduction of monarchical absolutism (1661, 1665, 1670), laid it down that the king, on being anointed, was in possession of certain divine attributes, and certainly had the right to call priests and tax the church. In Sweden, much the same line of thinking is to be found in the works of Carl Lundius (1638–1715), practising lawyer and professor at Uppsala, who is particularly interesting in that he combined a hardheaded approach to everyday legal proceedings with a romantic Rudbeckianism, an uncompromisingly theocratic conception of the ultimate authority in law-making, and great ability in arguing his case with opponents: Sandblad, H., 1967.

During this period, however, the most widespread acceptance of the importance of monarchy came from those who like Linnaeus knew their Bible well, and had a realistic commonsense conception of family life. There were many in the Scandinavia of the time who would have subscribed wholeheartedly to the doctrines expounded so elegantly by Sir Robert Filmer in *Patriarcha* (1680), had they had the opportunity to read the book, one of the most interestingly articulate of them being Johannes Forsenius (d. 1705), regimental chaplain at Narva, who in his *Purpur principis* (1694) presented an eloquent justification for absolute monarchy by expounding the twenty-ninth chapter of the book of *Job*: Normann, C.-E., 1948.

IV.i.5.10:

Uppsala ms. 49. Swedish and Latin. The only Latin is the 'Rex Sveciæ' in the title. The three points after Scheffer would seem to indicate that Linnaeus also had others in mind.

Adolf Frederick (1710–1771) was the son of the prince-bishop of Lübeck, his hereditary claim to the Swedish throne consisting of a great-grandmother on his mother's side, one of the daughters of Charles IX (1550–1611): IV.i.5.8. In 1725 his cousin Karl Friedrich (1700–1739) duke of Holstein-Gottorp, son of Charles XII's elder sister Hedvig Sophia (1681–1708) had married Anne (1708–1728), daughter of Peter the Great (1672–1725). Their son Karl Peter Ulrich (1728–1762) therefore had a strong claim both to the throne of Sweden and to that of Russia. On June 21st 1739, at almost exactly the same time as Malcom Sinclair (1690–1739) was murdered by Russian agents in Silesia (IV.i.3.5; IV.iii.2.8) and the Hats were beginning to push in earnest for war with Russia, Adolf Frederick was appointed guardian of the recently-orphaned duke. Sweden finally declared war on Russia on July 28th 1741, so precipitating the palace revolution which brought the duke's aunt Elizabeth (1709–1762) to the Russian throne (II.iv.2.7), and opening the way for the Russian occupation of Finland (II.ii.3; IV.i.4.3).

It was through this complicated web of events that Adolf Frederick emerged as king of Sweden. The empress was childless, and decided to adopt the young duke as her heir – partly as a result of her attachment to her dead sister, partly in order to break her nephew's links with Sweden. During the autumn of 1742, therefore, at the height of the war, the future emperor Peter III was despatched by his guardian to St. Petersburg, where he was received into the Russian Orthodox church and began his grooming for his future role. Having the upper hand in Finland, and wanting to prevent the emergence of a Nordic union through the acceptance of the crown prince of Denmark as heir to the Swedish throne, the empress then offered to restore Finland to Sweden if the future czar's former guardian was accepted in Stockholm as heir to the throne. Given the disastrous military situation this was an opportunity not to be missed, and on June 23rd 1743 all four estates of the realm decided to accept the offer. Adolf Frederick was duly elected heir to the throne, with right of inheritance for his male descendants, signing an agreement to this effect in Hamburg on September 3rd 1743, and moving his official residence to Stockholm some six weeks later.

Adolf Frederick was good-natured and courteous rather than forceful or determined. When he succeeded on the death of Frederick I in 1751, a council and parliament dominated by the Hats got him to sign a royal assurance which severely limited his constitutional significance, leaving him with little more than the power to confer knighthoods and create peers. As Linnaeus notes, the badly-timed and ill-organized effort made by the court party under the leadership of Eric Brahe (1722–1756) to secure more power for the crown ended in complete disaster: Brahe and those who assisted him were publicly executed (III.i.1.4; IV.i.2.5), parliament circumvented the need for the royal signature on official documents by authorizing the use of a rubber stamp, one of the most aggressive of the Hats Nils Palmstierna (1696–1766) was given a free hand as chairman of a commission for the searching out of anti-government conspiracies, and used it with a vengeance, not least in his capacity as chancellor of the university of Lund. By the end of the 1750s it looked as though the monarchy was on the point of being done away with altogether.

The tide turned for a variety of reasons, the most important being the changes which took place in the fortunes of the political parties – the temporary emergence of an influential splinter-group and the return to power of the Caps. On January 5th 1761 the disgruntled Hat Carl Fredrik Pechlin (1720–1796), with the collusion of the Hat speaker of the House of Lords and the French ambassador, forced the resignation from the council of the Hats Anders Johan von Höpken (1712-1789), Carl Fredrik Scheffer (1715–1786) and Nils Palmstierna, all of whom, in their different ways, had made a point of opposing the crown, as, indeed, Pechlin had: IV.i.2.8. Although Scheffer subsequently returned to the council, he was forced out again, together with Claes Ekeblad (1708–1771), someone else who had recently been at odds with the king (IV.i.3.4), when the Caps gained the upper hand in 1765: IV.i.2.11.

These party-political patterns were complemented by developments within the royal family itself. In 1744 Adolf Frederick married Lovisa Ulrika (1720–1782), sister of Frederick the Great, a strong-willed and strong-minded woman who during the early 1760s managed to lay the foundations of somethirg resembling a new court party. In 1746 she gave birth to the future Gustav III (d. 1792). Ten years later the Hats arranged for Scheffer, an expert in the theoretical elaboration of their own interpretation of the constitution, to be appointed his tutor. By 1766 the prince was thinking for himself, however, and the political state of the country was beginning to make it obvious that the constitutional arrangement of 1751 needed revision. Even Scheffer began to admit this, and there is evidence that Linnaeus' attitude to party politics also began to change at about this time: III.i.1.4; IV.i.2.8; IV.i.2.11. Daniel Tilas (1712–1772), who had rejoiced at the Cap victory of 1765, was saying a year later that the whole country had been demoralized by the so-called liberties of the constitution, that both the main parties were equally culpable, that the law of talion was the only one which still appeared to command general respect: Lagerroth, F., 1915; 1937; Tilas, D., 1974, 287–293; Fahlbeek, E., 1915/1916.

Linnaeus is certainly justified in emphasizing that Adolf Frederick's career as a monarch was unusual in that it was much more the creation of international circumstances and party politics than it was of heredity and personal fulfilment. If these jottings can be said to have a basic theme, it is that the set-backs of the 1750s were being countered by the political tendencies of the 1760s, and this too was certainly a justified interpretation of contemporary developments.

IV.i.5.11:

London ms. 15, no. 24, Swedish and Latin; Uppsala ms. 89, Swedish.

Linnaeus was uncharacteristically careless when he copied out this case. The London version begins quite correctly: "Ziegler, wife of the mayor of Vasa, ill-natured and spiteful but handsome and randy ...". In the Uppsala copy this becomes: "Ziegler, mayor of Vasa, had a handsome but ill-natured and randy wife". The third sentence is also reformulated, the phrase: "for life, but pardoned" being deleted. It has, therefore, been thought advisable to correct the opening sentence.

The confusion is all the more remarkable, since after visiting Vasa in the August of 1732, Linnaeus made the following note: "The mayor, Elias Grizell, an extraordinarily dashing person, treated me with exemplary respect; I was entertained here better than anywhere else in Österbotten. His wife, beautiful not only facially but also in character, is half-sister to Ziervogel, wife of the former dean of Roslagen". He must, at that time, have gathered this information directly from the mayor's wife: Linné, C. von, 1913, 204–207, 237.

Elias Grizell (1687–1750), son of the vicar of Sorunda, just south of Stockholm, became mayor of Norrtälje in 1722, and was appointed to the same post in Vasa on December 6th 1728: Aspelin, H.E.M., 1892/1897, II: 491; Millqvist, V., 1911, 127. In 1728 he married the recently-widowed Gertrud Ziegler (1702–1759), daughter of the Stockholm sugar-baker and court confectioner Joachim Christoph Ziegler (1671–1734) and his wife Gertrud Lang, who had previously been married to Christian Fredric Ziervogel (1660–1696), also a court confectioner, their daughter Brita Justina (1688–1751) having married Lars Balck (1680–1730), dean of Börstil, which is situated in Roslagen: Fant, J.E., 1842, 231–232.

Gertrud Ziegler had first married in 1721 – Johan Strandberg (d. 1727), by whom she had two children, a boy Johan Joachim, born in 1722, who later found employment with the Swedish East India Company, and a daughter Margareta Juliana, born in 1724, who later married a businessman in Åbo and moved to Russia. She had two sons with Elias Grizell, both of whom became soldiers: *Svenska ättartal* XI (1896) 479–483.

Linnaeus' opinion of the mayor's wife quite obviously changed over the years. It looks as though the first sentence on her was passed before the Brahe trouble blew up in the summer of 1756 (III.i.1.4; IV.i.2.5). After that date, the queen, Lovisa Ulrika (1720–1782), would in any case have had less scope for intervening in the normal course of the law: IV.i.5.10.

The implication of the case-history seems to be that the queen's intervention was unjustified, that it was only right that it should have been overruled by the subsequent course of events. It has been placed here as an example of the humanizing role of the monarchy.

IV.i.5.12:

London ms. 3; Uppsala ms. 75. Swedish. The London text is very similar to the first section of its Uppsala counterpart. It is evident from the handwriting that the material concerning Buckingham was added to the Uppsala version, apparently at a somewhat later date.

Linnaeus' interest in the general history of the house of Stuart was probably aroused by the careers of the two pretenders, James Francis Edward Stuart (1688–1766), son of James II (1633–1701), who made unsuccessful attempts to take the British throne in 1708 and 1715, and his son Charles Edward Stuart (1720–1788), who re-enacted his father's failures in 1745. All three attempts impinged in one way or another on the course of Swedish politics: IV.i.4.1; IV.i.4.2; III.iv.3.18.

Mary queen of Scots (1542–1587), faced with insurrection in her own country, fled to England on May 16th 1568, and was taken under the protection of her cousin Elizabeth I (1533–1603), who offered to mediate between her and her people, and in the first instance hoped to restore her to her throne. She was given into the custody of George Talbot, Earl of Shrewsbury (1528–1590), who had his residence at Tutbury castle in Staffordshire. Since Elizabeth, a protestant, was unmarried, there was always the possibility that Mary, a Roman catholic, might inherit the English throne. Mary was implicated in *three* main Roman catholic plots for bringing this about: that devised by the Florentine banker Roberto di Ridolfi (1531–1612) in 1570, that facilitated by the Spanish agent Francis Throckmorton (1554–1584) in 1583, and that instigated by her romantic admirer Anthony Babington (1561–1586) in 1586. She was pardoned after the first, set under the harsh governance of the puritan Amyas Paulet (1536–1588) after the second, and only executed after the third, on February 8th 1587.

James (1566–1625), the only son of Mary and her consort and cousin Henry Stuart, lord Darnley (1546–1567), was born on June 19th 1566, and after the enforced abdication of his mother, crowned king of Scotland on July 29th 1567. It is apparent from the letters James wrote to Elizabeth concerning his mother's execution, that although it had the undeniable advantage of rendering his own status as king unequivocal, he sincerely wished to save her life: not on account of filial affection, which he could hardly pretend to feel for a woman he could not remember ever having seen, but on account of his political predicament, – her execution was bound to reflect on his status and honour, and would be in flagrant violation of the law of divine right, in accordance with which sovereigns are accountable to God alone: Akrigg, G.P.V., 1984, 79–87.

James, the sixth of Scotland, became James I of England on the death of Elizabeth in 1603. After the civil war (1642/1648), his son Charles I (1600–1649) was tried on the authority of parliament for offences against the people of England, and beheaded at Whitehall in London on January 30th 1649. In the speech he delivered from the scaffold, he took his enforced connivance in the death of Thomas Wentworth, earl of Strafford (1593–1641) as the immediate occasion of his own end: "God forbid that I should be so ill a Christian, as not to say that Gods Judgements are just upon me: Many times he does pay Justice by an unjust Sentence, that is ordinary; I will only say this, That an unjust Sentence (Strafford) that I suffered for to take effect, is punished now, by an unjust Sentence upon me". Charles I, 1649, 6–7.

George Villiers, duke of Buckingham (1592–1628) married Katherine Manners (1600–1649) in 1620, and despite his being well-known as something of a ladies' man, there are no grounds for maintaining that during the unsuccessful courting-expedition to Madrid (February–October 1623) he was ever seriously unfaithful to his wife. The object of the expedition was to further the effectiveness of James I as the peacemaker of Europe by arranging for his eldest son Charles to marry the daughter of Philip III (d. 1621), infanta Maria Anna (1606–1646), who subsequently entered into a happy marriage with the emperor Ferdinand III (1608–1657). Despite Charles' actually falling in love with his proposed bride, his suit was unsuccessful on account of what were regarded in Spain as his unorthodox religious beliefs and his quite clearly inept courting techniques: Lockyer, R., 1984, 125–167; Gregg, P., 1981, 73–92.

Linnaeus is almost certainly retailing the view of Edward Hyde, earl of Clarendon (1609–1674) concerning the worsening of the relationship between James I and Buckingham during the early months of 1625: 1849, I: 14, 15, 33. There is, however, little evidence that it was justified, and even less supporting the persistent rumour that Buckingham and his mother Mary Beaumont (1570–1632) brought about the king's death by poisoning him. Given the political climate of the time, it was probably inevitable that their efforts to restore his health by herbal remedies disapproved of by the royal physicians, should have given rise to the accusation that they had purposely brought on his final illness: Fuller, T., 1845, V: 568–571; *Calendar of State Papers and Manuscripts relating to English Affairs existing in the Archives and Collections of Venice, and other Libraries of Northern Italy* (ed. A.B. Hinds, London, 1927) vol. XXVIII: 44, no. 94 (February 13th 1648).

IV.i.5.13:

Uppsala ms. 171b. Swedish. The word "alive" is inserted after "exhibiting", "eight" is changed into "nine" o'clock.

Linnaeus evidently heard of these events when he was working with Johan Burman (1707–1780), supervisor of the Pharmaceutical Garden in Amsterdam, between 1735 and 1737. When the garden was founded in 1682, one of its most energetic patrons was Jan Huydecoper (1625–1704), the mayor of the city, and it was his son Jan Elias Huydecoper (1669–1744), infantry lieutenant 1686, captain in the civil guard 1704, who took part in the attempted kidnapping in 1706, and who told Linnaeus about it thirty years later: Kuijlen, J., 1983, 20–24; Elias, J.E., 1963, II: 700–701; Polak, M.S., 1987, 213–216, nos. 147–162; Fries, Th.M., I: 215–226.

Official preparations began on November 23rd 1706, when the State Council passed a resolution authorizing the freebooter Pieter van Guethem (d. 1709) to carry it out. Marlborough's victory at Ramillies on May 23rd 1706 had forced the French onto the defensive (IV.i.4.1), and opened up the prospect of liberating the Southern Netherlands. This audacious undertaking was conceived of as a follow-up, and was meticulously planned: Van Guethem's troop consisted of sixteen officers and fourteen dragoons, well-supplied with hard currency, all of them French-speaking and in possession of genuine French passports, several of them experienced in horse-trading for the French army; six staging-posts were established, each with six horses, as well as five extra posts with a horse for the prisoner's chaise.

Three groups of ten men rode Pariswards from Cambrai by three different routes, took up lodgings in Sèvres, St. Cloud and Boulogne, and spent a few days reconnoitring the road from Versailles to Paris, especially the bridge at Sèvres where it crosses the Seine. The kidnapping was planned to take place just beyond the bridge, at about dusk on Thursday March 24th 1707. A scout was posted on the Versailles side, the main body of Van Guethem's troopers on the Paris side, just to the north of the road. At about eight o'clock there was a whistle from the scout – a coach and six, bearing the royal arms, footmen dressed in the royal livery, preceded by a horseman with a torch, was approaching from Versailles. Soon after it had crossed the bridge, it was surrounded. The occupant of the coach, Jacques-Louis, marquis of Béringhen (1651–1723), knight of the Order of the Holy Ghost, master of the king's horse, was hurried off north through the Bois de Boulogne.

Van Guethem was careless enough not to see to it that the personnel did not get away. The scout was captured crossing the bridge. By nine o'clock Louis XIV was fully informed of what had happened. Messengers were despatched to river-crossings, frontier-posts, and to the ports of the north-east.

Van Guethem's party reached the first staging-post during the night, and at dawn the following day crossed the Oise near Creil by pontoon. At the second post, near Remy, just west of Campiègne, the prisoner was transferred to a chaise. The Somme, near Ham, was reached on the Friday evening, and the party, which now consisted of the prisoner and three officers, rested for the night. They crossed the river early next morning, and were making for the next post when they were apprehended by the local sheriff and his troopers. Van Guethem addressed Béringhen. "Sir, you are no longer my prisoner, I am yours". Since he had official approval for the operation from the State Council and had been careful not to violate accepted conventions, he and his officers were treated as prisoners of war, as pawns in subsequent diplomatic exchanges: Veenendaal, A.J., 1975; *Dictionnaire de Biographie Française* VI: 20.

In the manuscript, this case comes after that concerning the Parisienne (III.ii.3.2), and Linnaeus indicates that the two are "analogous", that is to say, that they both illustrate the fact that the will of God overrides human objectives. It has been placed here, since it may be taken as illustrative of Linnaeus' attitude to republican politics.

IV.i.5.14:

Uppsala ms. 47. Latin and Swedish.

This is evidently a reference to the way in which Peter the Great (1672–1725) is reputed to have dealt with the revolt of the Streltsy, the conservative imperial guard regiments who marched on Moscow during the summer of 1698 in protest against his westernizing reforms. The executions took place at Preobrazhenskoe on October 23rd, and according to Johann Georg Korb (d.c. 1720), secretary of the Austrian legation at the time, the emperor himself beheaded some of the condemned: Korb, J.B., 1700; Massie, R.K., 1982, 259.

Peter III (1728–1762) was Peter the Great's grandson, the son of Anne (1708–1728), wife of Karl Friedrich (1700–1739) duke of Holstein-Gottorp, his daughter by his second wife Catherine I (1684–1727): IV.i.5.10. On August 21st 1741 he was married to Sophia Augusta Frederika of Anhalt-Zerbst (1729–1796), the future empress Catherine II. He succeeded to the imperial throne on the death of his aunt the empress Elizabeth on December 25th 1761, and there is every likelihood that his wife arranged for him to be done away with soon after he had been forced to draft a letter of abdication on June 29th 1762. On July 6th news of his death was sent to the empress by his gaoler Aleksei Orlov, who reported that he had expired after a scuffle with prince Fedor Bariatinskii. He was also said to have been done away with by Girgorii

Orlov, who was certainly not in the vicinity. In the official announcement put out on July 7th it is stated that he died of colic following a haemorrhoidal attack, but various rumours concerning the precise circumstances of his death were in circulation: *Merkwürdige Lebensgeschichte Peter des Dritten* (Frankfurt and Leipzig, 1762); Marche, C.F.S. de, *Nouveau Mémoires, ou Anecdotes du regne et du détronement de Pierre III* (Berlin and Dresden, 1765).

The descendant of the pardoned prisoner mentioned by Linnaeus was probably the gaoler Aleksei Orlov (1737–1808), member of the landed gentry of the province of Tver, who entered the imperial guards regiment in 1749, and played a leading role in getting Peter III to sign the abdication document: Alexander, J.T., 1989, 10–16.

IV.i.5.15:

London ms. 17, no. 46a, Latin; Uppsala ms. 137, Swedish. In the second section, Linnaeus writes "Leyonhufwud" for Stenbock. cf. II.iv.2.7.

Empress Anne (1693–1740) of Russia, daughter of Peter the Great's elder brother Ivan V (1666–1696), died childless on October 28th 1740, the throne passing to Ivan VI (1740–1764) son of her niece Anne duchess of Brunswick (1718–1746), who was only ten weeks old when his great-aunt passed away. Anne had appointed her favourite Ernst Johann Biron (1690–1772) regent, but after only three weeks he was replaced by Burkhard Münnich (1683–1767), who exercized authority under the regency of the emperor's mother: IV.i.3.5.

Elizabeth (1709–1762) was Peter the Great's daughter by his second wife Catherine I (1684–1727). During the War of Austrian Succession (1740/1748), France attempted to counteract Russia's support for Maria Theresa by encouraging Sweden to think of regaining the Baltic provinces, and by helping Elizabeth to remove the pro-Austrian regency council and have herself proclaimed empress: II.iv.2.7. The activities of her physician Jean Herman Lestocq (1697–1767) played an important part in the realization of this policy.

Lestocq was born in Hanover of French parents. In 1713 he was appointed physician to Peter the Great, and on the emperor's death Catherine I appointed him physician to her daughter Elizabeth. When Peter II died on January 30th 1730 he suggested to his twenty-one year-old charge that she should claim the throne, advice which was turned down at the time, but accepted, as Linnaeus notes, when it was repeated eleven years later. It was evidently Lestocq who accompanied Elizabeth to the guards' barracks on the evening of November 24th 1741, and so set the palace revolution going.

He remained in Elizabeth's favour until 1748, when his enemies at court, notably count Alexis Bestuzhev (1693–1768), changed her opinion of him by spreading ill-founded rumours. He was tried, whipped and exiled to the Volga and then to Archangel, not returning to court until after the death of Elizabeth in 1761. Catherine II granted him a handsome pension, but he retired from the court and during his last years abandoned himself, "à une malpropeté dégoûtante, qui augmenta ses infirmités". *Biografie universelle* 24: 320–321 (Paris, 1819).

IV.i.5.16:

Uppsala ms. 43, 43b. Swedish. In the seventh paragraph there is a gap after 'highness's', as if Linnaeus intended to fill in the name of someone else who was done away with.

During what is known as the Age of Liberty (1719–1772), the political parties in Sweden tended to reflect foreign interests, it being quite usual for Cap politicians to accept money made available through the Russian embassy, and for their Hat colleagues to milk its French counterpart in the same manner. Until the very end of his life, when he seems to have become increasingly aware of the importance of strengthening the monarchy, Linnaeus tended to move in Hat rather than Cap circles, and it is almost certainly from them that he drew the bulk of his information concerning what was going on in Russia. By and large he seems to have been pretty well informed.

The empress Elizabeth (1709–1761) had appointed Peter III (1728–1762) her successor soon after taking the throne (IV.i.5.10), and on August 21st 1745 had had him married to Sophia Augusta Frederika of Anhalt-Zerbst (1729–1796), the future Catherine II (IV.i.5.14). He was very probably the father of the son she gave birth to on October 1st 1754, the future Paul I (d. 1801), although in her memoirs she claimed otherwise. Stanislaus Poniatowski (1732–1798) the future Stanislaus II of Poland (1764–1795), arrived in St. Petersburg in the suite of the English ambassador Charles Hanbury Williams (1708–1759) in June 1755. He stayed there for four years before returning to Poland, was indeed "dashing, intelligent" and courteous, and certainly captured the emotional and intellectual commitment of Peter's wife. Soon after the birth of Paul, Peter had taken countess Elizabeth Vorontsova as his mistress, and she retained the position until his death: she was the niece of the vice-chancellor Michael Vorontsov, and evidently not only ugly, but also dull-witted and ill-natured.

In 1756 the empress Elizabeth suffered a stroke and her physical and mental condition began to decline. Since Peter was very much of a grown-up child, showing every sign of interrupted mental development, the suggestion was made that he should be bypassed in favour of his son. When the empress eventually passed away on December 25th 1761, however, there was no open opposition to his succession. This arose during the next few months as a result of his policies, since despite his childishness he soon set about altering the conduct of national affairs: the advisory cabinet was replaced by a military council, peace was made with Prussia, war declared on Denmark, far-reaching legislation was passed concerning trade, the peasantry and the nobility, and he was even rumoured to be contemplating the conversion of the country to Lutheranism. Catherine took her chance at the end of June, winning over the guards and requiring her husband to draft a letter of abdication. He was put under house-arrest at Ropsha, and just over a week later was dead (IV.i.5.14): Raeff, M., 1970; Mandariaga, I. de, 1981.

Augustus III, who had been king of Poland since 1733, died on October 5th 1763. A fortnight later Catherine II advocated the election of Stanislaus Poniatowski as "the individual most convenient for our common interests", and sent forty-thousand Russian troops into the country. The empress's candidate was elected on September 7th and crowned on November 25th. About ten per cent of the Polish population, the so-called dissidents, were not Roman Catholic but Protestant or Greek Orthodox, and in 1767 they appealed to the new monarch and to the Russian empress to see to it that their liberties were protected. The Roman Catholic church was, of course, opposed to this.

Had Peter III not reversed Russian policy and made peace with Frederick the Great on April 24th 1762, there was every likelihood that Prussia would have been defeated. After the conclusion of the general peace in February 1763, France therefore set about attempting to check Russian ambitions by championing Roman Catholic policy in Poland, and by heightening Turkey's awareness of a potential clash of interests in the Crimea and the Balkans. On October 6th 1768, when the Russian minister refused to give guarantees for the withdrawal of his country's troops from Poland, the grand sultan Mustafa III (1717–1773) declared war. Regardless of the truth or untruth of Linnaeus' story concerning the holy hermit and Mahomet, the situation was by no means lacking in high comedy, for while the Turks issued a manifesto declaring the Polish dissenters traitors to their church and religion, Catherine II appealed to Europe at large, "against the common enemy of the Christian name". Hostilities began with a Moslem sultan declaring himself the champion of Roman Catholicism and a deist czarina leading a crusade for Orthodoxy and Protestantism.

The outcome was the first partition of Poland – Russia and Prussia taking over large sections of it in February 1772, Austria moving in during the following August, the Polish parliament endorsing the situation on August 18th 1773, and a series of heavy defeats and extensive territorial losses for Turkey, finalized by the humiliating treaty of Kuchuk-Kainardja on July 21st 1774: Stavrianos, L.S., 1958.

IV.i.5.17:

Uppsala ms. 7, 7b. Swedish and Latin. In the last paragraph Linnaeus writes that Peter III made peace with "Russia".

Since Silesia had become Protestant at the Reformation, since in 1537 the duke of Liegnitz had guaranteed the inheritance of his duchy, together with Brieg, Jägerndorf and Wohlau, to the elector of Brandenburg, since the whole area had remained very uneasy under Austrian rule, at least since the outbreak of the Thirty Years' War, Frederick the Great had some justification for wresting it from the empress during the campaigns of 1740/1741. In 1744 Frederick began a second war in anticipation of a counter-attack for regaining the conquest, and by means of it strengthened his hold on the area still further. During the Seven Years' War the "three whores" – the empresses Elizabeth (1709–1761) and Maria Theresa (1717–1780) and the marquise de Pompadour (1721–1764), mistress of Louis XV, the governments of Russia, Austria, France, Sweden and Saxony, were ranged against Prussia and England. Silesia was repeatedly overrun by Russian and Austrian troops, and Frederick's ultimate expulsion seemed only a question of time. By the concluding Treaty of Paris, however, he was confirmed in undiminished possession of the territory; Naudé, A., 1884.

The coupling of the famous complaint concerning the unholy trinity of whores with the observation concerning Frederick's relationship with his wife is a typically Linnean stroke – it is in fact the case that Frederick's father forced him to marry Elisabeth Christine, princess of Braunschweig-Bevern (1715–1797), at the pleasure palace of Salzdahlum near Wolfenbüttel, on June 12th 1733, and that he never slept with her. What is more, this extraordinary state of affairs was evidently coupled in his mind with keeping his word. On September 4th 1732 he wrote as follows of his forthcoming marriage to his friend Grumbkow: "Marriage is the initiation of maturity, and as soon as I am married I shall be master in my own house and my wife will have the ordering of nothing. I shall marry as a gentleman should, that is, I shall allow madam to do what she finds best, and for my part do what I want to. I shall keep my word, I shall marry all right, but watch out for what happens afterwards: good day my lady, and fare you well!" ADB 6: 34–36.

The medical history of the empress Elizabeth is particularly well documented, and provides little justification for what Linnaeus heard about it through Ivan Ostermann (1725–1811), Russian ambassador in Stockholm from 1760 until 1774: SBL 28: 410. The physical and mental health of the empress began to decline in 1756, when she suffered her first stroke (IV.i.5.16). On November 17th 1761 she had a bout of fever, but recovered after taking medication. On December 12th she began to vomit blood and phlegm. Her doctors – Moisey, Schilling and Cruz – decided to let blood, and after a few days she seemed to recover. Although she was feeling particularly well on December 20th, two days later the vomiting began again, and on December 23rd she made her confession and took holy communion. On the twenty-fourth she took extreme unction, and during the evening ordered the prayers for the dying to be read, repeating them herself after the priest. She died at four in the afternoon on December 25th: Soloviev, S.M., 1997, 41: 281–283.

Constantine Scepin (1727–1770) was born at Vyatka, now known as Kirov, a town on a tributary of the Volga some four hundred and eighty miles east-north-east of Moscow. A member of the St. Petersburg Academy arranged for him to be given a grant to study in the West, the natural sciences with particular reference to botany, and on July 18th 1753 he was matriculated to read medicine and botany at Leiden. One of Linnaeus' pupils, the Norwegian Peder Ascanius (1723–1803), who was also studying botany at Leiden at the time, wrote to him about him on September 5th 1754, as did Johan Burman (1706–1779) from Amsterdam on June 30th 1758: *Letters* II.i: 99; II.ii:122.

While he was studying at Leiden Scepin's patron in the Academy died, so that there seemed to be little prospect of his returning to his homeland to take up a career as an academician. The Medical College therefore stepped in and agreed with the Academy to continue his grant on the condition that he took up a career as a medical doctor. Scepin was re-matriculated at Leiden on September 10th 1755, and on May 19th 1758 finally qualified as a doctor with a thesis on vegetable acids: *Schediasma Chemico-Medicum Inaugurale de Acido Vegetabili* (University Library sign. 239 B2-19). In this work he attempts to define the general nature of acids, considers several of a vegetable kind, discusses their medical uses, and concludes with a series of botanical notes in which he criticizes certain aspects of Linnaeus' work: "I have taken it upon myself to reform some of the genera defined by Linnaeus, renaming one of them in memory of a very good man and of his country, someone who has made great contributions to the republic of botany" (pp. 21–22).

After taking his degree, he visited France and England and then sailed for home. When the ship put in at Copenhagen for a few hours he went ashore, stayed too long, and discovered when he returned to the harbour that it had already set sail. He set off in pursuit of it in a small boat, failed to catch up, and eventually found himself back in Copenhagen without money or documents. A professor at the University provided him with the wherewithal to make his way home through Sweden, and in the course of doing so he visited Linnaeus at Uppsala. Linnaeus presented him with several of his publications, and for several hours discussed matters with him in the University garden. It must have been then that they fell out, possibly on account of Scepin's having mentioned certain crucial aspects of his doctorate.

Scepin made his name in Russia as an energetic and talented practitioner and lecturer. After working for some years as a military surgeon he was appointed lecturer – first at the Moscow Surgical School and then at a similar institution in St. Petersburg. There appears to be no record in Russia of his having had a hand in the death of the Empress. His career ran into trouble once he had taken to drink, and came to an abrupt end at Kiev, where he died during an outbreak of the plague. Information concerning his Russian background and activities is to be obtained from Kuprianov, V.V., *K.I. Scepin, doctor medicini* (Moscow, 1953), and has been kindly supplied to me by Alexandra Bekasova, Institute for the History of Science and Technology, Russian

Academy of Sciences, St. Petersburg.

The Danes occupied the duchy of Holstein Gottorp in 1713 (IV.i.3.4), and the final outcome of the Great Northern War was that the Gottorp portions of Schleswig were annexed by the Danish crown (1721) – a situation which naturally rankled with the members of the ducal house. Adolf Frederick (1710–1771), accepted as heir to the Swedish throne in 1743, settled his differences with Denmark on this account in 1750 (IV.i.5.10). His charge Karl Peter Ulrich (1728–1762), however, accepted as heir to the Russian throne in 1742, simply bided his time, waiting for the opportunity to assert his claim to the territories (II.iv.2.7; IV.i.5.16). This came with the death of the empress Elizabeth and his accession as Peter III.

His first move was to make peace with Prussia, despite the fact that Russian troops were in control of East Prussia and Silesia, and Frederick the Great was on the point of capitulating. Feelers went out from St. Petersburg in the February of 1762, and on April 24th a peace treaty was signed, a secret clause of which stipulated that Frederick would help the Czar to regain Schleswig. His second move was to mobilize his troops, the great bulk of whom were in Pomerania. The guards regiments received marching orders on June 12th. Denmark had managed to remain neutral throughout the Seven Years' War, although on May 4th 1758 the minister for foreign affairs Johann Bernstorff (1712–1772) had entered into a treaty with France in accordance with which a Danish army corps of twenty-four thousand men was to be maintained in the south of the country until the end of the war, in order to secure Hamburg, Lübeck, Gottorp and Holstein against invasion. In 1761 a renowned French general, Claude Louis de Saint Germain (1707–1778) was put in charge of it. Denmark began to mobilize these forces at the end of March 1762, and the fleet was prepared for an attack on the Russian Baltic coast. Between July 9th and July 12th a Danish army of some thirty-thousand men, about half the number the czar had at his disposal, advanced via Trave into Mecklenburg. On July 17th the Danish advance parties reported back that the Russian advance had come to a halt. The news had come through of the death of Peter III on July 6th, and the reversal of his foreign policy by his consort (IV.i.5.16). In 1767 Catherine II resigned the Russian crown's claims to Schleswig Holstein in the name of her son Paul, who confirmed this action on coming of age in 1773; Bech, S.C., 1977, 385–396.

The predominantly chronological ordering of the subject matter in this presentation of Linnaeus' conception of political power harmonizes well with the generalizing remarks with which he concludes IV.i.5.16 and IV.i.5.17. The international struggle for power and influence through diplomacy and war is brought into a moral perspective by calling to mind, "the many thousands who have to bite the dust for such a trifling cause", and rejoicing that peace has "ensured the freedom of millions". In the note on man in the *System of Nature* (1766/1768 I: 30–31) Linnaeus comments as follows on the systematic transition involved here:

> *Politically*, ... man hates the truly pious and persecutes them for holding speculative opinions different from his own, exciting numberless broils, not in order to do good, unaware of any real purpose. He fritters away his precious and irrecoverable time, thinking little on immortal and eternal matters, regulating the affairs of his children and his children's children, constantly entertaining fresh projects; unmindful of his real condition, he builds palaces instead of preparing his grave, until at length death seizes him ....
>
> *Morally*, ... man in his *benighted* state is unteachable, wanton, sectarian ... when *transformed* he becomes attentive, chaste, urbane, unassuming, frugal, calm, frank, gentle, kindly, gracious, contented.

IV.ii:

Cf. IV.ii.1.1.

IV.ii.1.1:

Uppsala ms. 6, 4. Swedish and Latin. The main divisions have been somewhat revised in the translation.

Although there appears to be no evidence that Linnaeus had any direct knowledge of the writings of Thomas Hobbes, there is a remarkable similarity in tone and detail between this characterization of Catholicism and that in the fourth part of *Leviathan*, especially the final chapter:

> For, from the time that the Bishop of Rome had gotten to be acknowledged for Bishop Universall by pretence of Succession to St. Peter, their whole Hierarchy, or Kingdome of Darknesse, may be compared not unfitly to the *Kingdome of Fairies*; that is, to the old wives *Fables* in England, concerning *Ghosts* and *Spirits*, and the feats they play in the night. And if a man consider the originall of this great Ecclesiasticall Dominion, he will easily perceive, that the *Papacy*, is no other, than the *Ghost* of the deceased *Romane Empire*, sitting crowned upon the grave thereof: For so did the Papacy start up on a Sudden out of the Ruines of that Heathen Power.

Hobbes took issue with the internationalism, the universalist pretensions of the Roman church, because he saw them as a force destructive of established and effective civil order. Linnaeus is much less worried by this clearly negative aspect, much more appreciative of what has been achieved in the way of curbing dynastic and national ambitions by building on an age-old secular tradition of imperial power (IV.i.5.4).

As is apparent from the rest of the section on the Churches, he is fully aware that although co-ordinating personal, social and political objectives by universalizing them within the organization of a church is a step in the right direction, it is no guarantee of personal morality. It is not infrequently the case that what is punishable in the world of ordinary society and politics is also in evidence in the lives of the clergy.

IV.ii.1.2:

Uppsala ms. 56b; cf. III.iii.5.1. Linnaeus thought he was certain of the date, but left a gap for the year.

He was in fact also in error concerning the date: inquisitorial executions took place in Lisbon on June 21st 1744, September 26th 1745, October 16th 1746, September 24th 1747, October 20th 1748, November 16th 1749, November 8th 1750, September 24th 1752, May 19th 1754, September 20th 1761, October 2nd 1778: Vekene, E. van der, 1982/1983 I: 125–244. Cf. Samuel Chandler (1693–1766) *The History of the Persecution* (London, 1756); Archibald Bower (1686–1766) *Authentic memoirs concerning the Portuguese Inquisition* (London, 1761) xv + 528 pp.

The great Lisbon earthquake, in which it is estimated that some thirty-thousand people died, took place on November 1st 1755. It was a clear and serene morning, and at about nine o'clock a hollow thunder-like sound was heard, followed almost immediately by the upheaval, the whole thing lasting about three minutes. The churches were filled with their congregations, each becoming a huge catacomb entombing the hapless beings in its ruins. Nineteen further shocks took place in the neighbourhood, killing another thirty-thousand people.

Linnaeus bases his reasoning concerning the event on the date – it was All Saints' Day, the feast celebrating the conversion of the Roman Pantheon into a place of Christian worship, one of the days of the year when the Iberian Inquisition was wont to stage its public burnings: "In its full development the Auto-da-fé was an elaborate public solemnity, carefully devised to inspire awe for the mysterious authority of the Inquisition, and to impress the population with a wholesome abhorrence of heresy, by representing in so far as it could the tremendous drama of the Day of Judgement". Lea, H.C., 1906/1907, III: 209. He was fortunate in being able to do this, since it was only in 1753 that Sweden had switched from the Julian to the Gregorian calendar. Had it not done so, he might have felt obliged to regard the date as having been October 21st, in which case he could well have been reduced to extracting some significance from what we know of the life of St. Hilarion.

Linnaeus quite evidently had no objection to execution as such (II.iii.3.3; III.i.3.14; III.iv.3.18), only to its being carried out unjustly (IV.i.3.4; IV.i.4.3), and in certain instances he accepts it as being the direct expression of the will of God (IV.iv.1.3; IV.iv.1.7). This text makes it clear that what he objected to was the use made of execution by the Inquisition – the idea that faith, the gift of God, can be imposed on anyone's conscience by force: Torquemada, T., 1537. During the eighteenth century, and especially in Protestant countries such as Sweden and England, which had never experienced anything resembling the Spanish Inquisition, there was certainly a tendency to exaggerate the horrors of its activities. During his eighteen years as inquisitor-general, Thomas

Torquemada (1420–1498) had probably had about two thousand people burnt, but he was reputed to have authorized the execution of five times that number. Nevertheless, it is in fact the case that followers of the Spanish quietist Miguel de Molinos (1640–1697) were burnt by the Inquisition as late as 1781, and that the organization was not finally abolished in Spain until 1834.

Hundreds of articles, poems, pamphlets and books were published on the Lisbon disaster, most of them doing what Linnaeus did and interpreting it in terms of church politics. The great majority of the clergy of the Church of England held the view that the city had been destroyed because its inhabitants were Roman Catholics. Those who survived the disaster attributed their misfortune to the fact that they had tolerated a few Protestants in their midst, and in order to avoid a recurrence of the lethal tremors made absolutely certain that everyone in their community had in fact been baptised: Brooks, C.B., 1994.

During the first half of the eighteenth century much institutionalized religion had drawn support from physico-theology, from the idea that an all-knowing, all-powerful and all-beneficent God has revealed Himself in nature, the order and activity of which He has ordained for the benefit of mankind. It is perfectly understandable, therefore, that soon after receiving news of the Lisbon disaster, Voltaire should have written as follows to a friend: "Nature is very cruel. One would find it hard to imagine how the laws of movement cause such a frightful disaster in the best of possible worlds .... What a wretched gamble is the game of human life! What will the preachers say, especially if the palace of the Inquisition is still standing? I flatter myself that at least the reverend fathers inquisitors have been crushed like the others. That ought to teach men not to persecute each other, for while a few holy scoundrels burn a few fanatics, the earth swallows up one and all". November 24th 1755; Besterman, T., 1969, 366. By going on, in his famous poem on the subject, to call in question the cogency of a belief in a beneficent Providence, to accuse the physico-theologians of callousness and even cruelty, Voltaire undoubtedly overplayed his hand. He certainly laid himself open to the criticism of Rousseau, who a few months later wrote him a long letter defending optimism, Leibniz and Pope, asking him if he was intent on aggravating our miseries and proving that all is evil, drawing the distinction between physical and moral imperfection: Rousseau, J.J., 1967, no. 424bis, August 18th 1756.

The ironical twist Linnaeus gives to Voltaire's argument by turning it upside down, maintaining that in the earthquake God was wreaking vengeance on the whole community, revealing His abhorrence of the inquisitors, His sympathy for their victims, is in perfect accord with the general principles of his theodicy. In an earlier work on metaphysics Voltaire himself had reached the very sound conclusion that evil is a human and not a divine creation, that respect of the natural world as it is in itself, it is as pointless to say that God is just or unjust as to say that he is blue or square. Linnaeus too, far from finding God in nature, had simply "tracked His footsteps there", knowing only that "all things are born of Him", aware of Him most completely not as He is in Himself, but as the Essence of the things he knew through the senses, the "founder and builder of all that shimmers here before our eyes". As he saw it, his most immediate awareness of God derived not from contemplating nature, but from the constant demands being made upon him by his numinous awareness of Nemesis – and this certainly entitled him to pass such a judgement on the disaster that had taken place at Lisbon (*Introduction* I.ii.2; II.iii.2).

IV.ii.1.3:

London ms. 16, no. 45, Latin; Uppsala ms. 163, Swedish. Linnaeus headed this 'Davignon D'Amiens', and began the first line of the text by writing 'Davignon' and inserting 'Damiens' above.

Robert François Damiens (1715–1757) was born near Calais, and after a somewhat undisciplined childhood took on a whole series of jobs – domestic servant, apprentice to a locksmith, soldiering, assistant-cook to an abbot, batman to a Swiss officer, valet to a gentleman. In 1739 he married a fellow servant Elisabeth Molerienne and had a daughter with her, but they never settled down together. In 1756 he took employment with a diplomat, who left his house in his care when he was posted to St. Petersburg, only to discover when he returned that Damiens had disappeared and relieved him of a substantial amount of money.

At some time or another during this unsettled career he evidently became obsessed by the French king's having persuaded the parlement of Paris to register Clement XI's bull *Unigenitus Dei Filius*, deeply disturbed by the Gallican church's refusal to grant the sacraments to the Jansenists, convinced that the situation could only be resolved by the death of Louis XV.

On January 5th 1757, as the king was entering his carriage at Versailles, he managed to get close to him and wound him slightly with a knife he had recently purchased in St. Omer. It was not clear at first what was happening, but he made no attempt to escape, and was seized almost immediately by the king's bodyguard. He was tried and condemned, suffering a series of frightful tortures before being finished off as Linnaeus describes in the Place de Grève on March 28th 1757; DBF 10: 49–50.

IV.ii.1.4:

Uppsala ms. 40b. Swedish.

Johan Georg Lillienberg (1713–1798) came of a Småland family originating in Eksjö; for his family and personal contacts with Linnaeus, see Selling, O.H., 1960: 103, note 146. After a career as a diplomat, courtier, judge and Hat politician, he was appointed lord lieutenant in Uppsala in 1757 – in 1761 deciding a case concerning manure for the botanic garden in Linnaeus' favour: Fries, Th.M., 1903, II: 112. Henrik Benzelius (1689–1758) was the brother of Erik (IV.ii.2.14) and Jacob Benzelius (1683–1747), both of whom were his immediate predecessors as archbishop of Uppsala, to which office he was elected in 1747, after an academic career in Lund and a period in parliament as a Hat politician.

Given the date of Lillienberg's appointment in Uppsala and that of Benzelius' death, it looks as though this remark probably dates from 1757/1758.

The first powerful Christian centre in the Uppsala area was established at about the turn of the millenium in Sigtuna, some seventeen miles due south of the present cathedral. The pagan centre and seat of government was then Gamla Uppsala, some two and a half miles north of the present cathedral, where one can still see the three mounds dedicated to Odin, Thor and Freyr. Bishops of Sigtuna continued to be appointed until the middle of the twelfth century, when the centre of the diocese was moved to Gamla Uppsala, and a cathedral, much of which is preserved in the present church, was constructed on the site of the old heathen temple. In 1164 the diocese was elevated to an archbishopric. The present Uppsala was then known as Östra Aros, and in 1258 pope Alexander IV gave permission for the centre of the archdiocese to be moved there. Although work on the new cathedral began two years later, it was not until September 1270 that the pope's decision was confirmed by a parliament called at Söderköping by king Valdemar (IV.i.5.7). The remains of St. Erik, the patron saint of Sweden, were moved from Gamla Uppsala to the new site in 1273, but it took one hundred and seventy five years to complete the building, the consecration taking place during the Whitsun celebrations of 1435.

The first Swedish Mass, prepared in 1531 by the first Protestant archbishop of Uppsala Olaus Petri (1493–1552), was meant to replace the low or private Mass of the Roman or Latin rite. From 1541 onwards, however, it was replaced by the full choral or High Mass in Swedish, replete with sermon. In the Swedish church to this day the main Sunday service is known as the 'High Mass'. In accordance with the church regulations of 1686, it began at eight o'clock in the morning: Rodhe, E., 1917.

IV.ii.2.1:

Uppsala ms. 9. Swedish.

The authorization of mining concessions was regulated by a royal decree of September 10th 1723. The Nordberg mines were evidently part of the great Kopparberg complex near Falun in Dalecarlia. The year must have been 1761, when Easter Day fell on March 22nd.

Linnaeus evidently heard of this from Anders Philip Tidström (1723–1779), one of his favourite disciples, who was appointed reader in chemistry at Uppsala on February 22nd 1758. On account of his wit and good humour, he was always a welcome guest at Hammarby: III.i.3.4. Linnaeus probably also heard of what was going on in Falun through his wife's family, his father-in-law Johan Moræus (1672–1742) having been the town physician there.

IV.ii.2.2:

Uppsala ms. 180, 106. Swedish.

Cf. III.i.3.7. The case has also been included here on account of its illustrating an aspect of canon law.

IV.ii.2.3:

London ms. 16, no. 41; Uppsala ms. 197. Swedish.

Eric Melander (1682–1745) was appointed by the crown to the professorship of theology at Uppsala largely on account of his services to the crown. He often took the lead for the Caps in the university consistory, and his doing so in December 1743 gave rise to the only recorded instance of Linnaeus' having expressed both anger and indignation in a consistory debate.

The professorship of poetry had fallen vacant through the death of Arvid Arrhenius (1683–1743) on May 9th. On October 8th, with ten Caps and seven Hats in the consistory, the assistant professor of theology Lars Hydrén (1694–1789) was placed top of the list of candidates by fifteen votes to two, the Hat clergyman and librarian Olof Burman (d. 1759) being placed third. Burman complained that his very substantial qualifications for the post, which included the publication of schoolbooks and a dictionary of poetry, teaching the subject at a university level, extra unpaid work in the library etc., had been overlooked. In the consistory meeting of November 17th, at which Burman's complaint was debated, the Caps had the majority and the complaint was rejected. Burman objected that there had been no quorum at the meeting. A series of rancorous meetings followed at which procedural matters were discussed in an ever-worsening atmosphere. On December 5th a meeting was held at which the Caps happened to have a majority and Melander went onto the open offensive, claiming that Burman's supporters were backing him not because of his merits or qualifications, but because if he got the post they would be able to promote their protégé Olof Celsius (1716–1794) to his present position as librarian. This was too much for Linnaeus, who objected that: "the putting forward of such a suggestion was an unacceptable way of discrediting the legitimate considerations of one's opponents and bringing them into disrepute with the authorities". Melander climbed down, and agreed to the deletion of the offending passage from the minutes. The final outcome of the whole business was that the chair went to Hydrén: Annerstedt, C., 1877/1931, III-1: 163–166; Fries, Th.M., 1903, II: 209–210.

Linnaeus takes notice of Melander's final illness in a letter to his friend Abraham Bäck (1713–1795); he died on February 8th 1745. His case-history has been placed here as an instance of the close involvement of the university teaching of theology with intrigue and politics: *Letters* I.iv, no. 639 (February 8th 1745).

IV.ii.2.4:

London ms. 14, no. 22, Latin: forty-seven lines, by far the longest case-history in the manuscript. Uppsala ms. 198, Swedish and Latin: the Swedish editors (Linné, C. von, 1968, 161) were evidently unable to read the 'som horade till pengar även" inserted between the lines in the eighth paragraph.

Gabriel Mathesius (1705–1772), though quite clearly an intellectual lightweight, was appointed professor of the Greek language at Uppsala in 1737 and professor of theology in 1745. Linnaeus gives expression to his basic conception of him in a letter he wrote to the secretary of the Academy of Science on December 21st 1750: "I can't help evaluating the wisdom of Mathesius as being superior to my own stupidity. He's never allowed his body to go short of rest, whereas I've murdered mine. He's achieved all I've ever managed to do, without ever taxing his body or his soul". This case-history has therefore to be seen against the background of the treatment of diligence in III.iii.7.1 and the remark concerning lazy Jack and busy Tom in the characterization of Fortune (IV.iv.2.1).

In 1746 the university's natural history collection was deposited in a sealed room in Mathesius' house, and was still there in 1752, when he requested that it should be moved, and that he should be paid rent for having looked after it: Fries, Th.M., 1903, II: 128. In 1766 Linnaeus voted for Lars Hydrén (1694–1789) and against Mathesius' cathedral appointment, and like all the others who did so was fined fifty silver dollars by the crown. On May 9th 1766 he writes to his friend Abraham Bäck (1713–1795) that the only houses of members of the university destroyed in the great fire were those of Mathesius and Wallerius' widow (III.ii.2.4; III.iv.3.16). On May 22nd 1772 he also informs Bäck of Mathesius' death. In the course of time, therefore, Linnaeus evidently began to feel that justice had been done to Mathesius, and this case-history is probably best understood as an attempt to work out an acceptable rationale for the process by means of which this came about: *Letters* II: 168, no. 280; V: 143, no. 1064; 195, no. 1115.

As Linnaeus presents things, there is an exact balancing of what was done with the punishment meted out. The primary offence was the way in which he had acquired the professorship in 1737. Since Israel Nesselius (d. 1739) had been professor of Greek since 1716 and was often seen to be enjoying his tipple together with Mathesius, the general opinion was that he had got tired of his job and had passed it on to his drinking companion as a gesture of friendship. It was well-known that he had co-operated with Petrus Ullén (d. 1747), professor of theology, in lending support to Anders Norrelius (1679–1750) in his divorce proceedings against Margareta Benzelstierna (1708–1772), daughter of archbishop Erik Benzelius (1675–1743), but not that he had colluded in this way with Mathesius in order to get the wife's solicitor Johan Laurentz Springer (d. 1762) beaten up: III.iv.3.10; IV.ii.2.5; IV.ii.2.14. It could be regarded as only right, therefore, that nearly thirty years later Mathesius should not have got the cathedral provostship, because the professor of philosophy Carl Asp (1710–1782) took exception to the way in which he had helped Daniel Annerstedt (1721–1771) to a curacy which Asp thought should have gone to Johan Floderus (1721–1789). It is worth noting that although Linnaeus admits that "the position should have gone to Mathesius", and although he later maintained that he would "never have chosen to be his scourge", he had no compunction about voting in support of Asp's motion: IV.ii.2.6.

The secondary offence was the way in which he had behaved in respect of the Axbergs. Jochum Axberg (d. 1730), merchant and alderman in Uppsala, had a daughter Catharina Christina (1718–1792), who was engaged to Anders Ryman (1706–1766), postmaster in Stockholm. Probably at about the time that he was appointed to the professorship Mathesius began to court her, proceeded to drop her for her widowed mother, and then to drop the mother when the daughter's marriage to the courtier and councillor Gabriel Magnus Piper (1721–1777) ruined the family's finances. It could be regarded as only right, therefore, that some twenty years later, when Mathesius was wooing Miss Uddbom, daughter of an Uppsala innkeeper and less than half his age, she should have fallen in love with and been made pregnant by Adolf Stenhammer (1727–1798), rural dean in the parish of Västra Ed in the diocese of Linköping: Annerstedt, C., 1877/1931 III-1: 70–73; 166–169; 403–408.

IV.ii.2.5:

London ms. 16, no. 42, Swedish. Uppsala ms. 80. Swedish and Latin.

Petrus Ullén (1700–1747) took his master's degree at Uppsala in 1729, and in 1736, largely on account of Cap influence, was elected to the chair of logic and metaphysics. He had taken holy orders, and in 1740 he entered parliament, but he had never been in charge of a parish, and the Hats therefore had some justification for insisting on his resignation. During the last two years of his life he was the fifth of the professors of theology at Uppsala. In 1747 he allowed his name to go forward as candidate for the chancellorship of the university, despite the fact that his rival was none other than the crown prince Adolf Frederick (IV.i.5.10). He was evidently "a small man, hunchbacked and quick-eyed, who combined an outstanding brilliance of mind with a clear and solid judgement – greatly respected by all on account of his learning, and very highly thought of by the students": BL 21: 347–349.

The controversy mentioned here by Linnaeus was an extension of that first stirred up by Anders Knös (1721–1799) in 1742, when under the auspices of Johan Ihre (III.ii.2.2), he submitted an elaborately Wolffian thesis on "the principles and inter-connectedness of natural and revealed religion" in the faculty of philosophy and not in the faculty of theology: *Introduction* III.iii.3. Ullén, like Nils Wallerius (III.ii.2.4), regarded the general tenets and the methodology of Wolffianism as well-suited for defending orthodox Lutheranism. Ihre tended to be sceptical of this, and Knös was of the opinion that Wolffian natural theology was unable to give a satisfactory account of the atonement. In November 1745 Ihre delivered an oration on the relationship between reason and revelation, allowing the latter precedent in matters religious, suggesting that reason should be regarded as the sister rather than the handmaid of theology, since it cannot be regarded as at odds with revelation, and should certainly never be regarded as its enemy: 'Eundem in modum ubi ratio & revelatio inter se concenduntur', *Programmata Upsaliensis* IV (1745/1770). Soon afterwards he backed a second attempt by Knös to open up a university career for himself by submitting a thesis in the faculty of philosophy. Once again it was considered to be a matter for the theologians rather than the philosophers, and in the first instance Matthias Asp (1696–1763) and Ullén were given the task of reading it through. Frängsmyr, T., 1972, 110–129, 156–169.

Much dithering, letter-writing and gerrymandering ensued, the final outcome being that Knös withdrew his thesis and took up parish work, Ullén pined away and died, Ihre concentrated on political theory and philology, and in 1749 there was a royal decree severely limiting the extent to which the philosophers were allowed to deal with theological topics. Linnaeus himself was caught up in, and greatly disturbed by, these developments: *Introduction* III.iii.3; Annerstedt, C., 1877/1931, III-1: 130–135; 167–169; 184–192.

IV.ii.2.6:

Uppsala ms. 152, 152b. Swedish.

Daniel Annerstedt (1721–1771) came of a Småland family originating in Annerstad, a village some twenty miles west-north-west of Stenbrohult, his father Petrus Annerstadius (1678–1732) being the vicar of Femsjö, just over the border in Halland. Like Linnaeus, he was educated at Växjö School before being matriculated at Uppsala. Linnaeus was very much aware of him as a fellow Smålander (*Letters* I: 281, no. 160, December 25th 1753), and several of the others mentioned in this case-history also had a Småland background. Annerstedt's betrothed, for example, the daughter of fencing-master Christopher Porath (1689–1745), came of a family with many Småland connections (Anrep III: 220–221); his wife Ebba Christine Humble (1733–1796), whom he married on July 12th 1761, was the daughter of Gustav Adolf Humble (1674–1741), bishop of Växjö; his rival for the curacy Johan Floderus (1721–1789) was the son of Per Floderus (1683–1741), who came of a family originating in Lekaryd near Växjö and was the rural dean at Skatelöv.

Annerstedt was first taken on by Johan Ihre (III.ii.2.2) in 1754 to teach Swedish literature. He graduated in theology in 1760, and the following year became vicar of Börje just outside Uppsala. He sat in the parliament of 1769/1770, and as a reward for supporting the court party and the Hats, was appointed professor of theology on September 17th 1770. His main field of interest was modern languages; SBL 2: 31–33.

Annerstedt's original offence was to abandon Porath's daughter for bishop Humble's once he saw that it was to his social and professional advantage to do so. We have record of Porath's daughter, still unmarried, serving as university caterer during the chancellor's visit in May/June 1774: Annerstedt, C., 1877/1931 III-1: 505. His second offence was the ungrateful manner in which he treated Mathesius in 1769. He was rewarded with his later financial difficulties, and his dying early of consumption (*Letters* V: 212, no. 1132, December 19th 1773), evidently on May 12th, not May 1st 1771.

Gabriel Mathesius' main offences are dealt with elsewhere by Linnaeus (IV.ii.2.4). There may have been some justification for his refusing to go along with the way in which Carl Asp (1710–1782), professor of philosophy, was backing Johan Floderus, who was then a lecturer in Greek, but Linnaeus' general conclusion is that he deserved what he got when Asp and Julinschöld (III.iv.3.7) combined to prevent him from becoming provost, and Annerstedt did him out of his university teaching. Linnaeus would not willingly have contributed more than he did to his difficulties, since in 1740 Mathesius had backed his appointment to the chair of medicine: Fries, Th.M., 1903, I: Bil. 30.

IV.ii.2.7:

London ms. 17, no. 47; Uppsala ms. 141, 141b. Swedish. Cf. II.iv.2.5.

Jacob Flachsenius (1683–1733), who taught Linnaeus systematic theology at Växjö School from 1723 until 1725, was almost certainly the source of his information concerning this case (*Introduction* III.iii.3). The probability that this was so is increased by the fact that in the London manuscript it follows immediately after the Peter Schmidt case (III.i.1.5), and by Linnaeus' misdating of the events. When he first heard of them, he evidently thought they were a recent occurrence.

Various inaccuracies have crept in as a result of the re-telling, although the broad outlines of Linnaeus' account are sound enough. The case is fully documented in the Åbo diocesan records and in various published legal reports: Gillby, J., 1961, 75–76.

The murder did not take place at Christmas, but at the time of the Åbo Quarter Sessions in September 1704. It was discovered not as a result of evidence given by the farmhand, but almost immediately, the delay in the sentencing being the result of the defendants' having denied the charges. It was the manservant who first struck the commissioner and Peldan who finished him off. The executions did not take place in 1722, but in 1706/1707, and there were not six but five who suffered the death penalty – the girl, Margareta Eriksdotter being let off with a whipping, and not for having held the candle, but for having kept quiet about having had to wash the bloodied sheets.

Henrik Florinus the elder, vicar of Kimito and then of Pemar in the diocese of Åbo, died in 1705. Maria Eriksdotter Pihl, the wife of Henrik Florinus the younger, died in 1704. Jacob Peldan, who was studying at the university of Åbo, was the nephew of the younger Florinus, and was engaged to the maid, who like the wife was also a Pihl.

The vicar Henrik Florinus the younger, the student Jacob Peldan, his betrothed the maid Pihl and the manservant Johan, were broken on the wheel and beheaded outside the customs house in Åbo in May 1706. Christina Gardeling, the wife of the district commissioner, denied any part in the murders, but was eventually found guilty and executed outside Kimito church in 1707. She was the granddaughter of Isak Rothovius (1572–1652), the great reforming bishop of Åbo, initiator of the University (1640) and the Finnish translation of the Bible (1642).

The first four texts in this section on the Churches touch upon ecclesiastical organization as a universal or international institution, as the spiritualized counterpart of imperial power (IV.ii.1.1–IV.ii.1.4); the next six upon the general organization of the church within Swedish society, the effect of the legislation, the university training of the clergy. These last eight case-studies touch upon the actual activity of the clergy in the parishes and dioceses, and have been ranged roughly in accordance with rank (IV.ii.2.7–IV.ii.2.14).

IV.ii.2.8:

London ms. 15, nos. 27 and 28; Uppsala ms. 181. Swedish.

The pastor in Malung, a parish in Dalecarlia on the river Västerdal, some forty-five miles from the Norwegian frontier, visited by Linnaeus when he travelled through the province during the July and August of 1734, was Vidik Andersson Harkman (1675–1749), who had been appointed to the living on September 9th 1719, after graduating and taking his doctorate at Uppsala, and teaching at the grammar school in Västerås. He had trouble with the congregation and the authorities, not only on account of his having had the church extended without getting the requisite permission from the diocesan council, but also on account of his management of the church-accounts, which included selling excess corn on the open market instead of making it available to the poor.

In 1710 Vidik Harkman married Brita Lidman (1687–1759), whose restlessness seems to have been the outcome of her turbulent family background.

She was the daughter of an unsuccessful Stockholm ironmonger whose financial misfortunes resulted in his dying in prison. Her mother subsequently entered into two further marriages with clergymen, neither of which provided the family with much economic or social security. She provided Harkman with eight children, six of whom, ranging in age from twenty-two to seven, were still alive when Linnaeus visited Malung in 1734.

The daughter in question was the eldest child Catharina (1711–1751). In 1728 she married Johannes Jacobi Gezelius (1698–1767), who after studying theology at Uppsala had been appointed curate at Malung in 1727. In 1733 he was appointed vicar of Lima, a parish some twenty-seven miles upriver from Malung, and in 1746 vicar of Orsa, a parish some fifty miles to the north-east of Malung, just north of Mora. By 1735 the couple had lost all four of their children, none of whom had reached the age of three. In February 1738 the court in Lima imposed a fine on Catharina for "assault and abusive language". In December 1750 her husband petitioned for divorce on account of her "drinking, smoking, dancing in nothing more than a linen dress, and keeping company with farm-hands and their like", and on account of her having run him into debt to the tune of 3,412 copper dollars. In 1750 the school chaplain in Mora Lars Samuelsson Elfvius (1706–1759), who had already been in the courts for drunkenness (1744), rowdiness (1746) and beating his wife (1749), was accused by his wife of having committed adultery with Catharina Gezelius: information supplied by Kajsa Bondpä, Malungs Hembygdsförening, who has since published on the subject: Bondpä, K. 1998.

The sons in question were the two youngest children, Carl (b. 1726) and Jacob (1727–1748). On the Saturday evening of January 30th 1748 they fell out about a whip. Carl struck his brother on the nose, drawing blood, and then ran him through the body with his rapier. He staggered from the room into the kitchen, where he was seen by two of the servant girls, and died the following day. Carl fled to Norway in order to escape justice, changed his name to Hartman, and subsequently found employment at the Avesta ironworks. Not all the pastor's descendants turned out so badly, since his third son Widick (b. 1720), though involved in a divorce case, was the maternal grandfather of the famous poet-archbishop Johan Olof Wallin (1779–1839): Muncktell, J.F., 1844, II: 196; Hansson, G., 1990, II-2: 441–442; Sjögren, J., 1963; *Släkt och Hävd*, 1996, no. 2. 79; Fries, Th.M., 1903, I: 144–159.

The pastor in Kvikkjokk, a settlement in Lapland at the upper end of lake Saggat, some thirty-eight miles as the crow flies from the Norwegian border, visited twice by Linnaeus when he travelled through the province during the June and July of 1732, was Petrus Alstadius (1679–1740), who had been appointed to the living on February 9th 1710, after studying at Uppsala, taking holy orders in 1703, and teaching for a while in the school at Luleå. The parsonage in Kvikkjokk was so badly situated on an islet in the stream, so old and dilapidated, that at high water there was a very real danger of its being swept away. On one occasion, when it looked as though this was about to take place, Alstadius evidently went out onto the bridge in front of his residence with the Bible in his hand, exclaiming: "Since God and the king have set me here as pastor of this parish, the devil and the waters shall not drive me hence".

In 1710 Alstadius married Christina Groth (1691–1737), daughter of the previous incumbent Matthias Jonæ Groth (c. 1663–1712), a lively, strong-willed woman who was reputed to have betrayed and bullied her husband, and to have been deeply involved in Lapp superstitions. During her final illness, when she was lying bedridden in Jokkmokk, she was said to have sent for a well-known Lapp soothsayer in order to consult him concerning her health. Finding her on the point of death, he consumed several drams of brandy, and after giving expression to a series of Lapp oracles, concluded by telling her that her soul was about to "pass on over the marshes of Saggat". Her body was taken back to Kvikkjokk and on April 20th 1737 buried between those of two of her daughters.

The daughter mentioned by Linnaeus was Magdalena, born on March 23rd 1715, who on March 6th 1730 gave birth to an illegitimate daughter Rakel Persdotter Alstadius (d. 1814), the father being Per Ersson (1689–1749), born in the village of Tärendö in the parish of Övertorneå of mixed Lapp-Finnish parentage, who had recently settled in Purkijaure near Jokkmokk. Her mother's lover, whom Linnaeus met in Piteå when he visited the area in 1732, was Joakim Kock (1690–1763), born in Stockholm, quartermaster in the Västerbotten regiment, who fought at Poltava and escaped from Russian captivity in 1711. The mother was also involved with the regimental clerk Henrik Lindbaum (c. 1698–after 1726), born in Luleå, matriculated at Uppsala in 1717, who wrote verse, including a rhymed description of the Lapps, was dismissed from his regimental duties in 1725, and with the help of the pastor's wife managed to escape justice and flee to Norway; Bygdén, A.L., 1923/1928, II: 28–29, 102; Nordberg, A., 1928, II: 547; Steckzén, B., 1955: 179, no. 392; Linné, C. von, 1913: 235–238.

IV.ii.2.9:

Uppsala ms. 104. Swedish.

The pastor in question in Kristinehamn, a town in Värmland on the north-eastern shore of lake Vänern, was Stefan Muræus (d. 1675). His successor was Birger Carlberg (1640–1683), son of the wealthy Värmland industrialist Johan Börgesson Carlberg (1606–1676), brother of Johan Carlberg (1638–1701), mystic, pietist and hymn-writer, highly regarded as a preacher at the court of Charles XI, appointed dean of Gothenburg cathedral in 1679, but not bishop until after his brother's death (1689). The court chaplain's friend and fellow accuser was the mill-owner and mayor of Kristinehamn Olof Pehrsson (1627–1693), who had married Crispin Flygge's half-sister Emerentia in 1653, and therefore had good reason to raise doubts concerning the birth of his son. He was imprisoned together with Carlberg in 1671. It may be worth noting that Pehrsson was the great-grandfather of Gustav Crispin Jernfeltz (1718–1767), whom we have good reason to suspect as having been the murderer of Isaac Schleicher (1706–1740): III.iii.2.6.

Linnaeus may well have been informed concerning these cases by his patron Nils Esbjörnsson Reuterholm (1676–1756), lord lieutenant of Värmland, whom he visited in Arboga in the June of 1746, during the early stages of his tour of West Gothland: Uggla, A.Hj., 1961; Reuterholm, N., 1957; Fries, Th.M., 1903, I: 326.

Peter Flygge (d. 1648), a merchant from Lübeck, made a fortune after being put in charge of the customs procedures in the lake Vänern area. His son Crispin Flygge (1628–1673) succeeded his father in the customs service, was ennobled in 1654, and in 1658 was also put in charge of the customs in Scania, Halland and Blekinge. He married Sigrid Ekehielm (d. 1700) in 1671, and died two years later, the peerage lapsing with the death of his son Crispin (1674–1680): Anrep I: 838–839.

Since Muræus died two years after Peter Flygge, Linnaeus must be mistaken concerning the duration of the widow's hospitality. The hymn by Johan Carlberg which he refers to is no. 293 in *Then swenska psalmboken* (Stockholm, 1695), its title being *The vexation brought on by my miseries*, and the most relevant line:

Crushed by fateful misfortune in the final days of his life.

IV.ii.2.10:

London ms. 15, no. 23; 16, no. 39; Uppsala ms. 126. Swedish.

Jacob Boëthius (1647–1718) took his degree and holy orders at Uppsala in 1682, taught at the school in Västerås, and in 1693/1694 got a living and then a provostship at Mora in Dalecarlia. His general view of things was strictly Bible-based and pietistic, his manner of living intensely devout and highly regulated. He was disturbed by the way in which the church ordinances of 1686, promulgated in the name of the newly instituted monarchical absolutism, impinged upon his calling, and when the fifteen-year-old Charles XII came to the throne in 1697, saw it as his God-given duty to draw up a thundering condemnation of the manner in which the country was being run, and to send it to the king's tutor and privy councillor Nils Gyldenstolpe (1642–1709), requesting that he should pass it on to his ward. This attempt to further the realization of his religious convictions resulted in his being arrested, relieved of his office and condemned to death, eventually to his being reprieved and incarcerated for life in Nöteborg, the fortress commanding the point at which the river Neva flows out of lake Ladoga, which Sweden had acquired from Russia in 1617.

The driving force behind this needlessly harsh treatment was Carl Piper (1647–1716), secretary of state since 1689, recommended by the dying Charles XI to his son, immensely influential throughout the early years of the young king's reign.

In 1702 the Russians retake Nöteborg and rename it Schlüsselburg. In 1709 Piper is captured at Poltava and heads the Czar's triumphal procession of Swedish prisoners in Moscow. In 1710 Boëthius is released from Nöteborg and returns to Sweden, Piper being incarcerated in the prison to which he had condemned the clergyman, and dying there in 1716: Petersson, E.L., 1878.

Linnaeus' interest in and knowledge of the case was probably the result of his being acquainted with Jacob Jacobsson Boëthius (1690–1748), town physician in Gothenburg, whom he visited there during his West Gothland journey in July 1746, and with his namesake (1727–1784), also town physician in Gothenburg, who also figures in his correspondence: Linnaeus, C., 1747; *Letters* III: 15, no. 465; 290, no. 602.

IV.ii.2.11:

London ms. 11, no. 2; Uppsala ms. 130. Swedish and Latin. The six lines of the London version contain three corrections; in both versions Linnaeus writes 'Bilmark'.

Thomas Kihlmark (1677–1758) was not as young as Linnaeus suggests when the trouble blew up in the Västmanland regiment. In 1723, as the result of his having instigated the enquiry the colonel of the regiment Bernhard Reinhold von Dellwig (1679–1746), who came of a family originating in Estonia, was charged with the embezzlement of equipment. The court did in fact find him guilty of irregularity in the disposal of fodder, horses and baggage-trains, which he had sold for hard currency and accounted for in token money. On October 15th he was condemned to loss of life and paying a large sum in compensation, a sentence commuted by the crown on November 20th, on account of his formerly good record of service, to life-imprisonment in the fortress at Bohus. He was in fact released after having served a year in prison and allowed to return to Estonia, all his Swedish property having been sequestrated and sold: Anrep I: 574–576.

On leaving the army, Kihlmark was appointed vicar of Tensta, a parish some twelve miles north of Uppsala, the mediaeval church of which is particularly notable on account of a series of murals, dating from 1437, depicting the life of St. Bridget (III.i.2.3). The advowson of the living was the right of Eric Brahe (1722–1756), who joined the court party in 1751, and was executed on July 23rd 1756 for having plotted to secure more constitutional power for the crown (III.i.1.4). Kihlmark's fateful sermon was probably delivered during the early summer of 1756, before the trial of the conspirators had got under way. It looks as though the curate may well have tipped off Dellwig's son as to the possibility of Kihlmark's making incriminating comments on it in his sermons. The son was also Bernhard Reinhold, and also had a military career – mainly in Germany and Russia.

IV.ii.2.12:

London ms. 11, no. 1, Latin. Uppsala ms. 139, Swedish and Latin. Linnaeus gets the name of Gezelius ('Petræus') and of his parish ('Grenie') wrong, and seems either to have added the final sentence of the Uppsala version later, or to have written it out after re-dipping his quill.

Georg Gezelius (1668–1720) came of a distinguished clerical family, his father of the same name (1631–1684) having been vicar of Husby, just north of Hedemora in Dalecarlia, his uncle Johannes Gezelius (1615–1690) having been bishop of Åbo and an accomplished publicist, grammarian, linguist and Biblical scholar. Grangärde is in the diocese of Västerås, some ten miles north-north-west of Ludvika. When the crown replaced him there, probably about 1715, he was given a living at Vika near Falun, some thirty miles away, also in the diocese of Västerås. In 1748 his son Martin Gezelius (d. 1760), an inland revenue official at Jönköping, lost his job on account of irregularities in his accounting. His grandson Georg Gezelius (1736–1789) was one of Linnaeus' pupils at Uppsala, and it may well have been through him that he gathered the information included in this case-history, either directly, or through his friend Carl Fredrik Mennander (1712–1786), who knew the grandson on account of his being employed as tutor in the home of his first wife's family, the Paléens: *Letters* VI: 389, no. 1457, 15th September 1755; III.i.2.1.

Johan Fahlander (1685–1746) threw in his lot with the insurgants from Dalecarlia who during the early summer of 1743 marched down to Stockholm, only to be defeated on June 22nd by troops loyal to the government (IV.i.4.4). Since one of their objectives was to get the crown prince of Denmark accepted as heir to the Swedish throne, it was perhaps only fitting that Fahlander should have been incarcerated in a prison first constructed by the Danes. Since he had gained the living of Grangärde in the way that he had, it was only right that he should not have been allowed to find his final rest there.

IV.ii.2.13:

Uppsala ms. 168. Swedish. Linnaeus spells the name 'Sirenius', which strictly speaking is more correct, since the family originated from Siringe near Skärstad in Småland, some fifteen miles north-north-east of Jönköping.

The primary career of Jakob Serenius (1700–1776) was as a churchman, and it is on account of it, and on account of Linnaeus' speculation concerning its future development, that his case-history has been placed here. Matriculated at Uppsala in 1714, he took holy orders in 1722 and served for a while in what had formerly been his father's parish, Färentuna near Stockholm. The most decisive development in his career took place in 1723, when bishop Jesper Swedberg (1653–1735) of Skara selected him for the task of establishing a church for the Swedish Lutheran community in London, a project into which he threw himself with great energy for well over a decade. This English connection coloured the whole of his subsequent view of things. He was elected fellow of the Royal Society and developed a close friendship with Edmund Gibson (1669–1748), the learned and influential bishop of London, adviser to Walpole on ecclesiastical affairs, editor of the *Anglo-Saxon Chronicle* and of Camden's *Britannia*, author of the standard work on the legal rights and duties of the English clergy. Serenius found the cultural life of London stimulating in the highest degree, and he began to publish – *Examen Harmoniæ Religionis Lutheranæ et Anglicanæ* (Leiden, 1726), *Engelska Åkermannen och Fåraherden* (Stockholm, 1727), and the highly successful *Dictionarium Anglo-Svetico-Latinum* (Hamburg, 1734, Stockholm, 1744; Nyköping, 1757[2]). When he returned to Sweden in 1735 he was appointed to a living in Nyköping. He took his doctorate in theology at Uppsala in 1752, and in 1763 was appointed bishop of Strängnäs, the rival Hat candidate being the not very inspiring professor of theology at Uppsala Lars Hydrén (1694–1789), see IV.ii.2.4. In 1764, although the most recently-appointed of the Swedish bishops, he was in the running for appointment as archbishop, but withdrew his candidature, the position going to Magnus Beronius (1692–1775). In matters of church organization, Serenius made his mark by introducing the Anglican practice of confirmation. He made a name for himself as a powerful and effective preacher, and in his capacity as bishop of Strängnäs, of Södermanland as Linnaeus puts it, organized the financing and establishment of a lectureship in natural history – a tribute to his enthusiasm for the work of Linnaeus. When the Caps came to power in 1765, he took the opportunity to push through the setting up of a Bible Commission of which Linnaeus became a member: Hagberg, L., 1952.

Serenius' secondary but equally prominent career was that of Cap politician. He sat in every parliament between 1738 and 1772 except that of 1755/1756, his abrasive and bitter-tongued manner, his formidable skills as an orator and debater, constituting a constant harassment for the Hats. He did all he could to counter the increasingly hostile attitude which eventually gave rise to the Russian war of 1741/1743 (II.iv.2.7), and when Russian troops were moved into Nyköping in the December of 1743, he went out of his way to make their commanding officer, the Irishman Peter Lacy (1678–1751), feel at home, and to preach conciliatory sermons: "Do not hate them, for it is not of their own will that they are here; our government has summoned them, their own government has sent them. It can truly be said, therefore, that they have been sent by God to persuade us of our pride and arrogance, we being incapable of being humbled in any gentler manner". To a very considerable extent, this championing of the Cap cause was, of course, the natural corollary of Serenius' enthusiasm for things English: Cederbom, L.A., 1904.

Had Serenius' career not been so clearly determined by his English attachment and his Cap convictions (cf. III.iv.3.18; IV.ii.2.5), there was much in it which Linnaeus might have wholeheartedly approved of. What is more, taken as a whole it had also been remarkably successful. From the manner in which Linnaeus presents it, one gets the impression that he was looking for a major fault and a fitting punishment which were simply not forthcoming – hence the retailing of the relatively trivial information concerning the student brawl, the family tension and the Hildebrand affair.

Siwert's cellar was evidently the basement of the house owned by the family of that name, prominent at Uppsala as university bookbinders and booksellers since the early years of the seventeenth century: Annerstedt, C., 1877/1931, I: 371; II-2: 413; Bih. III: 53. Johan Hermansson (1679–1737) was Skyttean professor at the university, his daughter Elsa Maria (1719–1794), sister of the wife of Linnaeus' colleague Nils Rosén von Rosenstein (III.i.1.4) was Serenius' second wife, by whom he had no children, his first wife having been Ulrika, daughter of David Lund (1657–1729) bishop of Växjö, by whom he had a daughter Christina: BL 14: 206. The cavalry captain David Henrik Hildebrand (1712–1791) contested the will of his deceased brother, the gentleman of leisure Gotthard Henrik Hildebrand (1710–1761), during the parliament of 1760/1762, which was dominated by the Hats. Serenius' daughter was married to a Stockholm lawyer by the name of Ternell, and it was he who arranged the get-together for members of the house of peasantry – not far from the old parliament building on Riddarholm, that is, somewhere along the south-western shore of the island on which the old town of Stockholm stands; SBL 19: 35–37.

IV.ii.2.14:

London ms. 13, 14, 15, nos. 16, 21, 33, Swedish and Latin. Uppsala ms. 132, 94, Swedish and Latin.

Haquin Spegel (1645–1714), bishop of Skara 1685, of Linköping 1691, archbishop of Uppsala 1711, has a place in Swedish literature as a hymn-writer and as the author of a mighty epic of eleven thousand verses, in the style of Du Bartas and Anders Arrebo (1587–1637), celebrating the six days of the creation and the seventh day of rest and contemplation, the revelation of God's power and wisdom in the works of His hands – the division of light from darkness, of the heavens from the waters, of the earth from the seas, the bringing forth of night and day, of the seasons and of plants, of fish, birds and animals, of the image of Himself in man: *Guds werk och hwila* (1685, 1705[2], 1725[3], 1745[4]).

On August 24th 1705 Spegel's younger daughter Elisabeth (1688–1720) married Johannes Steuch (1676–1742), son of the Matthias Steuch (1644–1730) who was to succeed Spegel as archbishop in 1714. Johannes was appointed professor at Uppsala in 1710, and twenty years later, after remarrying in 1724, succeeded his father as archbishop. His sister, Anna Christina, married the Skyttean professor Johan Hermansson (1679–1737): III.iv.3.7; IV.ii.2.13; Anrep IV: 152–153.

Spegel's elder daughter Margareta (1682–1760) became engaged to Joachim von Düben (1671–1730), who became a clerk in the royal chancellery in 1694, seems to have been sent on a spying mission to Aachen in 1706, was captured in Russia shortly after Poltava, and did not return from captivity until 1719, when she married him: Anrep I: 636.

The father of the Erik Benzelius (1675–1743) who "violated Düben's right" in Spegel's palace was the Erik Benzelius (1632–1709) appointed archbishop in 1700. The younger Benzelius studied at Uppsala and then abroad for a number of years, before returning as university librarian in 1702. On June 16th 1703 he married Anna (1686–1766), daughter of the evangelical, hymn-writing Jesper Swedberg (1653–1735), bishop of Skara, and soon afterwards began to lecture in the university and to devote much time and energy to research into Swedish history. His affair with Düben's betrothed must therefore have taken place between about 1706 and 1719. In 1723 he was appointed professor of theology, in 1726 bishop of Gothenburg, in 1731 bishop of Linköping. Although he was appointed archbishop just prior to his death, he never actually assumed the office; Anrep I: 149.

His wife "violated his right" with Adolf Mörner (1705–1766) of Morlanda, a village on the west coast near Lysekil, north of Gothenburg, who came up to Uppsala as a student in 1717, left for Paris in 1726, married Agneta Ribbing (1715–1776) in 1734 (III.ii.1.2), and began his career as a lord lieutenant in the county of Stockholm in 1750: Anrep II: 960.

Their daughter Margareta Benzelia (1708–1772) married the university librarian Andreas Norrelius (1679–1750) on December 29th 1726 (III.iv.3.10), and became notorious on account of the wealth of scandalous information made public in court prior to the granting of a divorce in 1733 (IV.ii.2.4). When Linnaeus first arrived in Uppsala in the September of 1728 he attended the lectures of Olof Rudbeck the younger (1660–1740), and subsequently lodged in his home, being employed there as tutor to his son (II.i.2.2). It is quite likely, therefore, that he was also in the house when Rudbeck took Benzelius' daughter in, and when Nils Rosén von Rosenstein (1706–1773), Linnaeus' rival for a permanent position at the university (*Introduction* I.ii.3) was found in her bed. Norrelius had travelled abroad between the August of 1716 and the November of 1719, and the publication which so incensed Rudbeck was probably the *Schediasma de avibus* (Amsterdam, 1720), an analysis of the prescriptions concerning edible and inedible animals in the eleventh chapter of *Leviticus*.

It looks as though the child she gave birth to when she claimed that the father was Johan Gerdesskröld (1698–1768), president of the court of appeal (III.ii.3.4), was the same as the forger who was extradited from Denmark and hanged in Stockholm. The Russian officers she entertained, presumably in Nyköping and Norrköping during the autumn of 1743 (IV.ii.2.13), were under the command of the Scottish Jacobite Francis Edward James Keith (1696–1758) and the Irishman Peter Lacy (1678–1751). She died in Copenhagen on December 27th 1772.

It is clear from this case-history that in Linnaeus' day, the bishops and archbishops of the Swedish church were selected for reasons other than their assiduity in observing the seventh commandment: cf. III.iii.8.1, no. 6. In fact if it were not for the higher clergy so in prominence here, the material might be better classified under III.ii or even III.i.

On the piglets and the porker, see Walther, H., 1963/1986, no. 26060; cf. III.ii.4.1; IV.i.4.2; IV.iv.3.2; IV.iv.3.3.

IV.iii:

Cf. IV.iii.1.1.

IV.iii.1.1:

London ms. 9, 11, 15; Stockholm ms. 1; Uppsala ms. 26b, 31; cf. title-page. Latin.

Ovid *Ars amatoria* I: 640. The quotation is still to be seen inscribed over the door of Linnaeus' bedroom at Hammarby. In the London manuscript (p. 15) it is the subtitle under the main heading of 'Nemesis'. The playfulness of it is apparent enough from its original context, since Ovid is giving advice on how to win over the fair sex: "Call as witnesses what gods you please. Jupiter from on high laughs at the perjuries of lovers, and bids the winds of Aeolus carry them unfulfilled away .... Since it is expedient that there should be gods, let us grant their existence, let us pour incense and wine on their ancient hearths. They have not retired to rest, moreover, gone to sleep as it were and lost all interest: live irreproachably, God is near; render back what has been given into your keeping; let duty keep her covenant; shun evil, keep your hands free of blood; be wise, avoid trouble, play with none but the women; keep faith save for this one deceitfulness – deceive the deceivers, who for the most part are an unscrupulous bunch; let them fall into the snare they have laid".

The earnestness of it is apparent in the context it is given in the note on man in the twelfth edition of the *System of Nature* (1766/1768): 30, and in the prominence it is given in the *Nemesis Divina* papers. Society (III.i.1–III.iv.3), politics (IV.i.1–IV.i.5) and the church (IV.ii.1–IV.ii.2) constitute the foundations of an ethical life which finds its fulfilment in an awareness of God.

IV.iii.1.2:

Uppsala ms. 3b. Swedish and Latin. Cf. IV.i.1.2, Virgil *Eclogues* I: 71–72, for the final Latin quotation.

Gustav Fredrik Gyllenborg (1731–1808), the author of the verses quoted here, came of a family well-known to Linnaeus: see his father's case-history III.iii.6.5. His uncle, Carl Gyllenborg (1679–1746), was one of the founders and leading lights of the Hat party, chancellor of the university of Uppsala from 1739 until 1746, and founder of the zoological museum there. In 1750 he himself became rector of the university of Lund, and on March 4th 1756, after the failure of Brahe's attempt to secure more constitutional power for the crown (III.i.1.4; IV.i.5.10), the Hat party had him appointed tutor to crown prince Gustav (1746–1792), a position he held until the prince reached his maturity on April 5th 1762. He was given the title of chamberlain in 1762, and after the prince succeeded to the throne in 1771 held various state sinecures. In 1802 he began to write his autobiography, completing it up to 1775: the manuscript, which is now in the university library at Uppsala, was published in 1885: SBL 17: 546–549.

Like Linnaeus, he discovered the stoic writers of antiquity at an early age, and this led him to take a decidedly negative attitude to much of the paraphernalia of orthodox Christian belief. His Hat commitment involved a natural openness to French influences, and he responded readily to the writings of Boileau and Rousseau, finding in the former the guidelines for the development of his satirical bent, and in the latter an intellectual justification for his instinctive dislike of town life and his abhorrence of political manoeuvring. He was at odds with many members of his party in regarding agriculture rather than industry as the true basis of a healthy society – a view broadened and deepened by his love of the Swedish countryside, especially the whole environment of the family estate at Skenäs, situated as it is on one of the smaller lakes in Södermanland, where he responded to the seasons in a manner reminiscent of Thomson, and wrote much of his finest poetry.

In 1753 the Stockholm civil servant, freemason and amateur poet Carl Fredrik Eckleff (1723–1786) founded a 'Society for mental Regeneration', a sort of literary academy which did in fact bring about a revolution in literary taste and style, largely on account of its managing to co-opt Gyllenborg, together with his friend Gustav Philip Creutz (1731–1785), and their mentor the poetess Hedvig Charlotta Nordenflycht (1718–1763). This group's first collection of poetry (1753/1756) was somewhat formal and derivative, but the second, from which Linnaeus takes the material quoted here, contains some of their very finest work: *Vitterhets arbeten* (2 pts., Stockholm, 1759/1762).

His first three sections are taken from a magisterial poem on *Human wretchedness*, the counterpart to a parallel work on *Human happiness* – a twin production which brings to mind *L'Allegro* and *Il Penseroso*, and as in

Milton's case, leaves no real doubt that in respect of depth and significance the palm is to be given to melancholy. Human life – birth, ageing, thraldom, poverty, war, wrong-doing, folly, greed, friendship, fame, virtue – is presented in its raw and naked immediacy, stripped of all the ameliorating and hope-engendering trappings of orthodox Christianity. Gyllenborg hedges his bets with the authorities by adding a footnote to the effect that here he is: "dealing with man simply in his natural state of wretchedness, not as enlightened by the one true religion", and rounds the whole thing off with stanzas which might be regarded as bearing some resemblance to Johnson's general conclusions in *The Vanity of Human Wishes* (1749):

> If you would travel well along the darkling way,
> Desire not in this life, from yourself or others,
> That part of perfection which nature cannot yield.
> Turn from the night of your sorrows, from glimpses of pleasure,
> And with a patient virtue set your unseeing gaze
>   On the end unknown.

For the passages quoted, see stanzas 36, 31, 32 (*Vitterhets arbeten* II: 149, 148).

The last two sections of the material quoted by Linnaeus are taken from *The World contemned*, a powerful satire in heroic couplets, originally based on Juvenal and Boileau, which Gyllenborg considered to be his masterpiece. It appeared in its original form in the first collection of poetry published by the group – the shepherd Celadon, at odds with the world, encumbered with a morality which causes him so much pain as the Mandevillian bees swarm about him, being quite clearly Gyllenborg's second self. In the revised version quoted by Linnaeus (*Vitterhets arbeten* II: 181 (lines 1–2), 180 (lines 3–5), 174 (lines 6–8), 175 (lines 9–10), 182 (lines 11–12)), it is developed into an elaborate and many-faceted condemnation of the whole contemporary social and political scene. It is now the young Lisidor, one cannot help calling to mind Molière's Alceste, who is confronted with the scandalously irresponsible manner in which the Hats had conducted the Pomeranian war (III.ii.4.3; IV.i.3.6), the economic turmoil and recurrent financial crises (III.iv.3.6; III.iv.3.7), the jobbing and gerrymandering of the place-seekers and empire-builders (IV.i.2.9; IV.i.2.10), the double-crossings and double-double-crossings of party politics (IV.i.2.8; IV.i.2.11). One can well understand that Linnaeus should have found the poem engaging and appealing, and why both he and Gyllenborg should have thrown in their lot with Gustav III after the coup of 1772.

There is a strong sense in Gyllenborg, as there is in Linnaeus, that the general corruption of manners they found about them was a betrayal of the intrinsic and traditional soundness of Swedish life – hence the much-quoted observation concerning the "three old Gothic words". All three, 'samvet', 'dygd' and 'mod' do in fact have Gothic roots, and in 'doughty' and 'mood' the last two at least also have a cognate form in modern English. The precise significance of the Swedish word for conscience, though etymologically it is the exact counterpart of the now obsolete English term 'inwit', has a somewhat complicated history, determined partly by the Bible (*Ecclesiastes* 10: 20; *Wisdom of Solomon* 17: 11; *Ecclesiasticus* 42: 18; *John* 8: 9; *Romans* 2: 15; *Hebrews* 10: 2, etc.), and partly by its Latin and German equivalents.

IV.iii.1.3:

Uppsala ms. 50, 50b. Swedish. The last phrase in point eleven was inserted later: cf. II.iv.3.1; IV.iv.2.1.

As has been pointed out in the *Introduction* (III.iii.3), where the significance of these points is considered at some length, they were almost certainly drawn up as a result of Linnaeus' brush with the faculty of theology at Uppsala in 1748: cf. IV.iv.4.6.

In Uppsala at this time, theologians such as Nils Wallerius, (1706–1764) tended to draw their philosophical armoury from Wolffianism: Frängsmyr, T., 1972. Linnaeus and his disciples were by no means convinced of the advisability of doing so, and Per Forsskål (1730–1763) put forward a highly critical analysis of many of its fundamental concepts, including the principles of sufficient reason and non-contradiction and the derivation of the idea of individual freedom: II.iv.2.3; III.ii.2.4.

It looks very much as though Linnaeus, like so many others in Scandinavia at that time, including Holberg (1748/1754, no. 465), equated naturalism with freethinking, and regarded it as a sort of English murrain, disseminated principally by Anthony Collins (1676–1729): O'Higgins, J., 1970; III.iv.3.18.

IV.iii.2.1:

Uppsala ms. 42, 42b. Latin and Swedish. Cf. II.i.1.2. The paragraphing has been left much as it is in the original.

As is usually the case with Linnaeus, this characterization of pride consists of assembling a list of Latin tags and Swedish Biblical quotations around certain leading themes, and then interspersing observations of his own, either in Latin or in Swedish. It looks as though he must have derived most of the tags from teaching books, probably those he used when he first learnt Latin at Växjö cathedral school.

On account of the cavalier manner in which he treats his basic material and his unreliability in indicating where he got it from, it is not easy to distinguish what has been assembled from what has been interspersed, but the following sources have come to light:

in the first two paragraphs all the sentences with the exception of no. 11

are in Latin, – no. 3 is Greek in ultimate origin, Homer *Odyssey* 23: 11, Sophocles *Antigone* 622, Euripides *Fragments* 436, edited Joshua Barnes (1654–1712) *Euripides quæ extant omnia* (Cambridge, 1694) 515, Lycurgus *Contra Leocratem*, edited Johann Jacob Reiske (1716–1774) *Oratorum Græci quæ supersunt* (Leipzig, 1770/1774) 8: 198, Latin by adoption, Publilius Syrus *Sententiæ* no. 671, and Cantabrigian in its familiar formulation, James Duport (1606–1679) *Gnomologia Homerica* (Cambridge, 1660) 282; no. 4 Publilius Syrus *Sententiæ* nos. 203, 649; no. 5 *Proverbs* 16: 18; no. 6 Virgil *Aeneid* 10: 284; no. 7 Horace *Epistles* I, 8: 17; no. 8 Seneca Ad Lucil. XIX.5;

the third paragraph is entirely in Swedish;

the fourth paragraph (cf. II.i.1.2), with the exception of the final triad of sentences is in Latin, and apart from no. 1, see appendix C, section 1, no. 5 Ovid *Ex Ponto* IV.iii.35 and no. 11 Persius *Satires* 4: 52, appears to be by Linnaeus himself; on Kierman, see IV.iii.2.6;

with the exception of the two concluding sentences the final paragraph is in Swedish; for no. 9, see Seneca N.Q. I præf. 6; no. 10 *De tranquillitate* xiii.2, cf. IV.iv.3.2, section 5.

IV.iii.2.2:

Uppsala ms. 5. Swedish.

The ten points of the law (III.iii.8.1), if not observed, can give rise to these six kinds of sin, for which there is retribution but no forgiveness or atonement.

It is worth noting that the only Biblical commandment not covered by Linnaeus' lists is the fourth, and that these sins are evidently ranged in order of decreasing seriousness.

IV.iii.2.3:

Uppsala ms. 27b. Swedish and Latin. Cf. IV.iv.2.1.

A sinful deed is the outcome not only of being unwary and not observing the law, but also of circumstances. Since to some extent it is therefore opportunity which makes the thief, an objective consideration of a situation in which evil intentions are frustrated may well justify the conclusion that "fortune has been granted by the grace of God".

Since the sins listed in the preceding text would appear to be ranged in order of decreasing seriousness, this passage constitutes a natural sequence.

Although the general significance of the event mentioned in the second sentence is evident enough, its historical context is not. It could be a reference to Haquin Spegel (1645–1714), archbishop of Uppsala: IV.ii.2.14. On sentence no. 5, see Ovid *Ex Ponto* IV.iii.49; on sentence no. 11, see III.iii.7.1.

IV.iii.2.4:

London ms. 31; Uppsala ms. 31b. Swedish, with three words of Latin. There is no clear paragraphing in the original.

With its rhetorical questions concerning the final framework of ethics, its picturing of the fruits of honest labour, its statements concerning the outcome for the individual, this cluster of quotations provides a fitting conclusion to what precedes it, an adequate introduction to the case-histories which follow.

On the triumphing of a just cause, see: *Proverbs* 12: 21, 24: 16; *Ezekiel* 18: 9.

IV.iii.2.5:

London ms. 26; Uppsala ms. 202. Swedish. Cf. IV.i.4.4.

This case-history of Christian Joachim Klingspor (1714–1778) has also been placed here as a particularly crass instance of inhumanity.

IV.iii.2.6:

Uppsala ms. 64, 64b. Swedish and Latin.

The fact that Gustav Kierman (1702–1766) was from Askersund, a small town just to the north of lake Vättern, made it all the more remarkable that he should have made the grade into the top rank of the merchant plutocracy of Stockholm, most of whom had family roots in much larger commercial centres. In his home town he evidently began his career as a pedlar or commercial traveller working for distributors or manufacturers in Borås, West Gothland, then a trading centre for foreign imports and an up-and-coming textile town.

He probably arrived in Stockholm during his late teens – being taken on by the 'godfearing' merchant Hans Henrik Björkman, whose widowed wife Margareta Gieding (1698–1750) he married on September 25th 1725. On April 24th 1753 he married Juliana Brandel (1712–1767) widow of the merchant Efraim Lothsack. It looks as though the cuckolded 'Forsberg' mentioned by Linnaeus may well have been the court chamberlain Peter Forssberg, husband of Maria Elisabeth Ahlström (IV.i.2.7), whose daughter Anna Elisabeth (1742–1768) married the Stockholm accountant Carl Hans Hauswolff (1736-1767) on April 22nd 1760: III.iii.2.4; SBL 16: 302.

He made his money by developing his interests in the textile industry, in the export of iron and in trading with the Levant. During the 1730s he found his natural political affiliation with the Hats, and continued to sit as one of the members for Stockholm in all the parliaments between 1738 and 1762, co-operating closely with his fellow Levant merchant Thomas Plomgren (1702–1754), taking a firm line on Brahe and his confederates in 1756, and being elected speaker of the house of burghers in 1755/1756 and 1760/1762: III.i.1.4; IV.i.2.4. He was elected political mayor of Stockholm in 1757 and commercial mayor for the period 1758/1765.

Kierman's daughter Ulrica (1734–1804) married Johan Diedric Duwall (1723–1801), a captain in the life guards, on July 17th 17571. In 1759 Kierman made over to his son-in-law the estate of Näsby near Taxinge, some twenty-eight miles west of Stockholm. In 1760 Duwall was co-opted onto the Hat secret committee, and in the 1761 appointments to the privy council, with Kierman's help, he just managed to pip the Cap Thure Gustaf Rudbeck (1714–1786) at the post. It was this success which brought about Kierman's downfall, for when the Caps came to power in 1765 Rudbeck took his revenge, and in no small measure: IV.i.2.11. Although Kierman died in prison on November 15th 1766, he was buried in the Riddarholm church in Stockholm. To a certain extent, Duwall managed to restore the family's fortunes, being ennobled in 1770: SBL 21: 120–124.

There is something almost ethical about the way in which Kierman addressed the bigwigs in the shipyard in 1756. There can be no doubt, however, that Linnaeus saw pride as his overriding motivation (IV.iii.2.1), and his attitude to his employer and his wives, as well as to Brahe and the inhabitants of Stockholm, as fully warranting the precipitation of his final punishment.

IV.iii.2.7:

London ms. 25; Uppsala ms. 70. Swedish.

Johan Buskagrius (1629–1692) was the son of a German merchant from Rostock who had settled at Falun in Dalecarlia. There is a possibility, therefore, that Linnaeus knew of him not only on account of his university career at Uppsala, but also through his wife's family, his father-in-law Johan Moraeus (1672–1742) having been town physician in Falun.

Buskagrius was matriculated at Uppsala in 1646, and on April 3rd 1652 performed the remarkable feat of defending a dissertation there in Hebrew. The university had high hopes of him, and for several years financed his studies abroad – under Johannes Buxtorf (1599–1664) at Basel, Johannes Hottinger (1620–1667) at Heidelberg, and Edward Pococke (1604–1691) at Oxford. On his return in 1660 he was appointed professor of oriental languages, and in 1661 married Kristina Simonia (d. 1706), daughter of Vilhelm Simonius, the Skyttean professor of rhetoric and political theory.

It is recorded in the minutes of the university consistory on February 4th 1671 that: "Buskagrius' case was discussed: since he has completely neglected his duties and sits alone in a melancholy state of mind, it was found necessary that he should be admonished, in the first instance by the faculty of philosophy, and then by the pro-vice-chancellor". The case was found to be hopeless, however, and on September 26th 1672 Uno Johannis Terserus (1642–1675) was appointed as his successor. The financing of his chair had always been somewhat irregular, involving the requisitioning of funds originally intended for the medical faculty, and it was not until 1685 that his pension was properly sorted out: BL 3: 112–113; SBL 6: 786–787; Lindroth, S., 1975, 220–230.

Buskagrius' exclamation, though made in the heat of the moment, is evidently to be seen as a violation of the second of the ten laws (III.iii.8.1) and as involving disdain for God (IV.iii.2.2).

IV.iii.2.8:

London ms. 7, 20 (Artedi); Uppsala ms. 161. Swedish. In the third paragraph, 'travels' is deleted and replaced by 'dispatched'.

Malcom Sinclair (1690–1739) came of a military family, his father Vilhelm Sinclair of Finnekumla, some twelve miles east-south-east of Borås in West Gothland, having held the rank of major-general under Charles XII and been commandant of Malmö. The young Sinclair joined the life guards, fought at Poltava, and was then captured at Perevolochna and interned in Kazan, some six hundred and fifty miles east of Moscow, not returning to Sweden until 1722. It was during this period that he knifed his fellow prisoner and formed his opinion of the Russians. Once he was back in Sweden and had been acquitted of murder, he was promoted to the rank of lieutenant and put on the pay-roll of the fifteenth company of the life guards, eventually being promoted to the rank of major in the Uppland infantry.

Between May 27th and July 5th 1738 Sinclair was prepared by the Hat council of state for a diplomatic mission to Constantinople, the long-term objective of which was to secure Turkish co-operation in an attack on Russia. Despite the briefing's being a matter of top security Michail Bestuzhev (1688–1760) the Russian ambassador got wind of it, and arranged for a portrait of Sinclair to be forwarded to Burkhard Christoph Münnich (1683–1767), the Russian prime minister: IV.i.3.5. Sinclair left Stockholm for the Turkish capital during the night of July 6/7th. Malmström, C.G., 1893/1901, II: 274, 363.

By the early April of 1739 Sinclair had fulfilled his mission and left Constantinople for Stockholm with letters from the Sultan; Münnich had prepared plans to have him ambushed and murdered and dispatched his agents. The two parties met near Naumburg in Silesia on June 17th. Sinclair, who was travelling together with a French merchant by the name of Couturier, was apprehended by a posse of six to eight Russian officers, taken from the chaise, led into a nearby copse, and shot – the troupe riding on into Dresden with the Frenchman, who was then released. The mutilated boly was discovered by a shepherd and interred. It was later exhumed by the Swedish authorities and transported to Stralsund, where it was given full ceremonial burial in the church of St. Nicholas. In 1909 the Swedish government arranged for a memorial to be erected on the site of the murder. News of what had happened contributed greatly to the build-up of tensions which eventually gave rise to the outbreak of war between Sweden and Russia in the July of 1741. It was noted in Stockholm that although both Austria and Russia had declared themselves to be shocked and appalled by what had happened, it had not been long before Küttler the leader of the expedition had been promoted to the rank of major, and Levitsky his second-in-command to that captain: BL 14: 271–275.

The event attracted great public interest in Sweden throughout the whole of the period of the war, and gave rise to some remarkable ballad literature, in which such military heroes of the past as Gustavus Adolphus, Banér and Charles XII converse with Sinclair in the Elysian fields or Valhalla: Axel Daniel Leenberg (c. 1705–1744) *Minnesrunor öfver ... Malcom Sinclair* (Stockholm, 1739); Anders Odel (1718–1773) *Hjeltarnas samtal med ... Malcom Sinclair* (Lund, 1741); Petter Momma (1711–1772) *Omständeling, Berättelse, om ... Malcom Sinclair* (Stockholm, 1741); Hörnström, J., 1943.

The outcome of the war tended to take the wind out of the sails of this sort of thing. Since the Sinclair case is not noted in the London manuscript, it could be that Linnaeus first took an interest in it not on account of its political connotations and the fateful comment concerning the Russians, but simply as an example of divine retribition for murder. The case of Peter Artedi (1705–1735), however, is noted in the London manuscript. It could well be the case, therefore, that Linnaeus' original intention was to give some prominence to the outcome of his comment concerning the Dutch, and that he later dropped the idea as being at odds with his general assessment of his character: "He gave me the impression of being modest and not over-hasty in forming an opinion, yet at the same time he seemed alert, determined and mature – a man of old-fashioned virtue and trust". Linnaeus, C., 1738, *Vita Petri Artedi*; III.iii.8.1 no. 8.

Both Linnaeus and Artedi had grown up in a rectory, been destined by their parents to read theology, and then switched to natural science. They both appreciated the central importance, in any field of learning, of developing a well-founded and efficient taxonomy. The friendship and mutual respect between them is fully apparent in Linnaeus' edition of Artedi's work on fish. They had first become acquainted when Linnaeus came up to Uppsala from Lund in 1728. In 1732 Artedi's research-grant ran out, and two years later his family supplied the funds for his travelling abroad to take his doctorate. Linnaeus left Stenbrohult in April 1735, and first met up with Artedi in Leiden on June 27th. On July 6th he visited him in Amsterdam, where he was lodging with a certain Hendrik Jutting, who lived in the Warmoesstraat near the Nieuwebrugsteeg, and where he had found employment with a German apothecary Albert Seba (1665–1736), who lived along the Haarlemmerdijk. It was almost certainly on this occasion that Artedi expressed his opinion of the Dutch. Linnaeus spent most of the rest of July and August in Leiden, and on September 13th began his employment with George Clifford (1685–1760) at *De Hartenkamp* near Haarlem.

Artedi was drowned during the early morning of September 28th, when returning to his lodgings from a festive gathering at Seba's place. Linnaeus heard three days later, and was soon involved in attempting to sort out his affairs. Since Seba had not paid Artedi his salary regularly, he was behind with his rent, and Jutting was threatening to confiscate all his manuscripts and realize on them. Linnaeus eventually got Clifford to purchase the manuscripts, and Seba to pay the cost of burying Artedi in a pauper's grave: Engel, H., 1950/1951, *Ons Amsterdam* 2de Jaargang, no. 4, April 1950: 146–151; SBL 2: 305–309.

IV.iii.3.1:

Uppsala ms. 37. Latin.

Seneca *Thyestes*: 388–403: the conclusion of the chorus preceding Thyestes' returning from banishment, accompanied by his three sons. The passage seems to have been quoted from memory, since it is by no means word-perfect.

IV.iii.3.2:

Uppsala ms. 59, 59b. Latin and Swedish. cf. III.i.3.1: *Ecclesiasticus* 30: 22 is also written out on 62b, and deleted.

The two opening sentences are in Latin, and are evidently Linnaeus' versions of *Ecclesiastes* 3: 12 and *Isaiah* 22: 13.

IV.iii.3.3:

Uppsala ms. 28. Swedish. Cf. III.ii.4.1.

It was as a result of his differences with Nils Rosén (1706–1773), his rival for a permanent academic position at Uppsala, that Linnaeus, in 1734, "put everything in the hands of God": *Introduction* I.ii.3; Fries, Th.M., I: 171–191; Malmeström, E., 1926, 69–80; 1964, 51–62.

IV.iii.3.4:

Uppsala ms. 13b. Swedish.

This is the only entry on this side of the sheet, on the other side of which is II.i.1.6 – the likening of man to a waxen candle, a comparison which is almost certainly influenced by certain passages in the Bible.

When Joseph's elder brothers became aware that Jacob their father loved him best, that Joseph's dreams foretold that they were destined to be beholden to him, their hatred was aroused and they plotted to kill him. As things turned out, they sold him into captivity in Egypt, where he rose in the Pharaoh's favour by interpreting his dreams correctly, was given control of the affairs of the country, and through the foresight of his policies saved the population from starvation. Unbeknown to his brothers, he was able to help them when they were sent by Jacob to buy corn in Egypt. There would appear to be no need for forgiveness when he first reveals himself to them: "Now do not be distressed or blame yourselves for selling me into slavery here; it was to save lives that God sent me ahead of you" (*Genesis* 45: 5). The brothers believe that it is only for their father's sake that Joseph has taken this attitude, and once Jacob is dead they beg him to forgive their crime, since they are "servants of their father's God" and therefore his "slaves". This moves Joseph to tears, and to the fuller revelation noted by Linnaeus: "Do not be afraid. Am I in the place of God? You meant to do me harm; but God meant to bring good out of it by preserving the lives of many people, as we see today" (*Genesis* 50: 19, 20).

Despite its involving the interpretation of dreams (II.iii–II.iv), the story of Joseph is essentially a matter of the family (III.i.–III.ii), of economic life (III.iv) and of the kingdom (IV.i.), concerned almost exclusively with God's guidance in everyday practical affairs. Unlike the story of Jacob, it involves no callings, visions or sacrifices (*Genesis* 46: 1–5). There is nothing of the priest or ritualist about Joseph, and despite his deep attachment to his family, very little of the Israelite as such. His message is truly universal: cf. IV.iv.4.6.

IV.iii.3.5:

Uppsala ms. 3. Swedish.

Olof von Dalin (1708–1763), the author of these stanzas, was the son of a clergyman in Halland. After studying at Lund he came to Stockholm as tutor to the Rålamb family, and through the opportunities which this opened up for him was soon at home in the polite society of the capital. He began his literary career by establishing a highly successful journal in the style of Addison and Steele (1732/1734), and soon branched out into writing lyrics, satires and even a Voltairean epic, as well as plays in the style of Molière and Holberg. He was appointed royal librarian in 1737, and after spending a couple of years in Paris, was commissioned to write the official history of Sweden, which at the time of his death he had completed to the beginning of the seventeenth century. He was appointed tutor to crown prince Gustav in 1750, and despite opposition from parliament on account of his advocating an increase in the constitutional power of the crown, kept his position at court until his death. He wrote with elegance and polish, pathos and wit in a great variety of forms, and succeeded in attuning the literary climate in Sweden to what was then fashionable in mainstream Europe: SBL 10: 50–65.

The 'Thoughts on God's providence' quoted here by Linnaeus, was one of Dalin's earliest poetic productions, and was evidently dictated to a friend when he was incapacitated by illness. It first appeared in print in 1736, and was often sung to a hymn-tune. The whole poem consists of twenty-two stanzas, those quoted being numbers 8, 19, 4, 20, 17, 22. Linnaeus put a line through what he had written, evidently intending to write it out again in a more satisfactory manner, but in the existing manuscript there is no evidence that he ever did so. Cf. IV.iv.4.4; IV.iv.4.5; Lamm, M., 1908, 141–146.

The poem is a fine if conventional statement of contemporary thinking in the tradition of Leibniz's theodicy – insight into the pre-established harmony of the two synchronized clocks, awareness of the imperfections of the natural and moral world, cause the poet to yearn for the grace of sharing more fully in the "counsel and secret decree" of the all-powerful Author: *Introduction* II.i.ii. It therefore rounds off rather neatly the ethical doctrine embodied in this section of the *Nemesis Divina*, initiated as it is by the admonition which finds its fulfilment in the reconciliation between Joseph and his brothers: IV.iii.1.1; IV.iii.3.4.

IV.iv:

Cf. IV.iv.1.3.

IV.iv.1.1:

Uppsala ms. 31. Swedish. Cf. London ms. 34, Uppsala ms. 1b, I.v, where the wording, punctuation and spelling are slightly different. Unidentified, but not likely to be by Linnaeus.

IV.iv.1.2:

Uppsala ms. 31. Swedish. Quoted without the Biblical reference. Cf. I.ii.

IV.iv.1.3:

London ms. 16; Uppsala ms. 14. Latin.

The London manuscript contains the heading with a list of numbered case-histories under it, but no attempt to define fate. The Uppsala text begins with the statement that: "Fate is the law of God . . . ", which is then corrected. For a parallel to the statement concerning the Devil and his trumpet, see IV.iv.2.1.

It is worth paying close attention to the reasoning here. Linnaeus begins by stating that fate is the judgement of God, and by making use of the parable finally concludes that the judgement of God is fate.

The parable, involving as it does a specific instance, has therefore confused the issue rather than clarifying it. The truth of the matter is, surely, that although fate can in principle be identified with God, recognition of the identification does not warrant our making any further assertions, either about God or about the specific manner in which He judges. It is the same with the ontological proof of God's existence, which although it rests on the necessary principle of God's being identified with being, does not in itself enable us to make any further assertions, either about God or about specific aspects of nature and man. Unfortunately, although Linnaeus seems to have sensed the truth of this when he turned his back on the Spinozists, Leibnizians and Wolffians in developing his essentially Baconian approach to the world of nature and man, he seems to have overlooked it in this particular connection.

As has been indicated in the *Introduction* (III.iii.3), Linnaeus is in fact much closer to orthodox Christian thinking on this point than he might appear to be when he argues in this manner. Taken as a whole, the *Nemesis Divina* papers certainly seem to imply that fate is most closely associated with the first person of the Trinity. Like the Father, it is essentially inaccessible to the human intellect: he often notes, for instance, that our minds are incapable of grasping the eternal law by which everything is controlled and ordered, that the decrees of fate can only be known at second hand, by contemplating the ways in which they are expressed in nature. It is indeed through nature that fate has its bearing on man, and therefore on the basis of nothing more than the contradictoriness of experience that one identifies fate with the judgement of God.

IV.iv.1.4:

London ms. 32; Uppsala ms. 29b, 32, 32b, 33b. Swedish.

Since this section has been assembled from some very untidy manuscript material, it may be of value to take note of the basic situation: the first two sentences are at the top of 29b, and are followed by eleven further quotations, most of them Biblical, and all of them deleted; the next eight quotations are on 32, with a considerable gap at the bottom of the page; the following eight sentences are on 32b, with a large gap between the sixth and the seventh; 33 contains eleven quotations, most of them relating to III.ii.2.1, and all of them deleted; the last four quotations are from the undeleted material on 33b.

These Biblical quotations are being used to back up the general proposition that human action, though rooted in the will (II.i.1.1) and subject to time and chance, has its ultimate foundation in God. Taken as a whole, therefore, they constitute a certain advance upon the position Linnaeus attempts to establish in IV.iv.1.3.

IV.iv.1.5:

Uppsala ms. 31. Latin. Cf. I.iii.

Claudius Claudianus (c. 365–c. 408), the last of the Roman poets, published his epic on the downfall of Rufinus, the unworthy minister of Arcadius, at Constantinople in 396.

IV.iv.1.6:

Uppsala ms. 156. Swedish. Cf. II.iv.1.6.

Johan Svensson Collin (1707–1766), who became Linnaeus' brother-in-law in 1737, had studied with him at Växjö school between 1716 and 1727.

The case has also been placed here on account of the final observation.

IV.iv.1.7:

Uppsala ms. 98. Swedish and Latin.

Linnaeus heard of this case from his colleague Mårten Strömer (1707–1770), a native of Örebro, who succeeded Anders Celsius (1701–1744) as professor of astronomy at Uppsala, and left the university in 1756 in order to take up a position as instructor at the Admiralty Training School in Karlskrona. He figures quite extensively in Linnaeus' correspondence, and they seem to have co-operated closely in faculty and university affairs, notably the backing of Carl Gustav Tessin (1695–1770) for the chancellorship in 1751: Fries, Th.M., 1903, II: 208.

It was almost certainly a concrete case such as this which Linnaeus had in mind when he argued as he did in IV.iv.1.3 concerning fate and the judgement of God. He was sympathetic toward Gustav III's "assuming almost complete sovereignty" in 1772 (IV.i.2.11), and in the Scandinavian political theory of the period he would have had no difficulty in finding justification for theocratic absolutism (IV.i.3.3; IV.i.5.9), but the religious conviction of a Divine Arbiter is by no means essential to the acceptance of the proposition that any organization can benefit from the recognition that it embodies an absolute authority.

In the case recounted by Strömer, the prisoner confirms the ultimate justice of the king's decision. Given the situation, however, it was by no means impossible that he should not have done so. The case in itself cannot be used to draw any conclusions concerning the necessarily just decisions of absolute monarchs. There is, therefore, a clear difference between the recognisably contingent infallibility of a monarch or a pope, and the identification of God and fate.

There are, therefore, distinct gradations of certainty in the material Linnaeus collected around the central theme of fate, and it is essential that they should be brought out by classifying it properly. Tracing the broad development of the conception of God's judgement in the various books of the Bible can be a help in this respect. In the early writings of the Old Testament the judge is appealed to as the defender of rights, Jehovah is invoked in order to obtain assistance for Israel, and when he grants judgement it is a saving act (*Exodus* 12: 12). Isaiah qualifies this conception, declaring Israel itself, and especially the corrupt wielders of power within it, to be under threat from God's judgement (*Isaiah* 1: 2–9), and Amos completely inverts it, proclaiming Israel as a whole as no less under God's judgement than other nations (*Amos* 9: 7). In the New Testament this universalization and humanization of the concept is continued in Christ's being faced with Jews determined to kill him, and declaring that: "the Father does not judge anyone, but has given full jurisdiction to the Son", by Paul's reminding us that by judging others we condemn ourselves, by Peter's rhetorical question concerning the ultimate end of judgement (*John* 5: 22; *Romans* 2: 1; 1 *Peter* 4: 17).

IV.iv.2.1:

London ms. 22; Uppsala ms. 63, 27b. Latin and Swedish. Cf. IV.iii.2.3.

The heading and the first paragraph are on p. 63. In the last section, there is a gap after "inherit".

On Linnaeus' interest in the inhabitants of the parish of Danmark and of Hammarby in particular, see II.i.1.8, III.i.3.13, III.ii.1.6, III.iii.6.3, III.iv.2.4; on Erland Carlsson Broman (1704–1757), III.iv.3.4; the Devil and hammering, IV.iv.1.3; Nils Kyronius (c. 1700–1784), III.iii.2.3; Haquin Spegel (1645–1714), IV.ii.2.14.

Much of the last three sections here is a medley of quotations and short observations, and as in the case of IV.iii.2.1, it is not easy to distinguish one from the other. The following sources have come to light: section 3, no. 3 Axel Olrik 'Odinsjægeren i Jylland', *Dania* VIII (1901) 163; no. 5 Seneca Ad Lucil. XIX.5; no. 6 Publilius Syrus *Sententiæ* no. 274; no. 7 *Proverbs* 10: 13, 26: 3; no. 8 James Duport (1606–1679) *Gnomologia Homerica* (Cambridge, 1660) 282; section 4, no. 5 Ovid *Ex Ponto* IV.iii.49; no. 9 Seneca *Phoenissae* 659, cf. IV.iv.3.2, section 4, no. 5; no. 11 see III.iii.7.1.

The significance of fate can be grasped in principle but not in fact (IV.iv.1.3). The facts of fortune are evident enough – in a matter of three minutes the Lisbon earthquake entombs congregations in the ruins of their churches (IV.ii.1.2), Joseph's brothers make money by selling him into captivity in Egypt (IV.iii.3.4), the lazy prosper and the diligent are ruined, the ignorant are fêted and the learned ignored. Linnaeus opens the *Nemesis Divina* by picturing the all too familiar scene:

> What you see is confusion, raised and marked by none.
> The fairest of lilies you see choked by weeds.

If observed closely and patiently, however, the scene itself can yield some sort of a foundation for hope. By and large, it is not infrequently the case that while probity, diffidence and gentleness bear wholesome fruit, jobbery, pride and ruthlessness do not. Long-term analysis of fortune will usually bring out the steadiness of the pace in the fickleness of its turns, the order in the confusion, the wisdom at work in human affairs. Although it can create no direct insight into the will of God, it can provide us with many instances of the working of His grace.

Although the Linnaeus of the *Nemesis Divina* papers gives remarkably little prominence to the New Testament and the life of Christ, he does urge the theologians to defuse the common heresies concerning the divinity, teachings and miracles of Christ, the nature of the incarnation and the atonement (IV.iii.1.3). There are good grounds for maintaining, moreover, that just as his conception of fate is most closely associated with the first person of the Christian Trinity, so his conception of fortune is most closely associated with the second (*Introduction* III.iii.3).

IV.iv.2.2:

Lachesis naturalis 1907: 242–243, Swedish; London ms. 15, no. 30b, Latin; Uppsala ms. 27, Latin.

The earliest extant version of this story would appear to be that of the Persian poet Jami (1414–1492), edited and translated by B.E. Cowell in the *Journal of the Asiatic Society of Bengal* I: 10–17 (Calcutta, 1860); cf. H. Brockhaus 'Gellert und Jami', *Zeitschrift der Deutschen morgenländischen Gesellschaft* XIV: 706–710 (Leipzig, 1860):

> One day spake Moses in his secret converse with God, "Oh thou all-merciful Lord of the world, open a window of wisdom to my heart, shew me thy justice under its guise of wrong". God answered, "While the light of truth is not in thee, thou hast no power to behold the mystery". Then Moses prayed, "O God, give me that light, leave me not exiled far away from truth's beams". "Then take thou thy station near yonder fountain, and watch there, as from ambush, the counsels of my power".
>
> Thither went the prophet, and sat down concealed. Drawing his foot beneath his garment, he waited what would be.
>
> Lo from the road there came a horseman, who stopped like the prophet Khizr by the fountain. He stripped off his clothes and plunged into the stream, he bathed and came in haste from the water. He put on his clothes and pursued his journey, wending his way to mansion and gardens; but he left behind on the ground a purse of gold, filled fuller with lucre than a miser's heart.
>
> And after him a stripling came by the road, and his eye, as he passed, fell on the purse; he glanced to right and left, but none was in sight; and he snatched it up and hastened to his home.
>
> Then again the prophet looked, and lo! a blind old man who tottered to the fountain, leaning on his staff. He stopped by its edge and performed his needful ablutions, and pilgrim-like bound on him the sacred robe of prayer.

Suddenly came up he who had left the purse, and left with it his wit and his senses too. – Up he came, and, when he found not the purse he sought, he hastened to make question of the blind old man. The old man answered in rude speech to the questioner, and in passion the horseman struck him with his sword and slew him.

When the prophet beheld this dreadful scene, he cried,"Oh thou whose throne is highest heaven, it was one man who stole the purse of gold, and another who bears the blow of the sword. Why to that the purse and to this the wound? This award, methinks, is wrong in the eye of reason or law".

Then came the Divine Voice, "Oh thou censurer of my ways, square not these doings of mine with thy rule? That young boy had once a father who worked for hire and so gained his bread; he wrought for that horseman and built him his house. Long he wrought in that house for hire, but ere he received his due, he fell down and died, and in that purse was the hire, which the youth carried away. Again, that blind old man in his young days of sight had spilt the blood of his murderer's father. The son by the law of retaliation slays him today, and gives him release from the price of blood in the day of retribution!"

The story, in almost exactly the same form in which Linnaeus tells it, is to be found in Addison's *Spectator* no. 237, Saturday, December 1, 1711: Addison, J., 1965, II: 423, where it is attributed to "Jewish Tradition". That this is the authentic Jewish version of it is confirmed by Ginzberg, L., 1946/1961 III: 135–136. It was a commonplace throughout the eighteenth century: Parnell, T., 1722, *The Hermit*; Gellert, C.F., 1746/1748 *Das Schicksal*; Voltaire 1747 c. 18; Friess, F.C., 1758, 51–52.

The fundamental Leibnizian tenet of sufficient reason, the abstract and general axiom that nothing can be as it is without there being a determining cause of its being so (*Theodicy* II §44), is completely compatible with the idea that fate can in principle be identified with God (IV.iv.1.3). We have seen, however, that acceptance of this idea does not warrant our claiming any insight into the judgements of God, and it could also be argued that the Leibnizian tenet is devoid of any bearing on our interpretation of contingent facts. It is worth noting, therefore, that in so far as this rabbinical story throws light upon fortune, it does so not by trading on the concept of physical causality, but by indicating the outcome of ethical decision-making – by suggesting that man's most meaningful insight into the nature of God derives not from his ability to understand or manipulate nature, but from his moral capabilities.

IV.iv.2.3:

Uppsala ms. 151. Swedish and Latin.

Johan von Asp (1708–1779) – town lawyer in Stockholm 1731, court steward 1739, agriculturalist, owner of the Salnecke estate in Uppland, married Juliana Appelroth (b. 1721) daughter of a Stockholm councillor 1744, ennobled 1758, bankrupt 1768 – came of a family which took its name from the farmstead of Aspby in the parish of Torsåker, some thirty miles up-river from Härösand. His father Petrus Jonæ Asp (1667–1726), prominent as a clergyman in Norrköping, well-known for the lead he took in improving the school-system in northern Sweden, married Elisabeth Steuch (1677–1758), daughter of the future archbishop Matthias Steuch (1644–1730) in 1695: IV.ii.2.14. His elder brother Matthias Asp (1696–1763) became professor of theology at Uppsala in 1745, his younger brother Carl Asp (1710–1782) professor of philosophy there in 1755: Anrep I: 97; SBL 2: 380–388; III.i.1.6; IV.ii.2.6.

His daughter Maria Elisabeth (1748–1770) married Frederic von Post (1726–1805) on September 25th 1766. His son Pehr Olof von Asp (1745–1808) was matriculated at Uppsala on October 17th 1757, passed his civil service examination in 1763, and got substantial promotion on July 12th 1769. He may have been rather poorly when Linnaeus was writing, but he went on to carve out a distinguished diplomatic career for himself, being appointed secretary to the embassy in London in 1770, and holding down similar jobs in various European capitals during the following decades, before being appointed ambassador to Constantinople (1791) and London (1795): SBL 2: 383–388.

Although one can find some justification for the way in which Linnaeus concludes his case-history of Marlborough (IV.i.4.1), it is rather more difficult in von Asp's case. It could be argued, moreover, that he would have done better not to enter into speculation of this kind: IV.iv.1.3; *Romans* 2: 1.

IV.iv.3.1:

London ms. 27; Uppsala ms. 26, 1b. Latin and Swedish.

Frederik Christian Friess (1722–1802), the author of the work referred to here by Linnaeus was born in Copenhagen, son of Floris Friess (1680–1750) a captain in the Royal Navy, and his wife Barbara Pontoppidan (1689–1743), a relative of the prominent pietist, theologian and historian Erik Pontoppidan (1698–1764), professor of theology (1738), bishop of Bergen (1748), vice-chancellor of the University of Copenhagen (1755). It is evident from Friess's book that Pontoppidan had a great influence upon his intellectual and religious development.

When Friess's father was pensioned off at the end of the Great Northern War (1719), the family was left in very straitened circumstances, and he had to be given into the care of a relative in Hatting, a village just outside Horsens in Jutland. He finished his studies at Horsens Grammar School in 1738, and went on to the University of Copenhagen, where he graduated in theology in 1743. In 1745 he was appointed to the position of catechist in Vejle, and the following year to a similar position in Kolding. It was in Kolding that he founded his society for "the furthering of various useful undertakings", a project clearly influenced by the Baconian programme of reform first outlined in *The Advancement of Learning* (1605) and the *New Atlantis* (1627), and the source of much of the material included in his book: *Introduction* III.i.4.

In 1754 he was appointed to an army chaplaincy based in Kolding, and in 1759, partly on the strength of his publication, parish priest at Hjarup with Vamdrup, villages close to Kolding, in the diocese of Ribe, a position he held until his retirement in 1787. His first wife was Sophie Dorothea Bang (1726–1757), daughter of the parish priest at Bredsten near Vejle, his second wife Margarethe Storm (c. 1738–1780), daughter of the parish priest at Egtved with Ødsted near Vejle: DBL, 9: 436–440; Ehrencron-Müller, H., 1924/1939, III: 126–127; VI: 321–336.

The bishop of Ribe, the pietist hymn-writer Hans Adolf Brorson (1694–1764), thought highly of Friess, in 1752 characterizing him as "a profound and edifying scholar, exemplary in his manner of living, diligent in his office as a teacher; married, but with a very limited income, in poor economic circumstances": *Kirkehistoriske Samlinger* 1884/1886 III.v. 713. He turned out to be an energetic and effective parish priest, a later official report on his work as such describing him as a man "of great insights and compassions".

The position of catechist had been created by the pietist king Christian VI in 1736 in order to improve religious instruction throughout the country, but the posts were not well funded, and as in Friess's case in Kolding often gave rise to friction with the regular clergy. The parish priest in Kolding was a certain Mathias Bakke, sympathetic toward the Zinzendorf movement, the Moravian Brethren, and in 1748 Friess wrote to the bishop accusing him of heresy. Ten years later, when Friess was established as priest in his own right, he prosecuted Bakke for the non-payment of services rendered when he was working with him as an army chaplain. Bakke's son-in-law Johannes Harbo (d. 1753) was in charge of the Danish school in Kolding, and when there was a public complaint against his teaching in 1748, Friess supported it. Four years later the bishop was also critical of the poor education being given to ordinary children in the town: Fyhn, J.J., 1848; Bruun, G., 1939; Jensen, F.E., 1949.

Surprisingly enough, Friess was also active as an agriculturalist, pioneering new farming methods during his first decade in Hjarup, and being awarded a gold medal for the work by the king. By and large, he was, like Linnaeus, quite clearly part of the enlightened and progressive movements of the time: Kock, L., 1914; Jensen, F.E., 1944.

The full title of the original edition of the work referred to by Linnaeus is: *Theologisk og Historisk Afhandling om Den guddommelige Giengiældelses Ret, Eller Jus Talionis Divinum*. Oplyst og stadfæstet med adskillig Exempler af Guds Ord og andre troværdige Historier, Udi det, Til adskillige nyttige Materiers Afhandling indrettede Koldingske Societæt, Af F.C. Friess. Societætets Forstander. Kjøbenhavn, 1758, xiv + 166 pp. It was translated into Swedish by the hymn-writer Olof Rönigk (c-1710–1780), headmaster of the Katarina School in Stockholm, member of the hymn-book commission (1763/1767), who in 1760 had published a Swedish translation of Erik Pontoppidan's pastoral letters. Apart from there being no mention of the Kolding Society on the title-page of the Swedish edition, it is the exact counterpart of the original. The work was published in Stockholm in 1763 by Lars Salvius (1706–1773), who also published most of Linnaeus' Swedish works.

Although Linnaeus observes that Friess only "has a little to say" on the subject of the "divine law of talion", he made good use of the book, notably in drawing up his own title-page. It seems quite likely, therefore, that the idea of preparing the *Nemesis Divina* papers for eventual publication, the whole business of re-writing them, of assembling the Uppsala manuscript, which Linnaeus carried out during the autumn of 1765, originated in his having come across Friess's book.

In the London ms. 27 we find the material used on the later title-page, as well as notes on eight of the case-histories subsequently acknowledged in the Uppsala papers as originating from Friess: III.i.1.3 *Onanists* (1758 ed. p. 97), III.iii.1.5 A Scanian farmhand (p. 43), III.iii.5.4 Erik Grubbe (p. 77), III.iii.5.6 King Christian II (p. 86), III.iv.3.9 A Danish sheriff (p. 125), IV.i.3.1 Slagheck (p. 86), IV.i.3.2 Jöran Persson (p. 88), IV.iv.3.4 Gallus (p. 109). We also find material on Tycho Brahe (p. 76), which may well have been the origin of a cryptic remark in II.i.1.3, as well as an account of Jupiter and the bees (p. 80) and a note on Matthias Kampfe, a sixteenth-century sexton of St. Peter's Freiberg (p. 121), not included in the Uppsala version.

It looks as though Linnaeus continued to browse through Friess while assembling the Uppsala material, although he makes no acknowledgement of this: see I.i the treatment of the father-son relationship (p. 152), II.iii.1.2 on angels (p. 105), III.i.2.4 on a woman's pride (p. 115), III.ii on parents and children (p. 116), III.iii.1.12 on Tordenskjold (p. 106), III.iii.2.3 on the oppression of widows (p. 129), III.iii.8.1 on God's law in nature (p. 39), IV.i.1.1 on the suffering of the nations (pp. 58, 159), IV.i.1.2 on political authority (p. 40), IV.i.5.2 on Adoni-bezek (p. 124), IV.i.5.9 on Belshazzar's debauchery (p. 134), IV.i.5.13 on Louis XIV's invulnerability (p. 140), IV.iii.1.1 on God's omniscience (pp. 109, 150), IV.iii.2.1 on the writing on the wall (p. 32), IV.iv.1.3 on hanging oneself (p. 84), IV.iv.2.2 on rabbinical wisdom (p. 51), IV.iv.3.3 on leaving things to God (p. 28).

Just as the individual and the community work together in Salomon's House to abstract well-founded axioms from their observation of nature, which are then applied in order to master it, so Friess and his society pooled their perceptions of human conduct in order to abstract worthy principles from the apparent chaos of fortune, and open up the possibility of improving the moral world about them. The outcome of their enterprise is quite sophisticated, and since it is explicitly based in experience and not simply a matter of *a priori* reasoning, it is not open to the sort of criticism levelled at IV.iv.1.3.

Friess distinguishes between ordinary cases of talion, in which it is, "bound in a certain manner to certain laws and persons" (vii–xiv, §5, p. 35), extraordinary cases, in which it is evident that it "derives from the Allhighest, who is bound by neither laws nor persons" (§6, p. 57), and finally puts forward, on the analogy of the ontological argument, the postulate of the all-pervading justice of the Almighty (§13, p. 149):

There is no need for me to give any elaborate proof when I say that to me God's right to retribution appears to be unexceptionably just, and worthy of unending praise. For dust though I am, I can say with all due respect that it flows from the very nature of God, that if He were not just, in talion as in all else, he would not be God: being just is both an actual reality and an essential property, and is to be found in all things as pertaining absolutely to the divine and most perfect Being.

The systematic structure is somewhat different when Linnaeus in his final edition of the *System of Nature* distinguishes the seven main facets to the understanding of man – the biological factors of his physiology, diet and pathology, the natural factors of his life in society, the more purely spiritual factors of his political, moral and theological awareness – but the two undertakings can certainly be regarded as complementary, and both yield a usable model for processing information concerning divine Nemesis: Linnaeus, C., 1735[12] (1766/1768) I: 28–32; *Introduction* I.iii.3.

IV.iv.3.2:

London ms. 9, 11, 12, 22 (8 Latin tags), 33; Stockholm ms. 1 (10 Latin tags); Uppsala ms. 26, 26b. The exceptional continuity here is evidence of the importance Linnaeus attached to the assembling of this material, the defining of this general category. There is no paragraphing in the original.

In respect of the unidentified quotations, the following sources have come to light here: section 1, no. 2 see IV.iii.2.2; section 2, no. 3 see III.iii.7.1; section 3, no. 1 see III.ii.4.1, IV.i.4.2, IV.ii.2.14, IV.iv.3.3; no. 4 see III.iii.5.2; section 4, no. 1 see III.iii.8.1 no. 8, IV.iii.2.8 (Artedi); no. 3 IV.iii.1.1; no. 5 see IV.iv.2.1, section 4; no. 6 Juvenal *Satires* 8: 121, see III.iii.5.1, III.iv.2.1; section 5, no. 1 Seneca *De tranquillitate* XIII.2, see IV.iii.2.1; no. 4 see I.iv; no. 5 see IV.i.3.1; section 6, no. 3 Seneca *De Beneficiis* II.31.3, Ad Lucil. XI.85.37 see II.i.3.1 no. 31, IV.iii.2.1; no. 4 Horace *Epistles* I, 8: 17, see IV.iii.2.1; no. 5 Seneca *De providentia* V.8; no. 6 Seneca N.Q. III.29.4; no. 7 Plutarch *Solon* VIII.24, see III.ii.4.1, cf. II.i.3.2 no. 33.

The case-histories contained in the *Nemesis Divina* constitute the purely empirical aspect of the basic tenets embodied in this assemblage of quotations. Speaking in general terms, therefore, one can say that the tenets have been extracted partly from experience and partly from Scripture (I.i), and give expression to the six-fold conviction (1) that God is just, (2) that this justice has implications (3) for our own ethical life, and (4) that it works itself out (5) through fate (6) in punishment.

IV.iv.3.3:

London ms. 15, 33; Uppsala ms. 28. Swedish and Latin. Cf. III.ii.4.1, where the same material is inserted as an anticipation in family life of the right common to all.

Linnaeus' observation concerning his own experience in 1734 (IV.iii.3.3) seems to have been squeezed in at the foot of the page, evidently after he had completed the rest of the text.

The working out of the concept of Nemesis divina (IV.iv.3.2) is an essentially theoretical or historical activity: humanity analyzes and categorizes its past in order to gain insight into the man-God relationship. Nemesis as it is characterized here is a living experience, something which is known existentially, a direct confrontation with fortune which can give rise either to despair (IV.iv.3.4) or to awareness of God's infinite mercy (IV.iv.3.5).

IV.iv.3.4:

London ms. 27; Uppsala ms. 146. Swedish. In both manuscripts Linnaeus notes that he came across this account of Gallus in Friess, F.C., 1763.

Friess (1758 ed., 109) presents the case as an example of talion working not by the ordinary means of evident laws, but by the extraordinary means of incomprehensible connections – there being no rational link between a nosebleed and avoiding murder: on related Danish superstitions, see Sixtus Aspach (1672–1739) *De variis superstitionibus* (Hafniæ, 1697), Clemens Oligeri Olivarius (d. 1785) *Superstitiones vulgares circa morituros* (Hafniæ, 1740).

In 1755 Soldau was in East Prussia, right on the Polish frontier. It is situated a hundred miles due south of Königsberg on the upper reaches of the river Soldau, which flows south into the Vistula not far from Warsaw. It is now in Poland and is known as Dzialdowo. Between 1664 and 1801 eight members of the Tschepius family of Soldau studied at the university of Königsberg: Erler, G., 1910/1917 III: 454.

In Linnaeus' version, the slight inaccuracies in respect of the panicking and the binding derive from Friess's rendering of the original newspaper account, which appeared in the *Københavnske Danske Post=Tidender* 1755, no. 26, Monday, March 31st, and reads as follows:

> Soldau in Prussia, March 6th. On February 26th last, something quite terrible happened in our town church. George Gallus, an elderly citizen and linen-weaver, eighty-eight years old, who had always lived in an extremely irregular manner, constantly falling out with everyone, arrived at church on the said Sunday, went into the vestry just before the second ringing, and in trembling tones asked the sexton to beg the senior priest Mr. Tschepius to come into the church, saying that he wanted to confess and take communion. The priest was surprised at the request; but he wanted to talk to him, and was soon ready to go in. While he was donning his vestments, however, his nose began to bleed, so that he was delayed for a while; in the

meantime the second ringing began. The godless creature feared that things were about to go amiss, *that it would be too late* to do what he intended; he therefore knelt down in front of the altar and knifed himself, first in the throat and then in the body. People came running forward terrified, thinking in the first instance that he was dead; but he soon came to himself again. The priest then came into the church and caught sight of the bloody scene, the reckless scoundrel casting a savage glance at him. He was questioned, *his wounds were bound up*, and he is now in prison. In the course of the enquiry he has admitted that his intention was to knife the priest first and then himself, and that that was why he had taken two knives along. The only reason for his hating this innocent man, is that he has constantly admonished him to improve his ways.

IV.iv.3.5:

Uppsala ms. 175. Swedish. Cf. III.i.1.1.

This case-history has also been placed here on account of its final paragraph. In the whole of the *Nemesis Divina*, there is only one other instance of anyone acknowledging the significance of what is happening to them, and in Cronhielm's case this did not take place until the very last moment of his life: III.iii.1.11.

Joseph remained "troubled in spirit", and there are no grounds for thinking that Linnaeus expected awareness of God's infinite mercy to give rise to any other state of mind: see the final observation in II.i.2.1.

IV.iv.4.1:

Uppsala ms. 39. Latin.

The clear and even lay-out in the manuscript would seem to indicate that this pastiche of quotations was well-prepared before being copied out. The sources are as follows: Virgil *Aeneid* III: 56; IV: 67; IV: 629; V: 22; V: 465; V: 625; V: 710; VI: 376; X: 438; X: 501; XI: 254; XI: 425–427.

IV.iv.4.2:

Uppsala ms. 40. Latin.

The sources here are: Horace *Odes* II no. 10: 21–24; 15–18; Seneca *Thyestes* 622. Horace is giving advice on the golden mean, Seneca fusing the conceptions of the spindle of the fates and the whirlwind of fortune.

IV.iv.4.3:

Uppsala ms. 18. Latin and Swedish.

The heading and the lines are centralized on the page, and are the only entry there. *Psalms* 91: 15–16.

IV.iv.4.4:

Uppsala ms. 31. Swedish.

Olof von Dalin (1708–1763) 'Thoughts on God's providence' (1736), stanzas 10, 4. Cf. IV.iii.3.5. Linnaeus slightly alters the first line of stanza ten.

IV.iv.4.5:

Uppsala ms. 5b. Swedish.

Olof von Dalin (1708–1763) 'Thoughts on God's providence' (1736), the final lines, stanza twenty-two. Cf. IV.iii.3.5. Linnaeus writes "grace" above "will".

IV.iv.4.6:

Uppsala ms. 5b. Swedish.

As in the case of the other reference to *Genesis* 50: 19, 20 (IV.iii.3.4), the context in the manuscript here could be of significance: at the top of the page, "Where the free child of the air speaks of his delight" (II.i.1.4), "To the unfortunate, all are hostile" (III.iii.1.2), and the three lines from von Dalin's poem (IV.iv.4.5), are followed half-way down by these words from Joseph's address to his brothers.

It is perhaps worth calling to mind that Thomas Mann's *Joseph und seine Brüder* (1933/1943) also concludes with a mulling over of these words:

> "But brothers, my dear old brothers!", he replied, bending towards them with outstretched arms. "Why are you speaking like this? It is as if you were afraid and wanted me to forgive you! Am I then as God is? Down there I am supposed to be as Pharaoh is, and although he is referred to as God, he is really a poor thing and rather endearing. If you ask for my forgiveness, it looks as though you have not really understood the whole of the story in which we are involved. I can hardly blame you. One can certainly be in a story without understanding it. Perhaps this is how it was meant to be, with me being open to punishment for being only too well aware of the game being played. Did you not hear from father himself, when he gave me his blessing, that with me it was only a game, and a way of showing approval? And when he summoned you, was he thinking of the trouble there has been between us? No, he kept quiet about it, for he was also in the game, God's game. It was under His protection, in a state of blatant immaturity that I incited you to evil, and God has of course turned it to good, for I have now provided for many and so matured somewhat. If there is to be any forgiveness between us, it is for me to ask it of you, for in order that everything should turn out as it has, you have had to play the evil ones".

The other reference to these words from Joseph's address has been used to throw light on Linnaeus' ethical doctrine, on the transition from simply "living irreproachably" to doing so because one is in fact aware that "God is near" (IV.iii). The reference here can be seen as rounding off his theological doctrine, the transition from a mere acknowledgement of the "mighty Ruler" (IV.iv.1.1) to thankfulness for His goodness (IV.iv.4.7), – a transition or progression in which awareness of the inscrutability of fate or the Father (IV.iv.1.3), and of the apparent paradox or contradictoriness of fortune or the Son (IV.iv.2.1), plays its part in coming to terms with both fate and fortune by deriving precepts from the past (IV.iv.3.2) and testing them within the present (IV.iv.3.3).

In traditional Christian language, therefore, this final section of the *Nemesis Divina* may be said to be concerned with Providence or the working of the Holy Spirit (*Introduction* III.iii.3).

It is apparent from Linnaeus' fourteen-point programme for the refutation of the freethinkers, that the doctrine of the Trinity was a matter of central concern to him, for over half the themes he urges the theologians to take up have a direct bearing upon it (IV.iii.1.3). Apart from the *Nemesis Divina* papers, however, we have little direct evidence of how he himself would have tackled the problem. When he notes, for example, that although the freethinkers believe in God: "they do not believe in the other two persons of the Godhead, for they say that nature provides no evidence at all of these two further persons", it is to be presumed that his refutation would have involved emphasizing the limited significance of the "economy of nature", and the fact that "contemplating it is but a foretaste of the bliss of Heaven" (*Introduction* I.iii.3; II.i.1). In order to confirm this, however, reference has to be made to sources which are only indirectly concerned with the point at issue. When he faults the freethinkers for maintaining "that a divine creator of this magnitude, not to be contained by the whole world, could never have been born of Mary", little direct light is thrown on his own view by the *Nemesis Divina* papers. In order to grasp it, one has to take into consideration the numinous sensualism which forms the focal point of his famous apotheosis of fate, nature and Providence in the *System of Nature*: "He is wholly and completely *Sense*, wholly and completely *Sight*, wholly and completely *Hearing*; and although both *Soul* and sense, He is also solely *Himself*" (*Introduction* I.ii.2).

Given the way in which he defines general categories throughout the whole of the *Nemesis Divina*, it is to be presumed that when he faults the freethinkers for maintaining that "the Holy Ghost was accepted at a certain council, but that a single vote would have decided the matter the other way", he would have begun his refutation by adducing the wealth of Scriptural evidence: St. Paul's having used the word "Lord" of the Spirit (2 *Corinthians* 3: 17), and having echoed St. John (6: 63) in speaking of Him as "the Spirit of life" and the "life-giver" (*Romans* 8: 2; 2 *Corinthians* 3: 6), our Lord's having said that the Spirit of truth that will bear witness to Him "proceeds from the Father" (*John* 15: 26), St. Peter's having stated that "no prophecy ever came by the will of man, but men spake from God, being moved by the Holy Spirit" (2 *Peter* 1: 21). Given the concern with historical accuracy so in evidence in the drawing up of the case-histories, one can well imagine him making the point that even the very earliest version of the Old Roman Creed, the prototype of that of the Apostles, includes a statement concerning belief "in the Holy Spirit"; that since the first of the oecumenical councils, that convened at Nicaea in 325, was preoccupied with Arianism, it is hardly surprising that in the creed it endorsed, the doctrine should elicit no more than the bare statement that: "We believe in the Holy Spirit". Kelly, J.N.D., 1972, 102, 369, 216.

It was the creed drawn up by the one hundred and fifty bishops constituting the 381 council of Constantinople, as a basis for discussion with the thirty-six Macedonian or Pneumatomachian bishops who were denying the Godhead of the Holy Spirit, which first came close to stating the fully orthodox doctrine: "We believe ... in the Holy Spirit, the Lord and life-giver, Who proceeds from the Father, Who with the Father and the Son is together worshipped and together glorified, Who spoke through the prophets". Once proper consideration had been paid to the Lord's statement that: "The Spirit of truth ... will glorify me, for he will take what is mine and make it known to you" (*John* 16: 13–14), and the doctrine of the double procession of the Holy Spirit had been added to what had been drawn up at Constantinople, and accepted at the councils of Toledo (589) and Hatfield (680), the logical coherence of Trinitarianism was complete. The one and indivisible Godhead was seen as presenting itself, severally and simultaneously, as Father, Son and Holy Spirit, three persons which though identical in substance and attributes, differ in the relation they bear to each other on account of their different modes of origin within their own divine substance – the Father-creator deriving His being from none, the Son deriving His being from the Father by generation, the Spirit deriving His being from the Father and the Son by procession. Swete, H.B., 1876.

The Lutheran reformation gave rise to a renewed commitment to this doctrine, evidence of which is to be found in the *Augsburg Confession* (1530) and *The Book of Concord* (1580). Linnaeus was probably first confronted with it when he was studying systematic theology under Jacob Flachsenius (1683–1733) at Växjö school, where the textbook used was based on the work of Matthias Hafenreffer (1561–1619), professor of theology at Tübingen and correspondent of Kepler: Ritter, A.M., 1981; Hafenreffer, M., 1603, bk. 1, loc. i, §4, 48–77; *Introduction* III.iii.3.

In this re-structured *Nemesis Divina*, fate, fortune and Providence, the three aspects of the final section, are presented as bearing out the Trinitarian doctrine Linnaeus was urging the theologians to concern themselves with – fate remaining inscrutable and deriving its significance from no created thing, fortune arising from the evident contradictions of experience, Providence proceeding from the awareness that the inscrutability and the apparent contradictoriness are in fact interdependent.

IV.iv.4.7:

Uppsala ms. 203. Swedish.
Clearly envisaged, by Linnaeus himself, as the final conclusion.

Appendix A

The Manuscripts

1. The London manuscript is in the keeping of The Linnean Society of London, Burlington House, Piccadilly, London, W1V 0LQ: signature Linn. Pat. Mss.

It consists of 30 folio pages, each approximately 32.5 by 20.25 cm, plus one folded folio-style sheet with slightly jagged edges and a small bottom corner cut off, 25 cm at its widest by 19.25 cm, that is, four further quarto pages.

It seems to have been written out and re-worked between the publication of the tenth edition of the *System of Nature* in 1758/1759 and the drawing up of the Uppsala manuscript during the autumn of 1765, since it contains a reference to an event which took place at the end of July 1765 (II.iii.2.4).

For the purposes of preparing this edition, these pages have been numbered as follows:

1. two Latin and six Swedish sentences, plus the Königsmarck, Freidenfelt (II.iv.2.4) and Ribbing (III.ii.1.2) cases;
2. the Jönköping (III.iii.5.3) and Byng (IV.i.4.2) cases;
3. the Leijel (III.ii.3.4) and Stuart (IV.i.5.12) cases;
4. blank;
5. blank;
6. the Sånnaböke case (III.ii.3.3);
7. a list of sixty-two cases, apparently divided into seven sections;
8. the Justinian case (IV.i.5.6);
9. some twenty Latin and Swedish quotations (IV.iv);
10. a reference to IV.iv.2.2 and a Latin sentence;
11. six numbered cases: Fahlander (IV.ii.2.12), Kihlmark (IV.ii.2.11), Netherwood (II.i.2.3), Jöransson (III.iv.3.2), Cicero (III.iii.4.2), Gylling (IV.i.2.9), plus Griffenfeldt (IV.i.3.3), and a few Latin and Swedish phrases;
12. seven numbered cases: Ziegler (IV.i.5.11), Horn (IV.i.3.7), the gardening man (IV.i.2.6), Brahe (III.i.1.4), Voigtländer (II.i.1.7), Urlander (III.i.3.11), Hallman (III.iii.4.3), plus Cederhielm (III.iv.3.15) and king Gustavus (IV.i.5.8), and a couple of deleted lines of Latin;
13. five numbered cases: Engberg (III.iii.4.4), Rogberg (IV.i.2.6), Benzelia (IV.ii.2.14), Cronhjort (see note III.ii.1.3), Görtz (IV.i.3.4), plus Skeckta (IV.i.2.3);
14. four numbered cases: Benzelia (IV.ii.2.14), Rosén, Benzelia (IV.ii.2.14), Mathesius (IV.ii.2.4);
15. the heading Nemesis, five lines of Latin and Swedish, and fifteen numbered cases: Boëthius (IV.ii.2.10), Ziegler (IV.i.5.11), Wrangel (III.ii.4.4), Münnich (IV.i.3.5), Malung (IV.ii.2.8), Kvikkjokk (IV.ii.2.8),

Westrin (III.iii.3.3), Ostermann (II.ii.4), Rabbinical story (IV.iv.2.2): not numbered, Kyronius (III.iii.2.3), Springer (IV.i.2.4), Benzelia (IV.ii.2.14), Celsius, Latin, Stobé (IV.i.1.4), Uggla (III.iii.6.4).

16. the heading Fate, and eight numbered cases: Bjelke, Boëthius (IV.ii.2.10), Krabbe (III.ii.4.2), Melander (IV.ii.2.3), Ullén (IV.ii.2.5), Canutius (III.i.3.8), Brahe (III.i.1.4), Damiens (IV.ii.1.3);
17. a reference in Latin to II.iv.3.4, and three numbered cases: Elizabeth (II.iv.2.7), Lucullus (II.iv.2.6) not numbered, Cronstedt (II.iv.3.5) not numbered, Caesar (II.iv.3.4) not numbered, Schmidt, Florinus (IV.ii.2.7);
18. thirty lines of deleted Latin on the physiological effects of psycho-somatic states, followed by the Clerck case of July 22/23 1765 (II.iii.2.4);
19. the heading Sin, followed by a whole page of Latin, specifying the concept under fifteen separate headings;
20. a list of ninety-six cases, apparently divided into seven sections, three sections of Latin and Swedish sentences, three references in Latin to Pliny, Aristotle and Artedi, and four numbered cases: Sohlberg (III.iv.2.3), Benzelia (IV.ii.2.14), Celsius, Parliament (IV.i.2.1), Uggla (III.iii.6.4) not numbered;
21. the heading Sin, twenty-three lines of Latin under the sub-heading Adultery (III.i.3.1), and four numbered cases: Blackwell (III.iv.3.18), Psilanderhielm (III.i.3.12), Lagerbladh (III.i.2.1), Hökerstedt (III.i.3.5);
22. the heading Folly, with fifteen lines in Latin and Swedish, involving reference to Pliny, five lines in Latin on Fortune (IV.iv.2.1);
23. the heading Wantonness, followed by a page of Latin, specifying the concept under seven separate headings;
24. twenty-seven lines of Latin on Onanism (III.i.1.3), thirty lines of Latin on Adultery (III.i.3.1);
25. six unnumbered cases: Klingspor (IV.i.4.4), Tavaste-bo (III.iii.1.10), Blackwell (III.iv.3.18), Dörnberg (III.i.3.3), Wallerius (III.ii.2.4), Buskagrius (IV.iii.2.7);
26. three unnumbered cases: Herkepaeus (IV.i.2.2), Klingspor (IV.i.4.11), German (II.iii.3.6);
27. a reference to the 1763 translation of F.C. Friess's work on Talion (IV.iv.5.1), followed by eight unnumbered cases; Scanian farmhand (III.iii.1.5), Grubbe (III.iii.5.4), Christian II (III.iii.5.6), Slagheck (IV.i.3.1), Persson (IV.i.3.2), Gallus (IV.iv.3.4), Danish sheriff (III.iv.5.9), Onanists (III.i.1.3), three references to material not clearly incorporated into the Uppsala text (IV.iv.3.1), and five deleted biblical quotations;
28. the Appelbom case (III.ii.4.3);
29. the numbering of the cases indicates that the fifteen (nos. 55–67) preceding those on this page are missing, which probably means that two pages of the manuscript have been lost; five numbered cases: farmer in Wallie

(III.ii.4.5), not numbered, Hauswolff (III.iii.2.4), Cronhielm (III.iii.1.11), Traveller (IV.i.1.3), Jönköping (III.iii.5.3), Nordenflycht (III.iv.3.13);

30. a list of the numbered cases, which indicates that the fifteen missing are, Mathesius, Voigtländer, Benzelius, Buddenbrock, Mathesius, Tiliander, Wallrave, Artedi, Nietzel, Kyronius, Lewenhaupt, Buddenbrock, Brahe, Stålsvärd, Blackwell; three numbered cases: Brålanda (III.i.3.14), Uppland (III.ii.2.5), Charles XII, not numbered, Mineworker (III.iii.1.4);
31. eleven Biblical quotations, followed by "live irreproachably" (IV.iii.2.4);
32. fourteen Biblical quotations (IV.iv.1.4);
33. twelve sentences, some Biblical (IV.iv.3.2, IV.iv.3.3);
34. the six lines, "O mighty Ruler", quoted as in I.v, not as in IV.iv.1.1, and five words of Latin.

2. The Stockholm manuscript is in the keeping of the Karolinska Institutets Bibliotek och Informationscentral, Solnavägen 1, Box 60201, 104 01 Stockholm: signature 44: 6: 4.

It consists of two folio pages, one containing some thirty-five quotations, about half of which are also to be found in the Uppsala ms. 26, 26b under the general heading of *Nemesis Divina* (IV.iv.3.2), the second containing fourteen lines in Latin on another subject, and five cases : Lindberg (III.ii.1.4), Yxkull (II.i.3.5), Brahe (III.i.1.4), Wallrave (III.iii.1.9) and Ekeblad (IV.i.3.6).

Since the Yxkull case does not occur in the London manuscript and includes a reference to an event which took place in January 1762, it looks as though this manuscript has to be dated 1762/1765. In lay-out and style it is very similar to the last four pages of the London manuscript.

3. The Uppsala manuscript is in the keeping of the Uppsala Universitetsbibliotek, Handsskrifts- och musikavdelningen, Box 510, Dag Hammarskjölds väg 1, 751 20 Uppsala: signature Cod. Ups. X 232.

When purchased by the University in 1845 it consisted of 203 octavo sheets, which had been numbered by their previous owner Dr. O.C. Ekman of Kalmar, who had acquired them in 1844. Sometime between 1845 and 1848 sheet no. 112 containing the Renhorn case went missing (IV.i.2.7). The whole manuscript is in Linnaeus' own hand, the only addition being a short note added by his son at the end of the Nordenflycht case on 100b (III.iv.3.13). Fifty-seven of the sheets have writing on both sides.

The great bulk of the manuscript was written out during the autumn of 1765 (*Introduction* I.i.2), although Linnaeus continued to add to it at least until the autumn of 1772 (IV.i.2.8, IV.i.2.11).

Dr. Ekman evidently numbered the sheets in the order in which they were when they came into his possession, and to a certain extent this may well reflect something of Linnaeus' own arrangement. By and large, the first sixty-three contain material related to the attempt to define general categories. The rest contain the case-histories, which originally may well have been arranged alphabetically, for if the present sequence is divided into seven sections – 1. 64–93; 2. 94–96; 3. 97–121; 4. 122–150; 5. 151–159; 6. 160–182; 7. 183–202, and these sections are then re-arranged in the order 2, 5, 4, 6, 7, 3, 1, the result approximates broadly to a general alphabetical sequence.

APPENDIX B

THE CURRENCY

The financial significance of the sums of money mentioned throughout the *Nemesis Divina* is apparent from the following facts concerning the Swedish currency of the time:

(1) between 1719 and 1776 the country's monetary system was based on the silver-standard of the rix-dollar, which contained 26.6973 grammes of the metal, and was the equivalent in:

    1719 of 3 silver-dollars, 9 copper-dollars, or 56 copper-marks;
    1745 of 3.44 silver-dollars, 10.33 copper-dollars, or 41.33 copper-marks;
    1755 of 3.41 silver-dollars, 10.25 copper-dollars, or 41 copper-marks;
    1765 of 7.25 silver-dollars, 21.75 copper-dollars, or 87 copper-marks;
    1768 of 3.5 silver-dollars, 10.5 copper-dollars, or 42 copper-marks;
    1775 of 5.79 silver-dollars, 17.31 copper-dollars, or 69.25 copper-marks;

(2) after 1654 the country also minted a gold-ducat, which contained 3.3966 grammes of the metal, but the value of it tended to fluctuate against that of silver; in 1741 it was worth two rix-dollars; one "tun" of gold was the equivalent of about 100,000 silver dollars (III.ii.3.4; III.iv.3.2; III.iv.3.3; III.iv.3.7; IV.iii.2.6; IV.iv.2.3).
(3) after 1745 there was a great deal of banknote inflation.

The actual significance of the amounts mentioned is apparent from the following facts concerning expenses, salaries and property-prices:

(1) the total cost of Linnaeus' one hundred and fifty one day journey to Lapland, May 12th–October 10th 1732, was 255 silver dollars, 211 of which were paid by the Royal Academy of Science;
(2) between 1745 and 1775, Linnaeus' average annual salary as a professor at Uppsala was 1,780 silver dollars; Dietrich Nietzel, keeper of the university's botanic garden from 1739 until 1756, began on an annual salary of 450 copper dollars plus a 50 dollar living allowance, and ended on a salary of 900 copper dollars; Lindstedt, the labourer appointed to assist him in 1744 was paid 60 copper dollars a year, the remuneration for this work being raised to 100 copper dollars a year in 1750;

(3) in the autumn of 1758 Linnaeus purchased the estate of Hammarby for 40,000 silver dollars, and the neighbouring hamlet of five homes and a croft for a further 40,000 silver dollars, 20,000 of which he raised on a mortgage: Fries, Th.M., 1903 I: 120, II: 103, 380.

APPENDIX C

THE NOTE ON MAN

Systema Naturae, 12th ed. 1766/1768 I: 28–32.

Man: "Know thyself". Solon's precept, once inscribed in golden letters in the temple of Diana, is the basis of wisdom:

Physiologically – as consisting of nerves interwoven with fibres, as a fragile machine, most perfect in youth, better endowed than other animals in respect of nearly all the faculties. "Man, intended for exercizing dominion over the whole animal creation, is cast naked on the naked ground, bewailing his lot, unable to use his hands or feet, incapable of acquiring any kind of knowledge without instruction, unable of his own accord to speak, walk or eat". *Pliny*. "The kind of life allotted us by nature is evident from its being ordained as an omen, that we should come into the world crying". *Seneca*. "It is humiliating to the pride of man, to consider the pitiable origin of this most arrogant of animals". *Pliny*.

"Dwell on what is your own!"

Dietetically – physical welfare and contentment having the possibility of yielding more happiness than the wealth of *Croesus*, the glory of *Solomon*, the pomp of *Alexander*. They are preserved by *moderation*, destroyed by excess, affected by variation, inhibited by what is unusual, furthered by what is familiar. The refinements of the kitchen create the glutton, pleasing by the pernicious means of fire and wine. "Hunger is satisfied with little, luxury demands abundance. Imagination yearns for vast quantities, and since nature is content with the little that is available, superfluity has to be sweated for". *Seneca*. Life will work out in the way that you live it.

"Take care of your health!"

Pathologically – the life of man resembling a bubble about to burst, the thread of a pendulum in the whirligig of time. "There is nothing on earth weaker than man". *Homer*. "There is no living thing which is more fragile, more subject to diseases, troubles and dangers. The whole of human life is but a span: half of it has to be spent in a state resembling death; then there is the age of infancy when there is no discretion, and the torment of living into old age, when the senses are blunted and the limbs become stiff, when there is a wasting away of the faculties of sight and hearing, of our ability to walk, of the teeth which enable us to nourish ourselves". *Pliny*. "Thus a considerable part of death has already occurred, all the past belonging to it. Nature will soon recall and bury all those you see before you, everyone you can conceive of as existing hereafter; for death summons all alike, whether the gods show favour or not". *Seneca*.

"Remember your mortality!"

Naturally – as the audacious miracle of nature, prince of animated beings, for whom nature has provided all: as one of the apes – weeping, laughing, singing, speaking, teachable, discerning, wondering and most wise; but weak and naked, unprovided with natural weapons, exposed to all the injuries of fortune, needful of assistance from others, of an anxious mind, solicitous of protection, continually complaining, changeable in temper, obstinate in hope, slow in the acquisition of wisdom. By condemning what is dead and gone, neglecting what is present, setting his mind on an uncertain future, man abuses the fleeting moments of life, which are most precious and irrevocable. In all ages, it is thus that the best of present time flies on for miserable mortals: poverty summoning some to the labour of their daily work, wealth incarcerating others in luxury, suffocating them with superfluities; ambition soliciting into its ever restless paths, acquisitions creating fear, the anxiety of possessing what has been striven for; some are condemned to solitude, others to having their doors continually crowded with visitors. One bewails having children, another mourns their loss; tears will sooner fail us than the sorrows causing them, which only oblivion can remove. "On every hand the ills we face outweigh the blessings: perils encompass us, we rush into the unknown, we are enraged without cause, like beasts we destroy without hating, we shift with the following winds, which simply waft us to destruction: the earth yawns wide, awaiting our death". *Seneca*. "Other animals come together in the face of what is alien, man suffers most from his own kind!" *Pliny*.

"Live irreproachably, God is near!"

Politically – for instead of holding to what is right, man is subject to public error, by which he is benighted from his very birth; it is this which determines how he is fed, fostered, educated and directed, which evaluates his honesty, fortitude, wisdom, morality and piety. Man is therefore ruled by opinion, and lives in accordance with custom rather than reason. Though set amid what is perishable, all being born to suffer, he can finally achieve strength of mind;

but he becomes infatuated with the smiles of Fortune, and while neglecting her real benefits, greedily looks for future enjoyment to her gaudy trifles. Driven to madness by envious snarlers, he hates the truly pious and persecutes them for holding speculative opinions different from his own, exciting numberless broils, not in order to do good, unaware of any real purpose. He fritters away his precious and irrecoverable time, thinking little on immortal and eternal matters, regulating the affairs of his children and his children's children, constantly entertaining fresh projects; unmindful of his real condition, he builds palaces instead of preparing his grave, until at length death seizes him in the midst of his undertakings, and his eyes being opened for the first time, he sees that all man's doings are vanity. "Living as if we were immortal, it is as mortals that we die". *Seneca.*

"Be a man of old-fashioned virtue and trust!"

Morally – since man combines willing what is right when prompted by an animated medullary substance, and pursuing pleasure when prompted by the body's being liable to impressions. In his *benighted* state he is unteachable, wanton, sectarian, ambitious, prodigal, querulous, devious, morose, ill-willed, malicious, greedy; when *transformed* he becomes attentive, chaste, urbane, unassuming, frugal, calm, frank, gentle, kindly, gracious, contented. "Common to all living beings are grief, extravagance, ambition, covetousness, as well as the desire to live and anxiety about the future". *Pliny.*

"Do what is good and be glad!"

Theologically – man being the final purpose of the creation; placed on the globe as the masterpiece of the works of Omnipotence, contemplating the world by virtue of sapient reason, forming conclusions by means of his senses, it is in His works that man recognizes the almighty Creator, the all-knowing, immeasurable and eternal *God*, learning to live morally under His rule, convinced of the complete justice of His Nemesis. Theology explains the rest of revelation. "Knowledge of God is to be obtained through the Creation and through Scripture". *Augustine.* "The cognition of *God* is by means of nature, the recognition by means of doctrine. Man alone experiences *God* through both the nature and the revelation of which He is the author". *Tertullian.* "Learn what God requires of you, what part you have been allotted in the human commonwealth." *Persius.*

"Remember thy Creator!"

Cicero *Tusculan Disputations* I.xxii.52; Pliny *Natural History* VII.i.2–4; Seneca *To Polybius: On Consolation* IV.iii.3; Pliny *Natural History* VII.vii.2; Persius *Satires* IV.52; Seneca *Epistles* XVI.viii.1, IV.xi.1; Walther, H.,

1963/1986, no. 32641; Homer *Odyssey* XVIII: 130; Pliny *Natural History* VII.i.5, L.167–168; Seneca *Epistles* I.ii.4; *Physical Enquiries* II.59.6; V.18.8–9; Martial *Epigrams* II.59.3; Pliny *Natural History* VII.v.8; Ovid *The Art of Love* I: 640; Seneca *On the brevity of life* III.iv.1; Pliny *Natural History* VII.i.5; *Ecclesiastes* 3: 12; Augustine *Soliloquies* I.3; Tertullian *Against Marcion* I.18, I.10; Persius *Satires* III.71–72; *Ecclesiastes* 12: 1.

APPENDIX D

OF GOD

Systema Naturae 12th ed. 1766/1768 I: 10–11.

Roused as I was, I saw from the back, as He went forth, the *everlasting, all-knowing, almighty God*, and I reeled! I tracked his footsteps throughout the field of nature, and I found in each of them, even in those I could scarcely make out, an infinite wisdom and power, an unfathomable perfection! I saw there how *animals* were sustained by plants, *plants* by the soil, the *soil* by the Earth; how night and day the *Earth* revolved about the Sun, which gave it life; how the *Sun* and the *Planets*, together with the fixed *Stars*, turned as it were upon their axis, in inconceivable numbers within infinite space; and that all was sustained within this empty nothingness by the incomprehensible *First Mover*, the *Being of all Beings*, the *Cause and Steersman of all Causes*, the Lord and Master of this World. To say that He is *Fate* is not to be mistaken, for everything hangs upon His finger; nor is it wrong to say that He is *Nature*, for all things are born of Him; and it is also right if we say that He is *Providence*, for everything happens in accordance with His will. He is wholly and completely *Sense*, wholly and completely *Sight*, wholly and completely *Hearing*; and although both *Soul* and sense, He is also solely *Himself*. No human conjecture can discover His *Form*; suffice it that He is a *Divine Being*, eternal and unchanging, neither created nor begotten, an *Essence* outside of which nothing made has being, which although It has founded and built all that shimmers here before our eyes, can Itself be seen only in thought; for so sublime a Majesty invests so holy a throne, that only the soul can have access to It.

BIBLIOGRAPHY

Addison, Joseph (1965). *The Spectator*. Edited by D.F. Bond. 5 Volumes. Oxford.
Afzelius, Adam ed. (1823). *Egenhändiga Anteckningar af Carl Linnæus. Med anmärkningar och tillägg*. Upsala.
Akrigg, George Philip Vernon (1984). *The Letters of King James VI and I*. Berkeley and London.
Album Studiosorum Academiæ Franekerensis (1968). Edited by S.J. Fockema Andreae and Th.J. Meijer. Franeker.
Alexander, Henry Gavin ed. (1956). *The Leibniz–Clarke Correspondence*. Manchester.
Alexander, John T. (1989). *Catherine the Great. Life and Legend*. Oxford.
Allgemeine Deutsche Biographie: cited as ADB. Edited by R. von Liliencron and others. 56 Volumes. Leipzig, 1875–1912.
Almquist, Jan Eric (1942). 'Karl IX och den Mosaiska Rätten'. *Lychnos*, 1–32.
Andersen, Hans Christian (1870). 'Hønse-Grethes Familie'. *Tre nye Eventyr og Historier*. Kjøbenhavn.
Anderson, Lorin (1976). 'Charles Bonnet's taxonomy and chain of being'. *Journal of the History of Ideas* XXXVII, no. 1, 45–58.
Annerstedt, Claes (1877–1931). *Upsala Universitets Historia*. 11 Volumes. Upsala and Stockholm.
Anrep, Johan Gabriel (1858–1864). *Svenska Adelns Ättar-Taflor*. 4 Volumes. Stockholm: cited as Anrep.
Anselm of Canterbury (1965). *Proslogion*. Translated by M.J. Charlesworth. Oxford.
Arcadius, Carl Ohlson (1888–1889, 1921–1922). *Anteckningar ur Wexjö Allmänna Läroverks Häfder*. 2 Parts. Växjö.
Arndt, Johann (1605; 1606–1610). *Vom wahren Christenthumb*. 4 Parts. Braunschweig and Magdeburg.
Arndt, Johann (1647–1648). *Fyra böcker om een sann christendom*. Translated by Stephanus Laurentii Muraeus (d. 1675). Stockholm.
Arndt, Johann (1695). *Fyra böcker om een sann christendom ... Nu på nytt återigen upplagde, med skiöna maginalia* (sic). 4 Parts. Stockholm.
Askmark, Ragnar (1943). *Svensk prästutbildning fram till år 1700*. Stockholm.
Aspelin, Henrik Emanuel (1892–1897). *Wasa Stads Historia*. 2 Volumes. Nikolaistad.
Auditorium Academiæ Franekerensis (1995). Edited by F. Postma and J. van Sluis. Leeuwarden.

Bacon, Francis (1605). *The Advancement of Learning*. London. Spedding ed. III: 253–491.
Bacon, Francis (1609). *De sapientia veterum*. Londini. Spedding ed. VI: 605–686.
Bacon, Francis (1623a). *De Dignitate et Augmentis Scientarum*. Londini. Spedding ed. I: 413–837.
Bacon, Francis (1623b). *Historia Vitæ et Mortis*. Londini. Spedding ed. II: 89–226.

Bacon, Francis (1625[8]). *The Essayes or Counsels ciuill and morall*. London. Spedding VI: 365–604.
Bacon, Francis (1857–1874). *The Works*. Edited by J. Spedding, R.L. Ellis, D.D. Heath. 14 Volumes. London.
Bäck, Abraham (1779). *Åminnelsetal öfver Hr. Arch. och Riddare Carl von Linné*. Stockholm.
Baker, Charles Henry Collins and Baker, Muriel Isabella (1949). *The Life and Circumstances of James Brydges First Duke of Chandos Patron of the Liberal Arts*. Oxford.
Barber, William Henry (1955). *Leibniz in France*. Oxford.
Bates, Brian (1983). *The Way of Wyrd: Tales of an Anglo-Saxon Sorcerer*. London.
Bayle, Pierre (1695–1697). *Dictionaire historique et critique*. 2 Parts. Rotterdam; Amsterdam, 1740[5].
Beard, Thomas (1597). *The Theatre of Gods Judgements: wherein is represented the admirable Justice of God against all notorious sinners, great and small, specially against the most eminent persons in the world. whose exorbitant power had broke through the barres of Divine and Humane Law*. London. Edited by Thomas Taylor, 1648[4].
Bech, Svend Cedergreen (1977). *Danmarks Historie. Oplysning og Tolerance*. København.
Beckman, Bjarne (1930). *Dalupproret 1743 och andra samtida rörelser inom allmogen och bondeståndet*. Göteborg.
Beckmann, Johann (1911). *Schwedische Reise in den Jahren 1765–1766*. Edited by Th.M. Fries. Upsala.
Behn, Aphra (1688). *Oroonoko: or, the Royal Slave. A True History*. London.
Bellman, Carl Michael (1791). *Fredmans Sånger: Skrifter II*. (1922–1957). 10 Volumes. Stockholm.
Bergendorff, Conrad John Immanuel (1928). *Olaus Petri and the Ecclesiastical Transformation in Sweden*. New York.
Bergius, Peter Jonas (1758). *Inträdes-tal, om Stockholm för 200 år sedan, och Stockholm nu för tiden, i anseende til handel och vetenskaper, särdeles den mediciniska*. Stockholm.
Besterman, Theodore (1969). *Voltaire*. Oxford.
Biblia (1703). *Thet är All then Heliga Skrift På Swensko*; *Efter Konung Carls then Tolftes Befalning*. Published by Henrik Keyser. Stockholm
Biblia (1778). *Thet är: All Then Heliga Skrift, Gamla och Nya Testamentsens, På Swensko: Med förriga accuratesta Editioner jämförd*. Published by J.G. Lange. Stockholm.
Bibliothèque raissonnée des ouvrages des savans de l'Europe (1747). Amsterdam.
Bicknell, Edward John (1957[3]). *The Thirty-Nine Articles*. Edited by H.J. Carpenter. London.
Biografiskt lexicon öfver namnkunnige svenske män: cited as BL. 23 Volumes 1835–1857; new edition 8 Volumes 1874–1876; new series 10 Volumes 1857–1907. Örebro and Stockholm.
Blicher, Steen Steensen (1824). *Brudstykker af en Landsbydegns Dagbog*. Edited by Emil Gigas. København (1905).

Blomqvist, Gunnar (1941). *Shacktavelslek och Sju vise mästare*. Stockholm.
Blunt, Wilfrid (1984). *The Compleat Naturalist. A Life of Linnaeus*. London, 1971[1].
Boerhaave, Herman (1709). *Oratio qua repurgatæ medicinæ facilis asseritur simplicitas*. Lugduni Batavorum.
Boerhaave, Herman (1715). *Sermo academicus de comparando certo in physicis*. Lugduni Batavorum.
Boerhaave, Herman (1718). *Sermo academicus de chemia suos errores expurgantæ*. Lugduni Batavorum.
Boerhaave, Herman (1983). *Orations*. Edited and translated by E. Kegel-Brinkgreve and A.M. Luyendijk-Elshout. Leiden.
Boerman, Albert Johan (1953). *Carolus Linnaeus als middelaar tussen Nederland en Zweden*. Dissertation. Utrecht.
Boerman, Albert Johan (1978). 'Linnaeus and the scientific relations between Holland and Sweden'. *Svenska Linné-Sällskapets Åsskrift*, 43–56.
Bogan, Zachary (1653). *A View of the Threats and Punishments Accorded in the Scriptures, Alphabetically composed. With some briefe Observations upon severall Texts*. Oxford.
Bondpä, Kajsa (1998). 'Dråp och horsbrott i Malungs prästgård'. *Skinnarebygd. Malungs Hembygdsförenings Årsbok*: 63–98.
Bots, Johannes Alphonsus Henricus (1972). *Tussen Descartes en Darwin. Geloof en Natuurwetenschap in de 18e Eeuw*. Assen.
Boyle, Robert (1688). *A Disquisition about the Final Causes of Natural Things*. London. Dutch translation by W. Sewel, Amsterdam, 1688.
Braw, Christian (1993). *Tro på Wasas tid*. Borås.
Bremekamp, Cornelis Elisa Bertus (1953). 'Linné's views on the hierarchy of taxonomic groups'. *Acta Botanica Neerlandica*, vol. 2 (2): 242–253.
Brenner, Elias (1690–1691). *Thesaurus nummorum Sveo-Gothicorum*. 2 Volumes. Stockholm.
Brilioth, Yngve (1943). 'Kyrka, gymnasium, läroverk', in *Minneskrift vid Växjö gymnasiums 300-års jubileum*. Växjö.
Broberg, Gunnar (1972–1974). 'Den unge Linné speglad i några hittills obeaktade dokument'. *Svenska Linné-Sällskapets Årsskrift*, 7–20.
Broberg, Gunnar (1975). *Homo Sapiens L. Studier i Carl von Linnés naturuppfattning och människolära*. Motala.
Broberg, Gunnar (1978). 'Linnaeus and Genesis: a Preliminary Survey'. *Svenska Linné-Sällskapets Årsskrift*, 30–42.
Brockhaus, Hermann (1860). 'Gellert und Jami'. *Zeitschrift der Deutschen morgenländischen Gesellschaft* XIV: 706–710.
Brolin, Per-Erik (1953). *Hattar och Mössor i Borgarståndet* 1760–1766. Uppsala.
Broocman, Carl Fredric (1760). *Beskrifning öfwer the i Oster=Götland befintelige Städer, Slott, Sokne=Kyrkor, Soknar, Säterier, Öfwer=Officers=Boställen, Jernbruk och Prestegårdar, med mera*. Norrköping.
Brooks, Charles B. (1994). *Disaster at Lisbon: the great earthquake of 1755*. Long Beach.
Brown, Beatrice Curtis (1935). *The Letters and Diplomatic Instructions of Queen Anne*. London.

Bruun, Georg (1939). *Den danske Skole og De Collinske Skoler i Kolding*. Kolding.

Bygdén, Anders Leonard (1923–1928). *Hernösands stifts herdaminne; bidrag till kännedomen om prästerskap och kyrkliga förhållanden till tiden omkring Luleå stifts utbrytning*. 4 Parts. Stockholm and Uppsala.

Cain, A.J. (1958). 'Logic and Memory in Linnaeus's system of taxonomy'. *Proceedings of the Linnean Society of London*, 144–163.

Cain, A.J. (1992). 'Was Linnaeus a Rosicrucian?'. *The Linnean*, vol. 8, no. 3 (August), 23–44.

Carlquist, Gunnar (1918). 'Karl XII's ungdom och första regeringsår', in *Karl XII. Till 200-årsdagen av hans död*. Edited by S.E. Bring. Stockholm.

Carpelan, Tor (1954–1958). *Ättartavlor för de på Finlands Riddarhus inskrivna ätterna*. 2 Parts. Helsingfors.

Cederbom, Lars August (1904). *Jacob Serenius i opposition mot Hattpartiet 1738–1766*. Skara.

Charles I (1649). *King Charles His Speech Made upon the Scaffold at Whitehall-Gate, Immediately before his Execution, On Tuesday the 30 of Jan. 1648*. London.

Childrey, Joshua (1660). *Britannia Baconia: or, The Natural Rarities of England, Scotland and Wales, According as they are to be found in every Shire*. London.

Cicero (1970). *De Legibus*. Translated by C.W. Keyes. Loeb edition XVI, 287–519. Cambridge, Mass. and London.

Clarke, Samuel (1646). *A Mirrour or Looking-Glass both for Saints, and Sinners, Held forth in some thousands of Examples; Wherein is presented, as Gods wonderful Mercies to the One, so his Severe Judgments against the Other*. 2 Volumes. London, 1671^4.

Clarke, Samuel (1706). *A Discourse concerning the Unchangeable Obligations of Natural Religion*. London.

Collins, Anthony (1713). *A Discourse of Free-Thinking, Occasion'd by the Rise and Growth of a Sect call'd Free-Thinkers*. London.

Corbett, Thomas (1739). *An Account of the Expeditions of the British Fleet to Sicily, In the Years 1718, 1719 and 1720. Under the Command of Sir George Byng, Bart.* London.

Crocker, Lester Gilbert (1959). *An Age of Crisis; man and world in eighteenth-century French thought*. Baltimore.

Croker, John Wilson (1824). *Letters to and from Henrietta, Countess of Suffolk*. 2 Volumes. London.

Curman, Sigurd and Roosval, Johnny (1950). *Sveriges Kyrkor. Uppland I*. Stockholm.

Dal, Nils Hufwedsson (1726). *Lord Verulams Oprichtiga Utlåtelser*. Stockholm.

Dal, Nils Hufwedsson (1729). *Lord Verulams Påminnelser, wid Förefallande måhl*. Stockholm.

Dal, Nils Hufwedsson (1736). *Riddarens Francis Bacons Betraktelser*. Stockholm.

Dalgren, Lars and Nygren, C.E. (1934, 1939). *Karlstads stads historia*. 2 Volumes. Karlstad.

Dalin, Olof von (1747–1761). *Svea rikes historia ifran des begynnelse til wåra tider*. 3 Parts. Stockholm.

Danielson, Hilding (1956). *Sverige och Frankrike 1736–1739*. Lund.

Dansk Biografisk Leksikon: cited as DBL. Edited by Svend Cedergreen Bech. 16 Volumes. København, 1979–1984.

Dee, John (1570). *The Mathematicall Praeface to The Elements of Geometrie of ... Evclide of Megara*. London. Edited by A.G. Debus, New York, 1975.

Dellaporta, Giovanni Battista (1558). *Magiæ natvralis, sive De miracvlis rervm natvralium*. Neapoli.

Dellner, Johan (1953). *Forsskåls filosofi*. Stockholm.

Derham, William (1714). *Astro-theology, or a Demonstration of the being and attributes of God, from a survey of the heavens*. London, 1750^{9}. Dutch translation, Leiden 1728, 1739^{2}; Swedish translation, Stockholm, 1735.

Descartes, René (1641). *Meditationes de Prima Philosophiæ*. Paris.

Descartes, René (1897–1910). *Oeuvres*. Edited by C. Adam and P. Tannery. 12 Parts. Paris.

Deutschbaltisches Biographisches Lexikon 1710–1960 (1970). Edited by W. Lenz. Köln and Wien.

Dewhurst, Kenneth (1958). 'Locke and Sydenham on the teaching of anatomy'. *Medical History*, 2, 1–12.

Dewhurst, Kenneth (1963). *John Locke, Physician and Philosopher*. London.

Dewhurst, Kenneth (1966). *Dr. Thomas Sydenham: His Life and Original Writings*. London.

Dijksterhuis, Eduard Jan (1970). *Simon Stevin. Science in the Netherlands around 1600*. The Hague.

Dictionnaire de Biographie Française: cited as DBF. Edited by R. D'Amat and others. 19 Volumes to date. Paris, 1933–1997.

Dixelius, Olof (1984–1985). 'Carl Gustav Tessin, Linné och naturen'. *Svenska Linné-Sälskapets Årsskrift* 89–107.

Duris, Pascal (1993). *Linné et la France, 1780–1850*. Genève.

Ehnmark, Erland (1941). 'Linnés Nemesis-Tankar och Svensk Folktro'. *Svenska Linné-Sällskapets Årsskrift*, XXIV, 29–63.

Ehnmark, Erland (1944). 'Dygden och Lyckan'. *Svenska Linné-Sällskapets Årsskrift*, XXVII, 81–105.

Ehnmark, Erland (1951–1952). 'Linnaeus and the problem of immortality'. *Kungliga Humanistiska Vetenskapssamfundets i Lund Årsberättelse*, IV, 63–93.

Ehrencron-Müller, Holger (1924–1939). *Forfatterlexikon omfattende Danmark, Norge og Island indtil 1814*. 12 Volumes. København.

Ekstedt, Olle (1988). *Sällsamheter i Småland*. Stockholm.

Elgenstierna, Gustav Magnus (1911–1942). *Svenska Släktkalendern*. 13 Parts. Stockholm.

Elgenstierna, Gustav Magnus (1925–1936). *Den introducerade Svenska Adelns Ättartavlor med tillägg och rättelser*. 9 Parts. Stockholm.

Elias, Johan Engelbert (1963). *De Vroedschap van Amsterdam 1578–1795*. 2 Parts. Amsterdam.

Elmgren, Sven Gabriel (1861). *Öfversigt af Finlands litteratur ifrån år 1542 till 1770*. Helsingfors.

Engel, Hendrik (1950–1951). 'Some Artedi documents in the Amsterdam archives'. *Svenska Linné-Sällskapets Årsskrift* XXXIII-XXXIV: 51–66.
Engelman, Jan (1747). *Het regt gebruik der natuurbeschouwingen, geschetst in eene Verhandeling over de Sneeuwfiguren*. Haarlem.
Englund, Peter (1988). *Poltava: berättelsen om en armés undergång*. Stockholm.
Englund, Peter (1992). *The Battle of Poltava*. Translated by P. Hale. London.
Ereshefsky, Marc (1994). 'Some Problems with the Linnaean Hierarchy'. *Philosophy of Science*, 61, 186–205.
Eriksson, Gunnar (1980–1981). 'Carl von Linné'. *Svenskt Biografiskt Lexikon*, 23, 700–715.
Eriksson, Gunnar (1984). 'Olof Rudbeck d.ä.'. *Lychnos*, 77–119.
Eriksson, Gunnar (1994). *The Atlantic Vision. Olaus Rudbeck and Baroque Science*. Canton, Mass.
Erler, Georg (1910–1917). *Die Matrikel der Universität Königsberg in Preussen*. 3 Volumes. Leipzig.

Fabricius, Knud Frederik Krog (1920). *Kongeloven, dens Tilblivelse og Plads i Samtidens natur- og arveretlige Udvikling*. København.
Fahlbeck, Erik (1915, 1916). 'Studier öfver frihetstidens politiska idéer'. *Statsvetenskaplig Tidskrift för politik, statistik, ekonomi* 18: 325–345; 19: 31–54, 104–121.
Fant, Johan Erik and Låstbom, August Teodor (1842–1845). *Upsala Ärkestifts Herdaminne*. 3 Parts. Uppsala.
Farmer, David Hugh (1984). *The Oxford Dictionary of Saints*. Oxford.
Filmer, Robert (1680). *Patriarcha and Other Political Works*. Edited by P. Laslett. Oxford (1949).
Flachsenius, Jacobus Henrici (1678). *Collegium Logicum, Exhibens I. Præcepta Logica ... II. Quæstiones controversas, inter celebriores Logicos, præcipue inter Peripateticos & Ramæos agitatas ... III. Axiomata seu Canones ... Quibus Præmissus est Tractatus utilissimus De Philosophia in Genere, cum Appendice De Variis Philosophorum sectis antiquis & recentioribus*. 3 Volumes. Aboæ.
Folter, Rolf J. de (1978). 'A newly discovered Oeconomia Animalis by Pieter Muis of Rotterdam'. *Janus*, LXV, 183–204.
Forsenius, Johannes (1694). *Purpur principis ex verbis Jobi capit. 29*. Dorpati.
Frängsmyr, Tore (1971–1972). 'Den gudomliga ekonomin. Religion och hushållning i 1700-talets Sverige'. *Lychnos*, 217–244.
Frängsmyr, Tore (1972). *Wolffianismens genombrott i Uppsala*. Uppsala.
Fredbärj, Telemak (1962). 'Johannes Moræus, Linnæi svärfader'. *Svenska Linné-Sällskapets Årsskrift*, XLV, 103–127.
Fredbärj, Telemak (1964). 'Ett Nyfunnet Manuskript till Fundamenta Botanica'. *Svenska Linné-Sällskapets Årsskrift*, XLVII, 5–15.
Fredbärj Telemak (1967). 'Morga i Nemesis divina'. *Svenska Linné-Sällskapets Årsskrift*, 50: 93–94.
Fredbärj, Telemak (1970–1971). 'Linné som djäkne och gymnasist'. *Svenska Linné-Sällskapet Årsskrift*, 13–35.
Fries, Elias Magnus (1848). *Carl von Linnés Anteckningar öfver Nemesis Divina. Inbjudningsskrift till morgondagens philosophiska Promotion från Upsala Uni-*

versitets Stiftelse den Sjuttiondesjunde af tillförordnad promotor Elias Fries. Upsala. Danish summary by P.F. Barfod (1811–1896), *Dansk Tidsskrift* (ed. J.F. Schouw) III (1849).

Fries, Elias Magnus (1852). *Botaniska utflygter.* 3 Parts. Stockholm, 1843, 1852, 1864.

Fries, Thore Magnus (1903). *Linné. Lefnadsteckning.* 2 Parts. Stockholm.

Friess, Frederik Christian (1758). *Theologisk og Historisk Afhandling om Den guddomelige Giengiældelses Ret, Eller Jus Talionis Divinum.* Kjøbenhavn.

Friess, Frederik Christian (1763). *Theologisk och Historisk Afhandling, om Jus Talionis Divinum, Eller Om Den Gudomeliga Wedergällnings=rätten.* Translated by Olof Rönigk. Stockholm.

Frondin, Elias (1740). *Dissertatio gradualis de B. Sigfrido, primo wexionensium episcopo.* Respondent: Andreas Hallenberg. Upsaliæ.

Fuller, Thomas (1845). *The Church History of Britain.* Edited by J.S. Brewer. 6 Volumes. Oxford.

Fyhn, Jens Jørgen (1848). *Efterretninger om Kjøbstaden Kolding.* København.

Garrison, James W. (1987). 'Newton on the Relation of Mathematics to Natural Philosophy'. *Journal of the History of Ideas*, 48, 609–627.

Gascoigne, George (1575). *Certayne notes of Instruction concerning the making of verse.* London.

Gaster, Moses (1924). *The Exempla of the Rabbis.* London and Leipzig.

Geffroy, Mathieu Auguste (1861). 'La Nemesis Divina, écrit inédit de Linné'. *Revue des Deux Mondes*, XXXI année, periode ii, tome 32, 178–195.

Geijer, Erik Gustav (1832–1836). *Svenska folkets historia.* 3 Volumes. Örebro.

Gellert, Christian Fürchtegott (1746–1748). *Fabeln und Erzählungen.* 2 Parts. Leipzig.

Geyer, Carl-Friedrich (1992). *Die Theodizee-Diskurs, Dokumentation, Transformation.* Stuttgart.

Gibbon, Edward (1776–1788). *The History of the Decline and Fall of the Roman Empire.* 6 Volumes. London.

Gibson, Reginald Walter (1950). *Francis Bacon: A Bibliography of his Works and of Baconiana to the year 1750.* Oxford.

Gillby, Johannes (1961). 'Nemesis Divina i Brålanda och Kimito'. *Svenska Linné-Sällskapets Årsskrift*, XLIV, 74–76.

Ginzberg, Louis (1946–1961). *The Legends of the Jews.* Translated by H. Szold. 7 Volumes. Philadelphia.

Gori, Giambattista (1972). *La fondazione dell'esperienza in 's Gravesande.* Firenze.

Goropius Becanus, Joannes (1569). *Origines Antwerpianæ, sive Cimmeriorum Becceselana novem libros complexa.* Antverpiæ.

Grant, Michael (1985). *The Roman Emperors.* London.

Gravesande, Willem Jacob 's (1720–1721). *Physices elementa mathematica, experimentis confirmata.* Lugduni Batavorum, 1748[4].

Green, Henry (1856). *Sir Isaac Newton's Views on Points of Trinitarian Doctrine.* London.

Gregg, Edward (1980). *Queen Anne.* London.

Gregg, Pauline (1981). *King Charles I*. London.
Gullander, Bertil (1971). *Linné på Gotland. Utdrag ur Carl Linnaeus' dagboksmanuskript från gotländske resan 1741*. Stockholm.
Günther, Louis (1889–1895). *Die Idee der Wiedervergeltung in der Geschichte und Philosophie des Strafrechts*. 3 Parts. Erlangen.

Hafenreffer, Matthias (1603). *Loci Theologici, Certa Methodo ac Ratione, in Tres Libros tributi*. Tubingæ. Francofvrti, 1615[5].
Hafenreffer, Matthias (1612). *Compendium doctrinæ cælestis ex locis theologicis*. Edited by P. Kenicius. Stockholmiæ. Holmiæ 1714[17]. Translated by J. Svedberg, Skara, 1714.
Hafenreffer, Matthias (1613). *Templum Ezechielis, sive in IX. postrema Prophetæ Capita Commentarius*. Tubingæ.
Haffenreffer, Matthias (1657). *Synopsis locorum theologicarum Matthiæ Haffenrefferi. Pro scholis inferioribus in inclyto regno Sveciæ*. Gothoburgi. Strengnesiæ, 1697[8].
Hagberg, Knut (1939). *Carl Linnæus*. Stockholm.
Hagberg, Knut (1940). *Carl Linnaeus. Ein grosses Leben aus dem Barock*. Hamburg.
Hagberg, Lars (1952). *Jacob Serenius' kyrkliga insats*. Stockholm.
Håkanson, Lennart (1982–1983). 'Nemesis Divina violata'. *Svenska Linné-Sällkapets Årsskrift*, 93–98.
Hammarsköld, Lorenzo (1821). *Historiska Anteckningar rörande Fortgangen och Utvecklingen Philosophiska Studium i Sverige, från de äldre till nyare tider*. Stockholm.
Hansson, Gösta (1991). *Västerås stifts herdaminne. Stifthistoriskt och stiftsbiografiskt uppslagsverk*. Edited by Gunnar Ekström. II: 2 1700-talet. Västerås.
Harris, Frances (1991). *A Passion for Government. The Life of Sarah, Duchess of Marlborough*. Oxford.
Hasted, Edward (1797–1801). *The History and Topographical Survey of the County of Kent*. 12 Volumes. Canterbury.
Head, Constance (1972). *Justinian II of Byzantium*. Madison. Milwaukee, London.
Hederich, Benjamin (1770). *Gründliches mythologisches Lexikon*. Edited by J.J. Schwaben. Leipzig.
Hedin, Sven (1808). *Minne af von Linné, fader och son*. 2 Parts. Stockholm.
Hedlund, Emil (1936). 'Johan Stensson Rothman. Levnadsteckning'. *Svenska Linné-Sällskapet Årsskrift*, XIX, 67–120.
Hellquist, Elof (1948[3]). *Svensk Etymologisk Ordbok*. Lund.
Hervey, James (1748). *Meditations among the Tombs and Reflections on a Flower Garden*. London, 1792[26].
Hesiod (1978). *Works & Days*. Edited and translated by M.L. West. Oxford.
Hildebrand, Karl-Gustav (1942). *Stormaktstidens fältpräster*. Stockholm.
Hillerdal, Gunnar (1992). *Schola Wexionensis. Med Rötter i Medeltiden. Växjö Gymnasium 350 år*. Växjö.
Hobbes, Thomas (1651). *Leviathan, or the Matter, forme & power of a commonwealth*. London.

Hofsten, Nils von (1935). 'Systema Naturae. Ett Tvåhundraårsminne'. *Svenska Linné-Sällskapets Årsskrift*, XVI11, 1–15.

Hofsten, Nils von (1958). 'Linnés Naturupfattning'. *Svenska Linné-Sällskapets Årsskrift*, XLI, 13–35.

Höjer, Torgny (1938). 'Christopher Springer och principilatsfrågan vid 1742–43 års riksdag'. *Studier och handlingar rörande Stockholms historia*. Edited by Nils Ahnlund. I: 49–105. Stockholm.

Holberg, Ludvig (1748–1754). *Epistler, Befattende Adskillige historiske, politiske, met physiske, moralske, philosophiske, Item Skiemtsomme Materie*. 5 Parts. Kiøbenhavn.

Holberg, Ludvig (1944–1954). *Epistler*. Udgivne med Kommentar af F.J. Billeskov Jansen. 8 Volumes. København.

Holmbäck, Åke Ernst Vilhelm (1928). 'Våra domarregler', in *Festschrift tillägnad Axel Hägerström*. Uppsala.

Holmes, Geoffrey (1987). *British Politics in the Age of Anne*. London.

Holst, Walfrid (1936). *Carl Gustav Tessin. En grand-seigneur från 18:e seklet*. Stockholm.

Holthausen, Carl Johan (1769). *Soldatens Skyldigheter, I Fred och I Fält*. Stockholm. Reprint Stockholm, 1973.

Hooker, Richard (1594–1662). *Of the Laws of Ecclesiastical Polity*. 8 Books. London.

Hörnström, Johannes (1943). *Anders Odel. En studie i frihetstidens litteratur- och kultur-historia*. Uppsala.

Hosum, Mogens Knudsen (1706). *Inauspicatus dies Jovis, auspice Jova illustratus*. Hafniæ.

Hult, Olof Torgny (1934). 'Om Linné och Den Osynliga Världen'. *Svenska Linné-Sällskapets Årsskrift*, XVII, 118–128.

Hulth, Johan Markus (1921). 'Uppsala Universitetsbiblioteks förvärv av Linnéanska original-manuskript'. *Uppsala Universitets Biblioteks Minneskrift 1621–1921*, 407–424. Uppsala.

Hultman, Frans Wilhelm (1870). 'Svenska aritmetikens historia: Georg Stjernhelm'. *Tidskrift för Matematik och Fysik*, 3, 49–95. Uppsala.

Hutchinson, Francis Ernest (1947). *Henry Vaughan: A Life and Interpretation*. Oxford, 1971[2].

Hyam, Roger and Pankhurst, Richard (1995). *Plants and their Names*. Oxford.

Hyde, Edward (1849). *The History of the Rebellion and Civil Wars in England*. Edited by William Warburton. 7 Volumes. Oxford.

Hyltén-Cavallius, Gunnar (1897). *Kongl. Kronobergs Regementes Officerskår 1623–1896. Biografiska anteckningar*. Stockholm.

Hyltén-Cavallius, Gunnar Olof (1864–1868). *Wärend och wirdarne, ett försök i Svensk Ethnologi*. 2 Parts. Stockholm. 1921–1922[2], 1972[3].

Jackson, Benjamin Daydon (1915). 'Dietrich Nietzel (1703–1756)'. *The Gardeners' Chronicle* no. 1,487, Saturday, June 26, 1915: 353–354.

Jacobsen, Jens Peter (1876). *Fru Marie Grubbe. Interieurer fra det syttende Aarhundrede*. Kjøbenhavn. English translation Hanna Astrup-Larsen. New York (1917).

Jägerskiöld, Stig Axel Fridolf (1941). *Sanning och sägen om Karl XII's död.* Stockholm.

Jenks, Benjamin (1700). *Submission to the Righteousness of God. Or the Necessity of Trusting to a better Righteousness than our own, Opened and Defended, in a Plain and Practical Discourse upon Rom. X.3.* London, 1775[6].

Jensen, F. Elle (1944). *Pietismen i Jylland.* København.

Jensen, F. Elle (1949). 'Provst Mathias Bakke i Kolding'. *Vejle Amts Aarbog*: 187–202. Fredericia.

Johan Magnus (1554). *Historia ... De Omnibus Gothorvm Sveonvmqve Regibus.* Romæ.

Johnson, Samuel (1749). *The Vanity of Human Wishes. The Tenth Satire of Juvenal imitated.* London.

Johnson, Walter Gilbert (1945). 'Skriften om Paradis and Milton'. *The Journal of English and Germanic Philology*, XLIV, 263–269.

Kallinen, Maija (1995). *Change and Stability. Natural Philosophy at the Academy of Turku 1640–1713.* Helsinki.

Kant, Immanuel (1781). *Critik der reinen Vernunft.* Riga. 1787[2]

Kant, Immanuel (1793). *Die Religion innerhalb der Grenzen der blossen Vernunft.* Königsberg, 1794[2].

Kelly, John Norman Davidson (1972). *Early Christian Creeds.* London.

Kepler, Johannes (1619). *Harmonices Mundi.* Lincii Austriae. Translated by M. Caspar, München and Berlin, 1939.

Kepler, Johannes (1938–1988). *Gesammelte Werke.* Edited by Max Caspar and others. 20 Volumes. München.

Kienast, Dietmar (1990). *Römische Kaisertabelle. Grundzüge einer römischen Kaiser-chronologie.* Darmstadt.

Kittel, Gerhard (1933–1979). *Theologisches Wörterbuch zum Neuen Testament.* 10 Volumes. Stuttgart.

Kjaer, Severin (1904). *Erik Grubbe og hans tre Døttre, Anne Marie Grubbe, Marie Grubbe, Anne Grubbe.* København.

Klingspor, Gustav A. (1932). *Sanning och sägen om Sankt Sigfrid av Husaby och Växjö, 'som christnade Swirges land'.* Göteborg.

Knös, Andreas Olofsson (1742). *De principiis et nexu religionis revelatæ et naturalis.* Upsaliæ. Swedish translation by Johan Dellner, *Redogörelse för Högre allmänna läroverket i Nyköping*, 1938–1939, 9–33; 1941, 5–26. Nyköping.

Koch, Klaus ed. (1972). *Um das Prinzip der Vergeltung in Religion und Recht des Alten Testaments.* Darmstadt.

Kock, Ludvig (1914). *Oplysningstiden i den danske Kirke 1770–1800.* København.

Korb, Johann Georg (1700). *Diarium itineris in Muscoviam.* Viennæ Austriæ. English translation by C. MacDonnel. 2 Volumes. London (1863). Reprint New York, 1968.

Krijgz Articlar (1621). *Som fordom then stormechtigste furste och herre, herr Gustaff Adolph, then andre och store, ... hafwer låtit göra och författa.* Stockholm, 1697[15].

Krol, J.L.P.M. (1982). 'Linnaeus' verblijf op de Hartekamp in Heemstede'. *Het landgoed de Hartekamp in Heemstede.* Heemstede.

Kuijlen, Jos (1983). *Paradisus Batavus. Bibliografie van plantencatalogi van onder-*

wijstuinen, particuliere tuinen en kwekerscollecties, in de Noordelijke en Zuidelijke Nederlanden (1550–1839). Wageningen.

Kulturhistorisk Leksikon for Nordisk Middelalder fra vikingetid til reformations tid (1980[2]): cited as KLNM. Edited by J. Danstrup and others. 22 Volumes. Viborg.

Lagerqvist, Lars O. (1995). 'Äldre Vasatid'. *Myntninge i Sverige 995–1995. Numismatiska Meddelanden* XL: 125–178. Edited by K. Jonsson, U. Nordlind, I. Wiséhn. Stockholm.

Lagerroth, Fredrik (1915). *Frihetstidens författning. En studie i den svenska konstitutionalismens historia.* Stockholm.

Lagerroth, Fredrik (1937). 'En frihetstidens lärobok i gällande svensk statsrätt'. *Statsvetenskaplig Tidskrift för politik, statistik, ekonomi* 40: 185–211.

Lagus, Vilhelm (1891). *Album Studiosorum Academiæ Aboensis MDCXL-MDCCCXXVII.* Helsingfors.

Lamettrie, Julien Offray de (1748–1750). *Ouvrage de Pénélope, ou Machiavel en médecine.* 3 Parts. Genève (Berlin).

Lamm, Martin (1908). *Olof Dalin: en litteraturhistorisk undersökning af hans verk.* Uppsala.

Larson, James Lee (1967a). 'Linnaeus and the Natural Method'. *Isis*, 58, 304–320.

Larson, James Lee (1967b). 'Goethe and Linnaeus'. *Journal of the History of Ideas*, XXVIII, no. 1, 590–596.

Larson, James Lee (1971). *Reason and Experience. The Representation of Natural Order in the works of Carl von Linné.* Berkeley, Los Angeles, London.

Larsson, Lars-Olof (1991). *Växjö genom 1000 år.* Stockholm.

Larsson, Ludvig (1923). 'Kyrkoherde Linnaeus och Skåneprästerna', in *Stenbrohult i forntid och nutid.* Edited by G. Virdestam, 81–86. Diö.

Lea, Henry Charles (1906–1907). *History of the Inquisition in Spain.* 4 Volumes. New York.

Leibniz, Gottfried Wilhelm von (1710). *Essais de Theodicée.* Amsterdam. Latin translation, Frankfurt am Main, 1719.

Leibniz, Gottfried Wilhelm von (1875–1890). *Die philosophischen Schriften.* 7 Parts. Edited by C.J. Gerhardt. Berlin.

Lemke, Otto Wilhelm (1868). *Visby stifts herdaminne, efter mestadels otryckta källor utarbetadt.* Örebro.

Lempp, Otto (1910). *Das Problem der Theodicee in der Philosophie und Literatur des 18. Jahrhunderts.* Leipzig.

Lepenies, Wolf (1982). 'Linnaeus's *Nemesis divina* and the Concept of Divine Retaliation'. *Isis*, 73, 11–27.

Levertin, Oscar (1906). *Carl von Linné. Några kapitel av ett oavslutat arbete.* Stockholm, 1908[2].

Lewenhaupt, Adam Ludvig Carl (1920–1921). *Karl XII's officerare. Biografiska anteckningar.* 2 Volumes. Stockholm.

Lidén, Johan Hinric (1778). *Catalogus Disputationum, in Academiis et Gymnasiis Sveciæ.* Part 1. Upsaliæ.

Liljencrantz, Axel (1939–1940). 'Polhem och Grundandet av Sveriges Första Naturvetenskapliga Samfund'. *Lychnos* 289–308 (1939), 21–54 (1940).

Lindborg, Rolf (1965). *Descartes i Uppsala*. Uppsala.
Lindeboom, Gerrit Arie (1968). *Herman Boerhaave. The Man and his Work*. London.
Lindeboom, Gerrit Arie (1978). *Descartes and Medicine*. Amsterdam.
Linderhielm, Emmanuel (1962). *Pietismen och dess första tid i Sverige*. Stockholm.
Lindquist, David (1939). *Studier i den svenska andaktslitteraturen under stormaktstidevarvet*. Lund.
Lindroth, Sten Hjalmar (1975). *Svensk Lärdomshistoria. Stormaktstiden*. Stockholm.
Lindroth, Sten Hjalmar (1978). *Svensk Lärdomshistoria. Frihetstiden*. Stockholm.
Lindroth, Sten Hjalmar (1983). 'The Two Faces of Linnaeus', in *Linnaeus. The Man and His Work*. Edited by T. Frängsmyr, 1–62. Berkeley, Los Angeles, London.
Linnaeus, Carolus (1729). *Spolia Botanica*. Upsaliæ. Edited by J.E.E. Ährling, *Ungsdomskrifter*. 2 Parts, I: 503–105. Stockholm.
Linnaeus, Carolus (1731). *Dissertatio de Planta Sceptro Carolino*. Upsaliæ.
Linnaeus, Carolus (1732). *Lachesis Lapponica, or a Tour in Lapland 1732*. Edited by J.E. Smith. London, 1811; *Iter Lapponicum*. 1888. Edited by J.E.E. Ährling, *Ungsdomskrifter*. 2 Parts, II: 1–202. Stockholm.
Linnaeus, Carolus (1733). *Diæta Naturalis 1733. Linnés tankar om ett naturenligt levnadssätt*. Edited by A.Hj. Uggla. Uppsala, 1958.
Linnaeus, Carolus (1735a). *Systema Naturae 1735. Facsimile of the first Edition*. Edited by M.S.J. Engel-Ledeboer and H. Engel. Nieuwkoop, 1964. Dutch Classics on the History of Science VIII.
Linnaeus, Carolus (1735b). *Systema Naturae, sive Regna tria Naturæ systematice proposita per classes, ordines, genera & species*. Lugduni Batavorum, 1735; Stockholmiæ, 1740[2]; Halle, 1740[3]; Parisiis, 1744[4]; Halæ Magdeburgicæ, 1747[5]; Stockholmiæ, 1748[6]; Lipsiæ, 1748[7]; Stockholm, 1753[8]; Lugduni Batavorum, 1756[9]; Holmiæ, 1758–1759[10]; Lipsiæ, 1762[11]; Holmiæ, 1766–1768[12].
Linnaeus, Carolus (1736a). *Bibliotheca Botanica*. Amstelodami.
Linnaeus, Carolus (1736b). *Fundamenta Botanica*. Amstelodami.
Linnaeus, Carolus (1737a). *Flora Lapponica*. Amstelædami.
Linnaeus, Carolus (1737b). *Hortus Cliffortianus*. Amstelædami.
Linnaeus, Carolus (1738). *Petri Artedi Sveci, medici Ichthyologia sive opera omnia de piscibus scilicet*. Lugduni Batavorum.
Linnaeus, Carolus (1739). *Tal om märkwärdigheter uti insecterna*. Stockholm, 1752[3].
Linnaeus, Carolus (1744). *Oratio de Telluris habitabilis incremento*. Lugduni Batavorum.
Linnaeus, Carolus (1745). *Öländska och Gothländska Resa på Riksens högloфliga ständers befallning förrättad*. Stockholm and Upsala.
Linnaeus, Carolus (1746). *Fauna Svecica*. Stockholmiæ.
Linnaeus, Carolus (1747). *Wästgöta=Resa ... förrättad år 1746*. Stockholm. Edited by N. Beckman, Göteborg, 1928.
Linnaeus, Carolus (1748). *Specimen academicum de Curiositate Naturali*. Holmiæ.
Linnaeus, Carolus (1749). *Oeconomia Naturæ*. Upsaliæ.
Linnaeus, Carolus (1749–1769). *Amoenitates Academicæ*. 7 Parts. Holmiæ.
Linnaeus, Carolus (1750). 'Rön om slö-korn'. *Kungliga Svenska Vetenskaps-Akademien Handlingar*, XI, 179–185.

Linnaeus, Carolus (1751). *Skånska Resa, på Höga Öfwerhetens Befallning Förrättad.* Stockholm.

Linnaeus, Carolus (1759a). *Tal vid deras Kongl. Majesteters höga närvaro, hållit uti Upsala ... den 25 September 1759.* Upsala.

Linnaeus, Carolus (1759b). *Upsala Academiæ Rector, Carl Linnaeus, hälsar denna Academiens och Stadsens samtelige fäder och inbyggare så af högre som lägre stånd, 1759, decembr. 11.* Upsala.

Linnaeus, Carolus (1760). *Dissertatio academica de Politia Naturæ.* Upsaliæ.

Linnaeus, Carolus (1761^2). *Fauna Svecica.* Stockholmiæ.

Linné, Carl von (1766). *Clavis medicinæ duplex.* Stockholmiæ.

Linné, Carl von (1767). *Dissertatio academica Mundum invisibilem.* Upsaliæ.

Linné, Carl von (1792). *Prælectiones in Ordines naturales Plantarum.* Edited by P.D. Giseke. Hamburgi.

Linné, Carl von (1878). *Anteckningar öfver Nemesis Divina. Utgifne af Elias Fries och Th.M. Fries. Ny, omarbetad och tillökad upplaga.* Upsala.

Linné, Carl von (1878–1880). *Svenska Arbeten i urval och med noter utgifna.* 2 Parts. Edited by J.E.E. Ährling. Stockholm.

Linné, Carl von (1888a). *Iter Lapponicum* (1732), in *Ungdomsskrifter*, II: 1–202. Stockholm.

Linné, Carl von (1888b). *Ungdomsskrifter.* Edited by J.E.E. Ährling. 2 Parts. Stockholm.

Linné, Carl von (1905–1913). *Skrifter ... Utgifna af Kungl. Svenska Vetenskapsakademien.* 5 Parts. Edited by Th.M. Fries. Stockholm.

Linné, Carl von (1907). *Lachesis Naturalis.* Edited by A.O. Lindfors. Uppsala.

Linné, Carl von (1907–1943). *Bref och Skrivelser af och till Carl von Linné*: cited as *Letters.* Edited by Th.M. Fries, J.M. Hulth, A.Hj. Uggla. 10 Volumes. Stockholm, Upsala, Berlin.

Linné, Carl von (1913). *Iter Lapponicum. Andra Upplagen med Bilagor och Noter.* Edited by Th.M. Fries. Upsala.

Linné, Carl von (1919). *Linnæi Minnesbok.* Edited by Felix Bryk. Stockholm.

Linné, Carl von (1923). *Nemesis Divina. Utgiven och kommenterad av Knut Barr.* Stockholm.

Linné, Carl von (1929). *Ungdomsresor.* Edited by Knut Hagberg. Stockholm.

Linné, Carl von (1935). 'Linnés Almanacksanteckningar för år 1735: med inledning och förklaringar, ed. A.Hj. Uggla. *Svenska Linné-Sällskapets Årsskrift* 18: 134–148.

Linné, Carl von (1951). *Adonis Stenbrohultensis.* 1732. Edited by T. Fredbärj. Ekenäs.

Linné, Carl von (1953). *Iter dalekarlicum, jämte Utlandsresan Iter ad Exteros och Bergslagsresan Iter ad fodinas.* Edited by A.Hj. Uggla and others. Stockholm.

Linné, Carl von (1957). *Vita Caroli Linnæi.* Edited by E. Malmeström and A.Hj. Uggla. Uppsala.

Linné, Carl von (1960). *Nemesis Divina. Med kommentar och efterskrift utgiven av Knut Hagberg.* Stockholm.

Linné, Carl von (1961). *Inledning till Dietetiken.* Edited by A.Hj. Uggla and T. Fredbärj. Ekenäs.

Linné, Carl von (1968). *Nemesis Divina. Utgiven av Elis Malmeström och Telemak Fredbärj.* Stockholm.

Linné, Carl von (1981). *Nemesis Divina. Herausgegeben von Wolf Lepenies und Lars Gustafsson Hanser. Aus dem Lateinischen und Schwedischen übersetzt von Ruprecht Volz.* München and Wien.

Linné, Carl von (1996). *Nemesis Divina. Bezorgd en vertaald door Trudi de Vlaming-van Santen en Michael John Petry, ingeleid door J.M.M. de Valk.* Kampen.

Lockyer, Roger (1984). *Buckingham. The Life and Political Career of George Villiers, First Duke of Buckingham 1592–1628.* London.

Lovejoy, Arthur Oncken (1936). *The Great Chain of Being.* Cambridge, Mass.

Löwenhielm, Carl Gustav (1751). *Tal, om landt-skötsel, hållit för Kungl. svenska vetenskaps academien ... då han lade af sit derstädes förda praesidium den 19. Januarii, år 1751.* Stockholm.

Lundius, Carolus (1691). *Dissertatio de origine majestatis civilis.* Lundio.

Lundström, Herman (1902). 'Karl XII – Messias, en i utlandet omkring år 1718 omfattad trossats'. *Kyrkohistorisk Årsskrift*, 1, 1–18.

Luther, Martin (1959–1963). *Psalmen=Auslegung.* 3 Parts. Edited by E. Mühlhaupt. Göttingen.

Malmeström, Elis (1925). 'Linnés religionsfilosofiska betraktelser, i företal och inledningsord till Systema naturae'. *Kyrkohistorisk Åsskrift*, 25, 1–44.

Malmeström, Elis (1925–1926). 'Linnés Parentation över Andreas Neander'. *Svenska Linné-Sällskapets Årsskrift*, VIII, 97–118; IX, 45–60.

Malmeström, Elis (1926). *Carl von Linnés religiösa åskådning.* Dissertation, Uppsala, 12.5.1926. Stockholm.

Malmeström, Elis (1939). 'Frågor rörande Tro, Moral och Samhällsliv. Ett Linné-manuskript med Kommentarer'. *Svenska Linné-Sällskapets Årsskrift*, XXII, 59–76.

Malmeström, Elis (1942a). 'Carl von Linnés Kyrkohistoriska Ställning'. *Svenska Linné-Sällskapets Årsskrift*, XXV, 12–32.

Malmeström, Elis (1942b). 'Die religiöse Entwicklung und die Weltanschauung Carl von Linnés'. *Zeitschrift für systematische Theologie*, XIX, 31–58.

Malmeström, Elis (1954–1955). 'Carl von Linné och Isaac Newton'. *Svenska Linné-Sällskapets Årsskrift*, XXXVII–XXXVIII, 80–82.

Malmeström, Elis (1959). 'Det nytestamentliga inslaget i Carl von Linnés skrifter och åskådning'. *Från Småland och Hellas. Studier tillägnade Bror Olsson*, 265–278. Malmö.

Malmeström, Elis (1962). 'Linné och Katolicismen'. *Svenska Linné-Sällskapets Årsskrift*, XLV, 23–33.

Malmeström, Elis (1964a). *Carl von Linné. Geniets Kamp för Klarhet.* Stockholm.

Malmeström, Elis (1964b). 'Linnés Bruk av Bibelord i Nemesis Divina: en statistisk undersökning'. *Svenska Linné-Sällskapets Årsskrift*, XLVI, 42–51.

Malmström, Carl Gustav (1893–1901). *Sveriges Politiska Historia från Konung Karl XII's död till statshvälfningen 1772.* 6 Volumes. Stockholm.

Mandariaga, Isabel de (1981). *Russia in the Age of Catherine the Great.* London.

Mankell, Julius (1870). *Anteckningar rörande Finska arméens och Finlands krigshistoria.* Stockholm.

Manuel, Frank Edward (1974). *The Religion of Isaac Newton.* Oxford.

Manuel, Frank Edward (1980). *A Portrait of Isaac Newton.* London.

Martin, Martin (1716). *A Description of the Western Islands of Scotland*. London.

Massie, Robert Kinloch (1982). *Peter the Great. His Life and World*. London.

McAdoo, Henry Robert (1965). *The Spirit of Anglicanism: a survey of Anglican theological method in the seventeenth century*. New York.

McGuire, J.E. and Rattansi, P.M. (1966). 'Newton and the Pipes of Pan'. *Notes and Records of the Royal Society of London*, 21, 108–143.

McTurk, H.W. (1974). 'Sacral Kingship in Ancient Scandinavia'. *Saga Book of the Viking Society*, I, 139–169.

Mennander, Carl Fredrik (1748). *De mundo non senescente*. Respondent Otto Reinhold Nordstedt (d. 1750). Aboæ.

Migne, Jacques Paul (1886). *Patrologiæ Græco-Latinæ*. Tomus II. Parisiis.

Millqvist, Victor (1911). *Svenska riksdagens borgerstånd 1719–1866*. Stockholm.

Minnesskrift (1934), *ägnad 1734 års lag av jurister i Sverige och Finland den 13 december 1934, 200-årsdagen av riksens ständers beslut*. 3 Parts. Stockholm.

Mitchell, Bruce (1995). *An Invitation to Old English and Anglo-Saxon England*. Oxford.

Moser, Friedrich Carl von (1776). *Rettung der Ehre und Unschuld ... des weiland ... Georg Heinrichs, Freyherrn von Schlitz*. Hamburg.

Muncktell, Johan Fredrik (1843–1846). *Westerås stifts herdaminne*. 3 Parts. Upsala.

Munktell, Henrik (1936). 'Mose Lag och Svensk Rättsutveckling. Några Huvuddrag'. *Lychnos*, I, 131–150.

Murray, Harold James Ruthven (1913). *A History of Chess*. Oxford.

Naudé, Albert Heinrich Ferdinand (1884). *Politische Correspondenz Friedrich's des Grossen. Band 12, 1756*. 46 Volumes. Berlin (1879–1920).

Neckel, Gustav ed. (1927^2). *Edda, die Lieder des Codex regius*. Heidelberg.

Neuhaus, Gerd (1993). *Theodizee – Abbruch oder Anstoss des Glaubens*. Freiburg.

Newton, Isaac (1729). *Mathematical Principles of Natural Philosophy*. London, 1726^3. Translated by Andrew Motte. London. Edited by F. Cajori. 2 Parts. Berkeley, 1934.

Newton, Isaac (1737). *A Dissertation upon the Sacred Cubit of the Jews*. In *Miscellaneous Works of Mr. John Greaves*. Edited by T. Birch. 2 Volumes. II, 405–433. London.

Nichols, Margaret Ann (1966). *The Garden Tradition ... in Marvell and Milton*. Illinois.

Nieuwentijt, Bernard (1715). *Het Regt Gebruik der Werelt Beschouwingen*. Amsterdam, 1759^7.

Nieuwentijt, Bernard (1720). *Gronden van Zekerheid*. Amsterdam, 1754^3.

Nikula, Oscar (1972). *Åbo Stads historia*. Åbo.

Nilsson, Johan (1952–1960). *Västan om sjön Åsnen. Minnnen från södra Allbo*. 4 Parts. Torne and Växjö.

Nordberg, Albert (1928). *En gammal Norrbottensbygd. Anteckningar till Luleå sockens historia*. 2 Parts. Nederluleå.

Nordström, Johan (1924). *Georg Stiernhielms Filosofiska Fragment*. 2 Parts. Uppsala.

Nordström, Johan (1954–1955). 'Linné och Gronovius'. *Svenska Linné-Sällskapets Årsskrift*, XXXVII–XXXVIII, 7–22.

Normann, Carl-E. (1948). *Prästerskapet och det karolinska enväldet.* Stockholm and Lund.

Odel, Anders (1739). *Hjeltarnas samtal med den tappre och omistelige men på sin hemresa ifrån Constantinopel, i negden af Breslau den 19 Junii 1739 förrådeligen mordade svenska majoren vid Uplands regemente til fot, then väl borne herren herr Malcom Sinclair.* Lund.

Odel, Anders (1963). *Sinclairsvisan.* In *Sveriges litteratur. Del III Frihetstidens Litteratur.* Edited by L. Breitholtz. 75–80. Uppsala.

Ödmann, Samuel (1830). *Hågkomster från hembygden och skolan.* Upsala, 1925[6].

Olrik, Axel (1922). *Ragnarök, die Sagen vom Weltuntergang.* Berlin.

Olsson, Bror (1954–1955). 'Mundus Senescens. En skiss om tron på en åldrande värld i svenskt folkliv och litteratur'. *Lychnos*, 66–81.

O'Higgins, James S.J. (1970). *Anthony Collins. The Man and His Works.* The Hague.

Ooteghen, Jules van (1959). *Lucius Licinus Lucullus.* Bruxelles.

Östergren, P.A. (1902). *Till striden om 1734 års lagreform.* 2 Parts. Lund.

Ovid (1985). *Ars Amatoria.* Translated by J.H. Mozley, revised by G.P. Goold. Loeb edition. Cambridge, Mass., London.

Ovid (1994). *Metamorphoses.* Translated by F.J. Miller, revised by G.P. Goold. Loeb edition. Cambridge, Mass., London.

Paludan-Müller, Caspar Peter (1847). 'Er Kong Carl den Tolvte falden ved Snigmord? En historisk-kritisk Undersøgelse'. *Nyt historisk Tidsskrift* I: 1–128. Swedish translation by G. Swederus *Carl XII's död, efter äldre och nyare, till en del förr obegagnade källor.* Stockholm (1847).

Parnell, Thomas (1722). *Poems on Several Occasions.* Edited by A. Pope. London.

Patch, Howard Rollin (1927). *The Goddess Fortune in Mediaeval Literature.* Cambridge, Mass.

Pater, Cees de (1988). *Willem Jacob 's Gravesande. Welzijn, wijsbegeerte en wetenschap.* Baarn.

Pater, Cees de (1994). 'Willem Jacob 's Gravesande and Newton's Regulae Philosophandi'. *Lias*, 21, 257–294.

Payne, John (1793). *The Naval Commercial and General History of Great Britain.* 5 Volumes. London.

Pehrsson, Anna-Lena (1965). 'Nils Rosén von Rosenstein och Iatromekaniken'. *Svenska Linné-Sällskapets Årsskrift*, XIVIII, 26–59.

Pérez-Ramos, Antonio (1988). *Francis Bacon's Idea of Science and the Maker's Knowledge Tradition.* Oxford.

Petersson, Erik Lorentz (1878). *Prosten i Mora Jacob Boëthius. Lefnadsteckning mest efter otryckta källor.* Stockholm.

Petersson, Gunnar (1982). *Örter vid Linnés Råshult.* Älmhult.

Petersson, Oskar (1971). *Virestad. Tradition och historia.* Växjö.

Petré, Torsten (1958). 'Uppsala under merkantilismens och statskontrollens tidsskede 1619–1789'. *Uppsala stads historia.* Edited by Herbert Lundh. Part 3. Uppsala.

Petri, Olavus (1914–1917). *Samlade skrifter.* 4 Parts. Edited by B. Hesselman. Upsala.

Petry, Michael John (1979). *Nieuwentijt's Criticism of Spinoza*. Leiden.
Petry, Michael John (1994). 'Newton, Isaac'. *Theologische Realenzyklopädie*, XXIV, 422–429. Berlin.
Pettazzoni, Raffaele (1955). *L'omniscienza di Dio*. Torino. Translated by H.J. Rose. London (1956).
Plantinga, Alvin ed. (1968). *The Ontological Argument. From St. Anselm to Contemporary Philosophers*. London.
Pleijel, Hilding (1935). *Karolinsk kyrkofromhet, pietism och herrnhutism 1680–1772. Svenska kyrkans historia*. Volume 5. Stockholm.
Polak, M.S. (1987). *Inventaris van het archief van de familie Huydecoper 1459–1956*. Utrecht.
Polhem, Christopher (1951). *Nationalekonomiska och politiska Skrifter*. Edited by G. Lindeberg. Uppsala.
Polybios-Lexikon (1956–1975). Edited by Arno Mauersberger. 4 Parts. Berlin.
Pontoppidan, Erik Ludvigsen (1739–1741). *Marmora Danica selectiora sive Inscriptionum*. 2 Parts. Hafniæ.
Pontoppidan, Erik Ludvigsen (1741–1752). *Annales ecclesiæ Danicæ diplomatici*. 4 Volumes. Hafniæ.
Pontoppidan, Erik Ludvigsen (1758). *Sanheds Kraft til at overvinde den Atheistiske og Naturalistiske Vantroe*. Kjøbenhavn. Dutch translation by C.W.R. Scholten, Utrecht, 1767.
Porter, Roy (1977). *The Making of Geology*. Cambridge.
Posse, Johan August (1850). *Bidrag till svenska lagstiftnings historia ... till stadfästelsen af 1734 års lag*. Stockholm.
Prideaux, Mathias and John (1664). *An Easy Introduction for Reading all sorts of Histories*. Oxford.

Raeff, Marc (1970). 'The Domestic Policies of Peter III and his Overthrow'. *American Historical Review* 75: 1289–1310.
Ragueau, François (1577). *Leges Politicæ, ex Sacræ Jvrisprvdentiæ fontibvs havstæ*. Francofvrti ad Moenum.
Ragueau, François (1579). *Leges Politicæ, das ist, von allen bürgerlichen Satzungen oder Rechten, Erklärung, auss heiliger biblischer Schrifft gezogen*. Translated by Abraham Saur. Franckfurt am Mayn, 1582^2, 1588^3.
Ragueau, François (1597). *Leges Politicæ, das ist, von allen bürgerlichen Satzungen oder Rechten. Erklärung, auss heiliger biblischer Schrifft gezogen*. Translated by Abraham Saur, edited by Conrad Gerhard Saur. Franckfurt am Mayn.
Ragueau, François (1607). *Lex politica Dei, thet är: Gudz regementz ordning, ther vthi författade äre några regementz statuter och politiske rätter, vthaff then h. biblischa schrifft vth tagne, och til thes meere nytte effter thet Justinianiska sätte vnder wisse titlar stälte och förordnade, först igenom her Franciscum Raguelem vpå latijn samman dragit: sedhan igenom M. Abrahamum och Cunradum Saurium på tydsko wendt, och förmeerat. Och nu ... vpå swenskt tungomål vth satt, aff Hendrich Jönsson Careell*. Rostock; Stockholm, 1635^2, 1638^3, 1651^4.
Ramm, Axel (1907). *Linné om Småland: Några utdrag ur hans skrifter*. Göteborg.
Ranta, Raimo (1977). *Åbo stads historia*. Åbo.

Rathlef, Ernst Ludwig (1748–1750). *Akridotheologie; oder historische und theologische Betrachtungen über die Heuschrecken.* 2 Parts. Hannover. Dutch translation, Amsterdam, 1750.

Ravanel, Pierre (1660–1663). *Bibliotheca Sacra seu Thesaurus Scripturæ Canonicæ Amplissimus.* 3 Volumes. Genevæ.

Raynaud, Claudine (1976). 'The Garden of Eden in Marvell's Poetry'. *Cahiers Elisabéthains*, 10, 13–32.

Remgård, Arne (1968). *Carl Gustav Tessin och 1746–1747 års Riksdag.* Lund.

Reuterholm, Nils Esbjörnsson (1957). *Journal.* Edited by Sten Landahl. *Historiska håndlingar* 36: 2. Till trycket befordrade av Kungl. Samfundet för utgifvande af handskrifter rörande Skandinaviens historia. Stockholm.

Ridderstad, Carl Anton (1913–1920). *Östergötland.* 3 Parts in 4 Volumes. Stockholm.

Ritter, Adolf Martin (1981). 'The Dogma of Constantinople and its reception within the Churches of the Reformation'. *Irish Theological Quarterly* 48: 228–232.

Roberts, Michael (1967). *Essays in Swedish History.* London.

Roberts, Michael ed. (1973). *Sweden's Age of Greatness 1632–1718.* London and Basingstoke.

Roberts, Michael (1979). *The Swedish Imperial Experience 1560–1718.* Cambridge.

Roberts, Michael (1986). *The Age of Liberty. Sweden 1719–1772.* Cambridge.

Rod, Jakob (1972). *Dansk folkereligion i nyere tid.* København.

Rodhe, Edvard Magnus (1917). *Studier i den svenska reformations-tidens liturgiska tradition.* Uppsala.

Rogberg, Samuel (1770). *Historisk beskrifning om Småland i gemen, i synnerhet Kronobergs och Jönköpings lähner, ifrån äldsta til närwarande tid.* Edited Eric Ruda. Carlskrona.

Romanell, Patrick (1984). *John Locke and Medicine. A New Key to Locke.* New York.

Rosén, Nils (1738). *Compendium anatomicum, eller en kort beskrifning om de delar, af hwilka hela menniskans kropp består.* Stockholm.

Rothe, Caspar Peter (1745). *Peder Græve af Griffenfelds Liv og Levnet.* 2 Parts. Kjøbenhavn.

Rothe, Caspar Peter (1747–1750). *Peder Tordenskiolds omstændelige Livs og Heldte-Levnets Beskrivelse.* 3 Parts. Kjøbenhavn.

Rousseau, André-Michel (1960). 'En marge de *Candide*: Voltaire et l'affaire Byng'. *Revue de Littérature comparée* 34 April-Juin: 261–273.

Rousseau, Jean Jacques (1967). *Correspondance complète.* Edited by R.A. Leigh. IV: 1756–1757. Genève.

Rowse, Alfred Leslie (1956). *The Early Churchills. An English Family.* London.

Ruuth, Martti (1914). 'Karl XII i den mystiskt-separatistiska profetians ljus'. *Kyrkohistorisk Årsskrift*, 15, 434–448.

Rydbeck, Monica (1957). *Den helige Sigfrid. En rikets skyddspatron.* Växjö.

Sacheverell, William (1702). *An Account of the Isle of Man.* London. Edited by J.G. Cumming. Douglas (1859).

Sahlgren, Jöran (1922). 'Linné som predikant'. *Svenska Linné-Sällskapets Årsskrift*, V, 40–55.

Salter, Keith William (1964). *Thomas Traherne, Mystic and Poet.* London.

Sandblad, Henrik (1942). 'Politiska Prognostika om Johan III, Sigismund och Karl IX'. *Lychnos*, 87–96.

Sandblad, Henrik (1967). *En karolinsk nationsinspektor Juristen Carl Lundius (1638–1715)*. Göteborgs Nations Skriftserie 8. Uppsala.

Scandinavian Biographical Index (1994). Edited by Laureen Baillie. 4 Volumes. London, Melbourne, Munich, New Jersey.

Scepin, Constantine (1758). *Schediasma Chemico-Medicum Inaugurale de Acido Vegetabili*. Lugduni Batavorum.

Schefferus, Johannes (1673). *Lapponia Id est Regionis Lapponum et Gentis Nova et verissima descriptio*. Francofurti.

Scheutz, Nils Johan Wilhelm (1878–1880). *Biografiska anteckningar om lektorer vid Wexiö läroverk från Gymnasii stiftelse 1643 till närvarande tid*. 2 Parts. Wexiö.

Schröder, Johan Henric (1838). *Rektor Johannes Henrik Schröders rektorsprogram*. Upsala.

Scriver, Christian (1675). *Seelen=Schatz*. 5 Parts. Magdeburg and Leipzig, 1715^5.

Scriver, Christian (1697). *Ymnig tröste-skatt uti armod och fattigdom ... af tyska språket öfwersatt i thetta hårda sorge-åhr 1697*. Translated by Gabriel Salonius (d. 1729). Åbo.

Scriver, Christian (1723–1727). *Själlaskatt, i hvilken uppbyggeligen och trosteligen handlas om den menskliga själens höga värdighet*. Translated by Antonio Münchenberg, pastor in Borg and Löt, near Norrköping. 6 Parts. Norrköping.

Seaton, Ethel (1935). *Literary Relations of England and Scandinavia in the Seventeenth Century*. Oxford.

Sellberg, Erland (1979). *Filosofin och nyttan. 1. Petrus Ramus och ramismen*. Diss. Göteborg.

Selling, Olof H. (1960). 'Från Linnés Småländska Hembygd'. *Svenska Linné-Sällskapets Årsskrift* 43: 75–147.

Sidenvall, Fredrik (1992). *Det var en svensk som satt fången*. Strängnäs.

Sippel, Heinrich (1987). *Ein Streifzug durch Schlitzer Geschichte*. Schlitz.

Sjögren, Josef (1963). *Acta genealogica Malungensia*. Malung.

Sjögren, Karl Johan Vilhelm (1900–1909). *Förarbetena till Sveriges rikes lag 1686–1736*. 8 Parts. Upsala.

Skrede, Pähr O. (1987). 'Carl von Linné's antavla'. *Svenska Antavlor* II, 6: 316–322.

Sloan, Phillip R. (1976). 'The Buffon-Linnaeus Controversy'. *Isis*, 67, 356–375.

Soloviev, Sergei M. (1997). *History of Russia*. Volume 41 (1757–1761). Gulf Breeze, Florida.

Sparre, Sixten Knut Benjamin Axelsson (1930). *Kungl. Västmanlands regementes historia*. Part IV. *Biografiska anteckningar om officerare och vederlikar 1623–1779*. Stockholm.

Spegel, Haquin (1705). *Thet öpna Paradis*. Stockholm, 1725^2, 1745^3.

Spinoza, Benedict de (1670). *Tractatus Theologico-politicus*. Hamburgi (Amsterdam).

Spinoza, Benedict de (1677a). *Ethica*. Amsterdam.

Spinoza, Benedict de (1677b). *Tractatus politicus*. Amsterdam.

Stafleu, Frans Antonie (1971). *Linnaeus and the Linnaeans. The spreading of their ideas in systematic botany*. Utrecht.

Ståhle, Carl Ivar ed. (1968). 'Ur Domareregler (1540-talet)'. *Sveriges litteratur Del I*, 236–242.
Stanley, Eric Gerald (1975). *The Search for Anglo-Saxon Paganism*. Cambridge.
Stavrianos, Leften Stavros (1958). *The Balkans since 1453*. New York.
Steckzén, Bertil (1955). *Västerbottens Regementes Officerare till år 1841*. Umeå.
Stille, Arthur Gustav Henrik (1903). *Kriget i Skåne 1709–1710*. Stockholm.
Stjernman, Andreæ Antonii (1719). *Aboa Literata*. Holmiæ.
Strandberg, Carl Henric (1832). *Åbo Stifts Herdaminne, ifrån Reformationens början till närvarande tid*. Åbo.
Strandell, B. (1940). 'Alexander Blackwell – Fredrik I's livmedicus. Ett läkaröde'. *Svenska läkartidningen. Organ för Sveriges läkarförbund* 37: 1904–1923.
Stremninger, Gerhard (1992). *Gottes Güte und die Übel der Welt. Das Theodizee-problem*. Tübingen.
Strindberg, August (1907–1912). *En Blå Bok*. 4 Parts. Stockholm, 1918^2.
Strindberg, August (1948–1976). *Brev*. 15 Parts. Edited by T. Eklund. Stockholm.
Strömberg-Back, Kerstin (1963). *Lagen, rätten, läran. Politisk och kyrklig idédebatt i Sverige under Johan III's tid*. Lund.
Svenska Akademiens Ordbok (1893–). Lund.
Svenska Ättartal: tidskrift för svensk släktkunskap och släktforskning. 10 Volumes. Vadstena (1889–1908).
Svensk Uppslagsbok: cited as SU. Edited by G. Carlquist. 31 Volumes. Malmö (1929–1937).
Svenskt Biografiskt Handlexikon: cited as SBH. Edited by Herman Hofberg. 2 Volumes. Stockholm (1906).
Svenskt Biografiskt Lexikon: cited as SBL. Edited by B. Boethius, G. Nilzén and others. 29 Volumes to date. Stockholm (1918–1997).
Swan, Charles (1905). *Gesta Romanorum. Entertaining Moral Stories*. Edited by E.A. Baker. London.
Swartling, Birger (1909). *Georg Stiernhielm, hans Lif och Verksamhet*. Uppsala.
Sweriges Rijkes Landzlagh (1608). *Efter Carl then nijondes ... befalning, af Trycket utgången*. Stockholm.
Swete, Henry Barclay (1876). *On the History of the Doctrine of the Procession of the Holy Spirit from the Apostolic Age to the Death of Charlemagne*. Cambridge.

Taylor, Henry (1760). *An Essay on the Beauty of the Divine Oeconomy*. London.
Taylor, Jeremy (1651). *The Rule and Exercises of Holy Dying*. London.
Tennemar, Eskil (1981–1982). 'Hur dog drabanten Strutz?'. *Svenska Linné-Sälllskapets Årsskrift* 98–103.
Tersmeden, Carl (1912–1919). *Amiral Carl Tersmedens Memoarer*. Edited by N. Sjöberg and N. Erdmann. 5 Parts. Stockholm.
Thomson, George Malcolm (1979). *The First Churchill. The Life of John, 1st Duke of Marlborough*. London.
Thorschmid, Urban Gottlob (1765–1767). *Versuch einer vollständigen engelländischen Freydenker-Bibliothek*. 4 Volumes. Halle.
Thunander, Rudolf (1993). *Hovrätt i funktion. Göta Hovrätt och brottmålen 1635–1699*. Lund.

Thuronius, Andreas (1665). *Contemplationis philosophicæ De Scientiæ Naturalis Prooemio*. Aboæ.

Tilas, Daniel (1974). *Anteckningar och brev från riksdagen 1765–1766*. Edited by Olof Jägerskiöld. Stockholm.

Tornehed, Stig ed. (1992). *Carl von Linné Om Smålands naturalhistoria*. Commentary by Ingvar Christoffersson. Växjö.

Tornehed, Stig (1993). *Linné om Småland. Ur Öländska och Gotländska resan 1741 och Skånska resan 1749*. Växjö.

Torquemada, Tomás de (1537). *Compilacion de las instrucciones del Oficio de la Santa Inquisicion*. Granada; Madrid, 1627², 1667³.

Traherne, Thomas (1908). *Centuries of Meditations*. Edited by B. Dobell. London.

Tunstall, Brian (1928). *Admiral Byng and the Loss of Minorca*. London.

Turner, William (1697). *A Compleat History of the most Remarkable Providences, both of Judgment and Mercy, which have hapned in this Present Age ... To which is added, Whatever is Curious in the Works of Nature and Art ... Recommended as useful to Ministers in furnishing Topics of Reproof and Exhortation, and to Private Christians for their Closets and Families*. London.

Uggla, Arvid Hjalmar (1940). 'De Tidigaste Förbindelserna mellan Royal Society och Sverige', *Lychnos*, 302–324.

Uggla, Arvid Hjalmar (1961). 'Till Nemesis Divina-Kritiken'. *Svenska Linné-Sällskapets Årsskrift*, XLIV, 71–73.

Uggla, Arvid Hjalmar (1967). 'Om Linnés Nemesis Divina, i synnerhet de urprungliga Nemesis-Anteckningarna i London'. *Svenska Linné-Sällskapets Årsskrift*, L, 13–19.

Valentine, Alan (1970). *The British Establishment 1760–1784*. 2 Volumes. Oklahoma.

Varro, Marcus Terentius. *De gente populi Romani*. Now lost; see Augustine *De Civitate Dei*, book 18, chapter 8.

Vaughan, Henry (1650). *Silex Scintillans*. London.

Veenendaal, Augustus Johannes (1975). 'Pieter van Guethem, een Partizaan uit de Spaanse-Successieoorlog'. *Driekwart Eeuw Historisch Leven in Den Haag*, 202–216. 's-Gravenhage.

Vekene, Emil van der (1982–1983). *Bibliotheca Bibliographica Historiæ Sancta Inquisitionis*. 2 Volumes. Vaduz.

Vermeulen, Ben (1987). 'Theology and science: the case of Bernard Nieuwentijt's theo-logical positivism'. In *Science and imagination in XVIIIth-Century British Culture*. Edited by S. Rossi. 379–390. Milano.

Vickery, Roy (1995). *A Dictionary of Plant Lore*. Oxford.

Virdestam, Gotthard et al. (1921–1934). *Växjö stifts herdaminne*. 8 Parts. Växjö.

Virdestam, Gotthard (1924). 'Kyrkoherde Nic. Linnæus' släktanteckningar'. *Stenbrohult i forntid och nutid*, III, 77–89. Växjö.

Virdestam, Gotthard (1928). 'Linné och Stenbrohult'. *Svenska Linné-Sällskapets Årsskrift*, 7–40.

Virdestam, Gotthard (1930). *Småländska gestalter*. Älmhult.

Virdestam, Gotthard (1931). 'Kring några brev från Samuel Linnaeus'. *Svenska Linné-Sällskapets Årsskrift*, XIV, 115–125.

Voltaire, François-Marie Arouet (1733). *Letters concerning the English nation.* London.

Voltaire, François-Marie Arouet (1747). *Zadig ou la Destinée. Histoire orientale.* Londres (Amsterdam).

Voltaire, François-Marie Arouet (1759). *Candide, ou l'Optimisme, traduit de l'Allemand.* n.p. Edited René Pomeau. Oxford (1980).

Vries, Jan de (1970). *Altgermanische Religionsgeschichte.* 2 Volumes. Berlin.

Wade, Ira Owen (1958). *The Search for a New Voltaire.* Philadelphia.

Wal, Marijke J. van der (1995). *De moedertal centraal.* Den Haag.

Walker, Margaret (1985). 'Smith's acquisition of Linnaeus's library and herbarium'. *The Linnean*, I, no. 6, 16–19.

Wallenberg, Jacob (1928–1941). *Samlade Skrifter.* Edited by Nils Afzelius. 2 Volumes. Stockholm.

Wallerius, Johan Gottschalk (1758). *De geocosmo senescente.* Upsaliæ.

Wallerius, Nils (1750–1752). *Systema metaphysicum, omnis cognitionis humanæ fundamenta continens.* Stockholmiæ.

Wallerius, Nils (1752). *De usu et abusu physices in theologia naturalis.* Respondent L. Watlander. Upsaliæ (Holmiæ).

Wallerius, Nils (1754). *Compendium logicum.* Upsaliæ.

Wallerius, Nils (1756–1765). *Prænotiones theologicæ.* 6 Parts. Stockholmiæ.

Wallerström, Ingrid (1974). *Carl von Linné. Barndom, hem, skola. Kulturhistorisk skildring.* Göteborg.

Walther, Hans (1963–1986). *Proverbia Sententiaeque Latinitatis Medii Aevi: Lateinischer Sprichwörter und Sentenzen des Mittelalters in alphabetischer Anordnung.* 9 Volumes. Göttingen.

Wandal, Hans (1663–1672). *Juris regii anupeuthunou et solutissimi cum potestate summa, nulli nisi Deo soli, obnoxia, regibus christianis, e juris divini pandectis V. et N. Testam.* 6 Parts. Hauniæ.

Wanley, Nathaniel (1678). *The Wonders of the Little World; or, a General History of Man.* London. Edited by W. Johnston. 2 Volumes. London, 1806–1807.

Wasastjerna, Oskar (1879–1880). *Ättar-Taflor öfver den på Finlands Riddarhus introducerade adeln.* 3 Parts. Borgå.

Weber, Edmund (1978). *Johan Arndts Vier Bücher vom Wahren Christentum.* Hildesheim.

Weibull, Lauritz (1929). 'Carl XII's död'. *Scandia* 2: 229–274.

Werwing, Jonas (1746–1747). *Konung Sigismunds och Konung Carl den IX:des Historier.* Edited by Anders Anton von Stiernman. 3 Volumes. Stockholm.

Westfälische Lebensbilder (1930–1990). Edited by Aloys Bömer and others. 15 Volumes. Münster.

Whiston, William (1711). *An Historical Preface to Primitive Christianity Reviv'd.* London.

Whitelocke, Bulstrode (1772). *A Journal of the Swedish Ambassy, in the years M.DC.LIII and M.DC.LIV.* 2 Volumes. London.

Whitelocke, Bulstrode (1855). *A Journal of the Swedish Embassy*. 2 Parts. Edited by H. Reeve. London.
Widengren, Geo (1953²). *Religionens värld. Religionsfenomenologiska studier och översikter*. Stockholm.
Wieselgren, Oscar (1910). 'Anteckningar om Urban Hjärnes Bibliotek'. *Samlaren*, årgang 31. Uppsala.
Wieselgren, Peter Jonasson ed. (1831–1843). *Delagardiska Arkivet, eller handlingar ur grefliga delagardiska bibliotheket på Löberöd*. 20 Parts. Lund.
Wijkmark, Oscar Henning Vilhelm (1913). *Linné i bibelkommissionen. En episod*. Stockholm.
Wikman, Karl Robert Villehad (1964a). 'Carl von Linnés samling av småländska vidskepelser 1741'. *Svenska Linné-Sällskapets Årsskrift*, XLVII, 16–25.
Wikman, Karl Robert Villehad (1964b). 'Linnés Religion'. *Finsk Tidskrift*, 8, 411–416.
Wikman, Karl Robert Villehad (1968–1969). 'Superstitionerna i Lachesis-manuskriptet. Några Linné anteckningar jämte anmärkningar'. *Svenska Linné-Sällskapets Årsskrift*, 25–40.
Wikman, Karl Robert Villehad (1970). *Lachesis and Nemesis. Four chapters on the human condition in the writings of Carl Linnaeus*. Stockholm.
Wilson, Catherine (1995). *The Invisible World. Early Modern Philosophy and the Invention of the Microscope*. Princeton.
Wissowa, Georg (1894–1972). *Paulys Real-Encyclopädie der Classischen Altertumswissenschaft*. 34 Volumes. Stuttgart.
Wolff, Christian (1739–1741). *Theologia naturalis, methodo scientifica pertractata*. 2 Parts. Francofurti & Lipsiæ. Edited by J. École, 3 volumes, Hildesheim, 1978/1981.
Wrangel, Ewert (1897). *Sveriges Litterära Förbindelser med Holland särdeles under 1600-talet*. Lund. Dutch translation, Leiden, 1901.

Zedler, Johan Heinrich (1732–1754). *Grosses vollständiges Universal Lexicon Aller Wissenschaften und Künste*. 68 Volumes. Halle and Leipzig.
Zwergius, Detlev Gotthard (1754). *Det Siellandske Clerisie*. Kjøbenhavn.

INDEX TO THE TEXT

INDEX TO THE INTRODUCTION AND NOTES

ARCHIVES INTERNATIONALES D'HISTOIRE DES IDÉES

*

INTERNATIONAL ARCHIVES OF THE HISTORY OF IDEAS

137. Otto von Guericke: *The New (so-called Magdeburg) Experiments* [Experimenta Nova, Amsterdam 1672]. Translated and edited by M.G. Foley Ames. 1994 ISBN 0-7923-2399-8
138. R.H. Popkin and G.M. Weiner (eds.): *Jewish Christians and Cristian Jews*. From the Renaissance to the Enlightenment. 1994 ISBN 0-7923-2452-8
139. J.E. Force and R.H. Popkin (eds.): *The Books of Nature and Scripture*. Recent Essays on Natural Philosophy, Theology, and Biblical Criticism in the Netherlands of Spinoza's Time and the British Isles of Newton's Time. 1994 ISBN 0-7923-2467-6
140. P. Rattansi and A. Clericuzio (eds.): *Alchemy and Chemistry in the 16th and 17th Centuries*. 1994 ISBN 0-7923-2573-7
141. S. Jayne: *Plato in Renaissance England*. 1995 ISBN 0-7923-3060-9
142. A.P. Coudert: *Leibniz and the Kabbalah*. 1995 ISBN 0-7923-3114-1
143. M.H. Hoffheimer: *Eduard Gans and the Hegelian Philosophy of Law*. 1995 ISBN 0-7923-3114-1
144. J.R.M. Neto: *The Christianization of Pyrrhonism*. Scepticism and Faith in Pascal, Kierkegaard, and Shestov. 1995 ISBN 0-7923-3381-0
145. R.H. Popkin (ed.): *Scepticism in the History of Philosophy*. A Pan-American Dialogue. 1996 ISBN 0-7923-3769-7
146. M. de Baar, M. Löwensteyn, M. Monteiro and A.A. Sneller (eds.): *Choosing the Better Part*. Anna Maria van Schurman (1607–1678). 1995 ISBN 0-7923-3799-9
147. M. Degenaar: *Molyneux's Problem*. Three Centuries of Discussion on the Perception of Forms. 1996 ISBN 0-7923-3934-7
148. S. Berti, F. Charles-Daubert and R.H. Popkin (eds.): *Heterodoxy, Spinozism, and Free Thought in Early-Eighteenth-Century Europe*. Studies on the *Traité des trois imposteurs*. 1996 ISBN 0-7923-4192-9
149. G.K. Browning (ed.): *Hegel's* Phenomenology of Spirit: *A Reappraisal*. 1997 ISBN 0-7923-4480-4
150. G.A.J. Rogers, J.M. Vienne and Y.C. Zarka (eds.): *The Cambridge Platonists in Philosophical Context*. Politics, Metaphysics and Religion. 1997 ISBN 0-7923-4530-4
151. R.L. Williams: *The Letters of Dominique Chaix, Botanist-Curé*. 1997 ISBN 0-7923-4615-7
152. R.H. Popkin, E. de Olaso and G. Tonelli (eds.): *Scepticism in the Enlightenment*. 1997 ISBN 0-7923-4643-2
153. L. de la Forge. Translated and edited by D.M. Clarke: *Treatise on the Human Mind (1664)*. 1997 ISBN 0-7923-4778-1
154. S.P. Foster: *Melancholy Duty*. The Hume-Gibbon Attack on Christianity. 1997 ISBN 0-7923-4785-4
155. J. van der Zande and R.H. Popkin (eds.): *The Skeptical Tradition Around 1800*. Skepticism in Philosophy, Science, and Society. 1997 ISBN 0-7923-4846-X
156. P. Ferretti: *A Russian Advocate of Peace: Vasilii Malinovskii (1765–1814)*. 1997 ISBN 0-7923-4846-6
157. M. Goldish: *Judaism in the Theology of Sir Isaac Newton*. 1998 ISBN 0-7923-4996-2
158. A.P. Coudert, R.H. Popkin and G.M. Weiner (eds.): *Leibniz, Mysticism and Religion*. 1998 ISBN 0-7923-5223-8
159. B. Fridén: *Rousseau's Economic Philosophy*. Beyond the Market of Innocents. 1998 ISBN 0-7923-5270-X
160. C.F. Fowler O.P.: *Descartes on the Human Soul*. Philosophy and the Demands of Christian Doctrine. 1999 ISBN 0-7923-5473-7

ARCHIVES INTERNATIONALES D'HISTOIRE DES IDÉES

*

INTERNATIONAL ARCHIVES OF THE HISTORY OF IDEAS

161. J.E. Force and R.H. Popkin (eds.): *Newton and Religion.* Context, Nature and Influence. 1999 ISBN 0-7923-5744-2
162. J.V. Andreae: *Christianapolis.* Introduced and translated by E.H. Thompson. 1999 ISBN 0-7923-5745-0
163. A.P. Coudert, S. Hutton, R.H. Popkin and G.M. Weiner (eds.): *Judaeo-Christian Intellectual Culture in the Seventeeth Century.* A Celebration of the Library of Narcissus Marsh (1638–1713). 1999 ISBN 0-7923-5789-2
164. T. Verbeek (ed.): *Johannes Clauberg* and Cartesian Philosophy in the Seventeenth Century. 1999 ISBN 0-7923-5831-7
165. A. Fix: *Fallen Angels.* Balthasar Bekker, Spirit Belief, and Confessionalism in the Seventeenth Century Dutch Republic. 1999 ISBN 0-7923-5876-7
166. S. Brown (ed.): *The Young Leibniz and his Philosophy (1646–76).* 2000 ISBN 0-7923-5997-6
167. R. Ward: *The Life of Henry More.* Parts 1 and 2. 2000 ISBN 0-7923-6097-4
168. Z. Janowski: *Cartesian Theodicy.* Descartes' Quest for Certitude. 2000 ISBN 0-7923-6127-X
169. J.D. Popkin and R.H. Popkin (eds.): *The Abbé Grégoire and his World.* 2000 ISBN 0-7923-6247-0
170. C.G. Caffentzis: *Exciting the Industry of Mankind. George Berkeley's Philosophy of Money.* 2000 ISBN 0-7923-6297-7
171. A. Clericuzio: *Elements, Principles and Corpuscles.* A Study of Atomisms and Chemistry in the Seventeenth Century. 2001 ISBN 0-7923-6782-0
172. H. Hotson: *Paradise Postponed.* Johann Heinrich Alsted and the Birth of Calvinist Millenarianism. 2001 ISBN 0-7923-6787-1
173. M. Goldish and R.H. Popkin (eds.): *Millenarianism and Messianism in Early Modern European Culture.* Book I. Jewish Messianism in the Early Modern World. 2001 ISBN 0-7923-6850-9
174. K.A. Kottman (ed.): *Millenarianism and Messianism in Early Modern European Culture.* Book II. Catholic Millenarianism: From Savonarola to the Abbé Grégoire. 2001 ISBN 0-7923-6849-5
175. J.E. Force and R.H. Popkin (eds.): *Millenarianism and Messianism in Early Modern European Culture.* Book III. The Millenarian Turn: Millenarian Contexts of Science, Politics and Everyday Anglo-American Life in the Seventeenth and Eighteenth Centuries. 2001 ISBN 0-7923-6848-7
176. J.C. Laursen and R.H. Popkin (eds.): *Millenarianism and Messianism in Early Modern European Culture.* Book IV. Continental Millenarians: Protestants, Catholics, Heretics. 2001 ISBN 0-7923-6847-9
177. C. von Linné: *Nemesis Divina.* (edited and translated with explanatory notes by M.J. Petry). 2001 ISBN 0-7923-6820-7

KLUWER ACADEMIC PUBLISHERS – DORDRECHT / BOSTON / LONDON